Field Guide to the Carboniferous Sediments of the Shannon Basin, Western Ireland

T0176583

A Field Guide to the Carboniferous Sediments of the Shannon Basin, Western Ireland

Edited by

JAMES L. BEST & PAUL B. WIGNALL

WILEY Blackwell

This edition first published 2016
© 2016 by the International Association of Sedimentologists

Registered Office
John Wiley & Sons, Ltd, The Atrium, Southern Gate, Chichester, West Sussex, PO19 8SQ, UK

Editorial Offices
9600 Garsington Road, Oxford, OX4 2DQ, UK
The Atrium, Southern Gate, Chichester, West Sussex, PO19 8SQ, UK
111 River Street, Hoboken, NJ 07030-5774, USA

For details of our global editorial offices, for customer services and for information about
how to apply for permission to reuse the copyright material in this book please see our
website at www.wiley.com/wiley-blackwell.

Library of Congress Cataloging-in-Publication data applied for

ISBN: 9781119257127

A catalogue record for this book is available from the British Library.

Wiley also publishes its books in a variety of electronic formats. Some content that
appears in print may not be available in electronic books.

Cover image: James L. Best and Paul B. Wignall

Set in 9.25/11.5pt Times New Roman by SPi Global, Pondicherry, India
Printed and bound in Malaysia by Vivar Printing Sdn Bhd

1   2016

Contents

Contributors, vii

Acknowledgements, ix

About the Companion Website, xi

1 Introduction to the Field Guide, 1
Jim Best & Paul B. Wignall

2 The Shannon Basin: Structural Setting and Evolution, 16
John Graham

3 Basin Models, 35
Paul B. Wignall & Jim Best

4 Lower Carboniferous of the Shannon Basin Region, 48
Ian D. Somerville

5 Viséan Coral Biostromes and Karsts of the Burren, 79
Ian D. Somerville

6 The Clare Shales, 97
Paul B. Wignall, Ian D. Somerville & Karen Braithwaite

7 Architecture of a Distributive Submarine Fan: The Ross Sandstone
Formation, 112
David R. Pyles & Lorna J. Strachan

8 Evolving Depocentre and Slope: The Gull Island Formation, 174
Lorna J. Strachan & David R. Pyles

9   The Tullig and Kilkee Cyclothems in Southern County Clare, 240
    Jim Best, Paul B. Wignall, Eleanor J. Stirling, Eric Obrock &
    Alex Bryk

10 The Tullig and Kilkee Cyclothems of Northern County Clare, 329
    Paul B. Wignall, Jim Best, Jeff Peakall & Jessica Ross

11 The Younger Namurian Cyclothems around Spanish Point, 350
    Paul B. Wignall & Jim Best

Appendix: List of *GigaPan* Images, 361

References, 362

Index, 371

Contributors

Editors

Jim Best, *Jack and Richard Threet Chair in Sedimentary Geology, Departments of Geology, Geography & GIS, Mechanical Science and Engineering and Ven Te Chow Hydrosystems Laboratory, University of Illinois at Urbana-Champaign, 605 East Springfield Avenue, Champaign, IL 61820, USA.*

Paul B. Wignall, *School of Earth and Environment, University of Leeds, Leeds, West Yorkshire, LS2 9JT, UK.*

Contributors

Karen Braithwaite, *School of Earth and Environment, University of Leeds, Leeds, West Yorkshire, LS2 9JT, UK. now at: School of Veterinary Medicine and Science, University of Nottingham, College Road, Sutton Bonington, Loughborough, Leicestershire, LE12 5RD, UK.*

Alex Bryk, *Department of Geology, University of Illinois at Urbana-Champaign, 605 East Springfield Avenue, Champaign, IL 61820, USA; now at: Department of Earth and Planetary Science, University of California Berkeley, 307 McCone Hall, Berkeley, CA 94720, USA.*

John Graham, *Department of Geology, Museum Building, Trinity College Dublin, Dublin 2, Ireland.*

Eric Obrock, *Department of Geology, University of Illinois at Urbana-Champaign, 605 East Springfield Avenue, Champaign, IL 61820, USA; now at: ExxonMobil, 22777 Springwoods Village, Parkway, Spring, TX 77389, USA.*

Jeff Peakall, *School of Earth and Environment, University of Leeds, Leeds, West Yorkshire, LS2 9JT, UK.*

David R. Pyles, *EOG Resources, 600 17th Street Suite 1000N, Denver, CO 80202, USA. Formerly at: Chevron Centre of Research Excellence, Department of Geology and Geological Engineering, Colorado School of Mines, Golden, CO 80401, USA.*

Jessica Ross, *School of Earth and Environment, University of Leeds, Leeds, West Yorkshire, LS2 9JT, UK.; now at: Maersk Oil North Sea UK, Maersk House, Crawpeel Road, Aberdeen, AB12 2LG, UK.*

Lorna J. Strachan, *Earth Science Programme, School of Environment, University of Auckland, Auckland 1142, New Zealand.*

Eleanor J. Stirling, *BP Exploration Operating Company Ltd., Chertsey Road, Sunbury-on-Thames, Middlesex TW16 7LN, UK.*

Ian D. Somerville, *School of Earth Sciences, Science Centre West, University College Dublin, Belfield, Dublin 4, Ireland.*

Acknowledgements

We are very grateful to a whole host of people who have inspired and helped us in our research in Western Ireland over the past twenty-five years and also provided enthusiasm and assistance in completing this field guide.

Firstly, we are grateful to all of the contributors to this book in its long journey from inception to publication and thank them for their input, patience and dedication to producing a field guidebook to a region of globally-renowned and important geology that will hopefully be very widely used. We have also benefitted from visiting this area with many people over the years and are very grateful for the insights and expertise provided by Jeff Peakall (University of Leeds), Drew Phillips (Illinois State Geological Survey), Steve Marshak and Michael Stewart (University of Illinois), Jeff Nittrouer (Rice University) and Owen Sutcliffe (Neftex). We would also like to thank several generations of Leeds University and University of Illinois undergraduates and postgraduates who have worked in the region and greatly contributed to our understanding of the geology; especially Dan Bell, Alex Bryk, Karen Braithwaite, Rachael Dale, Heather Macdonald, Eric Obrock and Eleanor Stirling. We are thankful to Dan Bell for data that helped construct the geological map of Kilkee. We are also grateful for the help of the IAS Special Publications editors we have worked with – Ian Jarvis, Tom Stevens and Mark Bateman – for their encouragement and perseverance; and also to the IAS and the Jack and Richard Threet Chair in Sedimentary Geology at the University of Illinois for funding the final graphics compilation. We are indebted to Chris Simpson for his superb work on the final graphics that ensured the consistency of all illustrations in the guide and made them available for online download. Ian Francis, Kelvin Matthews, Delia Sandford and Radjan Lourde Selvanadin at Wiley Blackwell are thanked for their guidance and work in bringing this guide to publication.

We have also been incredibly fortunate to make lasting friends in County Clare in our many years working there and these people have provided a constant source of knowledge, help, good humour and true friendship. We are especially grateful to Patrick Blake and Patrick Egan (Liscannor), Orla and Mark Vaughan (Kilfenora Hostel and Vaughan's Pub, Kilfenora),

Brian Farrell (Burren Coaches, Ballyvaughan) and Geoff and Susanne Magee (Dolphinwatch, Carrigaholt) for their friendship and help.

Finally, we would like to dedicate this field guide to the memory of Trevor Elliott, who sadly passed away on 28th January 2013. Trevor was an inspirational field geologist who fostered the interest of many geologists, including us, in the Shannon Basin. His seminal work on many parts of the basin fill, his mentoring of colleagues in the region and his leadership of many industrial trips to these rocks have left a permanent imprint on many of us who were lucky enough to know and meet Trevor. We, and his many friends in this area of Western Eire, will sorely miss the sight of Trevor striding across the cliff top paths in his brightly coloured field clothes and discussing the geology with him over a Guinness in the evenings. His kindness and inspiration will remain with us for many years to come.

Jim Best and Paul B. Wignall,
January 2016

About the Companion Website

This book is accompanied by a companion website:

www.wiley.com/go/best/shannonbasin

The website includes:

- All figures from the field guide in PowerPoint format for use in teaching.
- GigaPan Images from various sites that are referred to in the field guide.
- All outcrop locations, as both kmz files and text files (UTM Zone 29U and OSI Grid co-ordinates).
- Web addresses for the Ordnance Survey of Ireland and Geological Survey of Ireland.
- Details of accommodation and travel in the region.

Chapter 1
Introduction to the Field Guide

JIM BEST & PAUL B. WIGNALL

1.1 The Aim of this Field Guide

This field guide provides a detailed account of the Carboniferous geology of the Shannon Basin, principally in County Clare and County Kerry, Western Ireland. This region has become a classic destination for field groups from across the world in the past 25 years due to its stunning exposures of a wide range of Carboniferous depositional environments – from carbonate platform to deep sea turbidites, from black shales to delta slope and from shallow marine environments to fluvial channels – that can be viewed on a wide range of spatial scales up to those of interest within hydrocarbon reservoir modelling. The region has become a testing ground for concepts within basin analysis and sequence stratigraphy and has been used as a source of outcrop analogues for many hydrocarbon reservoir studies across the globe. This guide provides a summary of both past work and ongoing debate on the interpretation of these exceptional outcrops, through description of the principal localities and their major features. We hope the guide will be valuable to both professional and amateur geologists, as well as a broader audience who want to know more about the rocks that form this beautiful landscape. This guide thus provides both an account of the deep-time evolution of this region in the Carboniferous some 320 million years ago, as well as setting the stage for a landscape that has a fascinating history of human settlement over the past 6000 years (Jones, 2004, 2007). The guide assumes a basic knowledge of geology but also includes some terminology associated with specific areas and topics, such as palaeontology and sequence stratigraphy.

A Field Guide to the Carboniferous Sediments of the Shannon Basin, Western Ireland, First Edition. Edited by James L. Best and Paul B. Wignall.
© 2016 International Association of Sedimentologists.
Published 2016 by John Wiley & Sons, Ltd.
Companion website: www.wiley.com/go/best/shannonbasin

Perhaps what strikes one most when walking over and examining the rocks described in this field guide is the incredible variety and wealth of superbly-preserved geological features present. These outcrops have provided the materials for a range of detailed research papers over a 60-year period and have led to this region being perhaps one of the most visited destinations by geological field parties in a global context; and for specialities that include sedimentology, palaeontology, structural geology, geophysics and reservoir geology. Yet, despite this extensive study, the Shannon Basin continues to reveal new features, to stimulate new interpretations and debate and there are many questions that remain to be answered about these sedimentary sequences. This field guide thus not only aims to provide details about many of the key localities, but also highlights areas of ongoing debate and discussion, in the hope that it will provide a synthesis and starting point for future study and teaching.

1.2 Background to the Area

The Carboniferous-age Shannon Basin encompasses an area within Western Ireland that includes the majority of County Clare and parts of County Kerry and County Limerick. The area lies along the Atlantic coast of Western Ireland (Fig. 1.2.1) and encompasses regions both to the north and south of the Shannon Estuary, a waterway whose trend has considerable geological importance as regards the formation and evolution of the Shannon Basin (Graham, Chapter 2). The topography of the area (Fig. 1.2.2) consists of higher terrain to the north in the limestone hills and karst terrain of the Burren and gently rolling hills of the Loop Head Peninsula that lies to the north of the Shannon Estuary, the mouth of Ireland's largest river. The Atlantic coast often possesses high and dramatic vertical cliffs that afford superb exposures within the Carboniferous sediments, but inland on the Loop Head Peninsula and in Counties Kerry and Limerick the exposures are far more limited. The area contains the world heritage area of the Burren in the north that comprises Lower Carboniferous limestones that also make up the Aran Islands (Fig. 1.2.1), and which exhibits superb geomorphology (Simms, 2006), natural history (D'Arcy & Hayward, 1991; Nelson, 2008) and archaeology (Jones, 2004, 2007). The region also contains the world famous Cliffs of Moher that display a section through part of the Carboniferous clastic basin fill, and the striking coastal scenery of the Loop Head Peninsula. The geology of the area allows access to a wide range of sediments that document the formation, fill and later deformation of the Carboniferous Shannon Basin, with many exposures being located along the western Irish Atlantic coast.

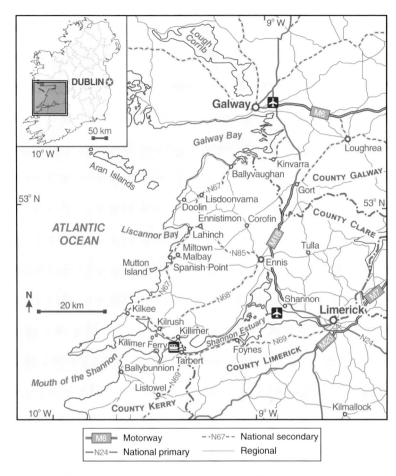

Fig. 1.2.1. Map of the region detailed in this field guide, with principal towns, roads and airports marked.

Tectonic deformation of the area occurred during the Variscan Orogeny, with compression from the south producing broadly west-east oriented fold structures, which decrease in their intensity further north. Thus, the limestones and sandstones of northern County Clare, that gently-dip at a few degrees to the south, are increasingly replaced by more intensely folded and faulted sediments further south on the Loop Head Peninsula and south of the Shannon Estuary, where pervasive pressure solution cleavage is also developed. Indeed, the degree of low-grade metamorphism also increases into the outcrops across the Shannon Estuary to the south.

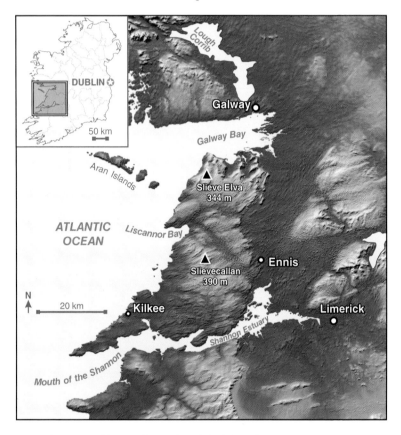

Fig. 1.2.2. Topographic map of the region, derived from Shuttle Radar Topography Mission (SRTM) data (from http://photojournal.jpl.nasa.gov/catalog/?IDNumber=PIA06672). Colour coding is directly related to topographic height, with green at the lower elevations, rising through yellow and tan, to white at the highest elevations.

1.3 Climate

The present-day climate in Western Ireland (Fig. 1.3.1) is mild, being dominated by the weather systems that track across the Atlantic and providing conditions that are seldom very cold but are often rainy. Weather conditions can range from large Atlantic storms that strike the coast and bring considerable rain, high seas and gale force winds, through to summer anticyclones that yield warm, dry periods. July is the warmest month (at Kilkee (Fig. 1.3.1), the average is 15.2°C (59.4°F) with January being the coldest (Kilkee average = 5.8°C or 42.4°F)). June and December are the

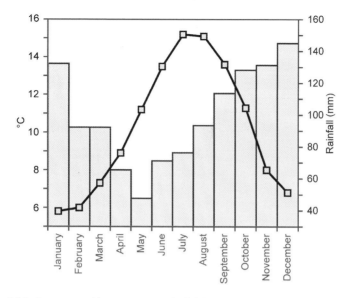

Fig. 1.3.1. Average monthly temperature and rainfall at Kilkee, County Clare (see location on Fig. 1.2.1).

driest and wettest months (at Kilkee, 71 and 154 mm rainfall, respectively). However, rainfall occurs in all months (Fig. 1.3.1) and thus it is always worthwhile packing waterproof clothes for fieldwork in the area, as well as sun-block and sunhats for the summer months. It is also worthy of note that at this latitude the winter days are short (*c.* 7.5 hrs between sunrise and sunset at the winter solstice) but you are rewarded by long days and light evenings in the summer (*c.* 17 hrs between sunrise and sunset at the summer solstice).

1.4 Accommodation, Travel and General Facilities

The field area can be reached easily by car or bus from Dublin (*c.* 4 hours by car) or visitors may also fly direct to either Shannon Airport, which is *c.* 1.5 hrs away from the town of Kilkee (Fig. 1.2.1), or Galway Airport that is *c.* 1.5 hrs away from Lahinch (Fig. 1.2.1). In-field transport along often small roads is easier by car or minibus. Larger coaches can negotiate the small roads that lead to the vast majority of the outcrops detailed herein, but it is essential to discuss the localities to be visited with the local coach operators. Most of the sites can be accessed easily and the local landowners are very gracious and obliging in allowing geologists to visit these localities. However, it is recommended that visitors ask permission for access to any areas where is it obvious that you are walking across private land and fields.

Tourism is important to the local economy in most of the coastal areas to be visited and consequently accommodation in the area is easily available through a host of options. A range of hotels is available in the larger towns, with excellent bed-and-breakfast guest houses also to be found in towns, villages and in the countryside. In addition, several hostels are present in the area (such as in Kilfenora, Doolin and Lahinch in northern County Clare) and these can cater for student groups. Besides these options, houses, cottages and caravans can be rented at many localities within the field guide area and the local tourist offices of Counties Clare, Kerry and Limerick provide a ready source of excellent information. Some useful sources of contact are given in the linked online website.

1.5 Safety

The field area is generally safe and many of the localities can be visited by large groups, with the extensive exposures providing easy access. However, a series of hazards, some potentially fatal, are present and should be borne in mind when planning visits to these localities. Some of the localities detailed herein are also weather dependent (slippery rocks, high winds, tides) and thus the visitor should always, at each outcrop, conduct a careful safety assessment of local conditions and hazards before proceeding to access the localities described in this field guide. A list of the principal hazards is given below but additional factors and hazards may be present at individual sites.

High cliffs: Some of the coastal outcrops are adjacent to high cliffs that call for great care, especially if the rocks are slippery when wet, as they often are, and when there are high winds, which may be stronger near the cliff edges. All such localities must be approached with extreme care and cliff edges should not be approached due to potential undercutting and instability of any overhanging ledges.

Slippery rocks: One of the most frequent hazards in this region is slippery rocks, caused by either rain or sea-water. Slippery surfaces are often worse in intertidal areas where seaweed and water can make the outcrops truly treacherous. Extreme care should be taken in this regard, with the limestones also providing sharp surfaces on which cuts can be sustained easily during any falls.

Loose rocks: Some localities possess high outcrops and cliffs that have loose rocks overhead and thus, as always, the wearing of hardhats is essential at these localities.

Tides: The western coast of Ireland has large tides (at Kilkee, the tidal range is between *c*. 2 and 5 m), with the Shannon Estuary being macrotidal and having the largest tidal range on the western Irish coast. Care is thus required at some localities to avoid being cut-off by the tide, with some intertidal localities also being tidally-restricted and covered at high tides. Visitors to the area should thus consult the appropriate tide timetables. Outcrops that are tidally-restricted are highlighted in the text.

Small roads: Access to many of these localities is along small country roads that are frequently narrow, with room only for one-way traffic with occasional passing places. Visitors should be aware of the care needed in driving along such roads, the fact that the soft, grassy edges of the tracks may often be unstable or adjacent to ditches and that cars drive on the left in Ireland. Driving in this region, with the geology and geomorphology making for superb viewing, is enjoyable, but caution is required in negotiating some of these roads. Additionally, at some localities vehicle parking is very restricted and visitors should ensure that they do not block any access points or passing places.

Large waves: The Atlantic Ocean produces some enormous waves that bombard the Irish coast during storms. However, even on calm days between storms, large waves can and do impact the coast and thus great care should also be taken not to access exposures near the water's edge on days when rough seas may produce large waves. Again, care is called for in assessing the sea conditions on arrival at the outcrop and how these may change during the duration of the field visit.

Farm animals and electric fences: Visitors should be wary of farm animals when walking across any fields, taking care to close gates that have been opened and to not disturb these animals. Electric fences are in use in some cattle fields and visitors should be aware of these wires. As always, please leave the countryside untouched and do not leave any litter behind.

1.6 Map Coverage

The Ordnance Survey of Ireland (OSI) has produced a range of topographic maps that cover this region and the link to the OSI website is given in the online website supporting this field guide. The Discovery series of maps are at a 1:50 000 scale and provide an excellent resource for use in the field identifying site access and roads. A series of sheets must be used to encompass the region detailed in this field guide (Fig. 1.6.1), with sheets 51, 52, 57, 58, 63, 64 and 65 covering the main regions of coastal exposure. The OSI also sell an electronic version of this map series that can be downloaded and used on a variety of handheld GPS devices. Additionally, the OSI sell a range of other maps, including historic 25 inch to the mile

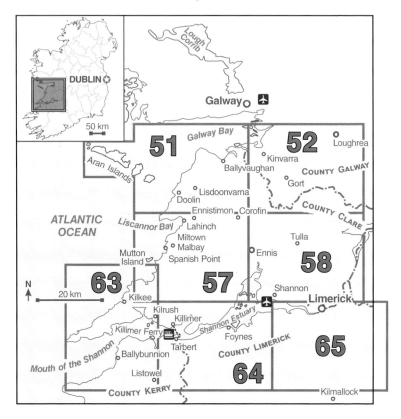

Fig. 1.6.1. Ordnance Survey of Ireland (OSI) 1:50 000 map coverage of the area (maps 51, 52, 57, 58, 63, 64, 65).

(1:2534) maps that were produced between 1897 and 1913. Although old, the detail of these maps along the coast is superb and they are a very useful resource for mapping in the region. Google Earth images are also generally excellent for the region and can be well used for locating outcrops. A kmz file for all of the principal localities and stops referred to in this guide is available from the linked online website.

In this field guide, eastings and northings are given for all sites as UTM (Universal Transverse Mercator; Zone 29U) co-ordinates, with six-digit eastings and seven-digit northings. Additionally, in the text all locations are given in Irish Grid co-ordinates that will enable use of the 1:50 000 maps; these grid references are given as map number and then a six figure easting and northing. The second and third numbers in these Irish Grid co-ordinates are those given as larger numbers on the 1:50 000 maps: thus a grid reference of Map 63, 087516 m E, 160225 m N should be read as

087516 m E, 160225 m N on the maps. For instance, the hostel at Kilfenora is at UTM 29U 485278 m E, 5871180 m N and its 1:50 000 map reference is Map 51 118167m E, 193910 m N. On the printed 1:50 000 map, readers may wish to simplify this to a six figure grid reference (i.e. for the above location Map 51 181 m E, 939 m N), but the full six figure eastings and northings are given for those using these Irish grid references with a GPS. Conversions to latitude/longitude, the Irish Grid (IG) and ITM Irish Transverse Mercator (ITM) are available using a free application on the OSI website (see resources link in the online website for this guide book; IG and ITM are used on OSI maps).

Geological maps of the region are published by the Geological Survey of Ireland (GSI) and an excellent source of information on their publications is given on the GSI website (see links in Resources in the online website for this guide book). Of most relevance to this field guide, a series of 1:100 000 bedrock geology maps with accompanying booklets have been produced by the OSI that discuss the geology of the Shannon Estuary region (Sleeman & Pracht, 1999), Galway Bay (Pracht *et al.*, 2004) and Dingle Bay (Pracht, 1996). In addition, several very useful books have been written that present broad treatments of the geology of Ireland and some of the localities in this region, including the general textbook by Woodcock & Strachan (2012), the general guide to the geology of the Burren by McNamara & Hennessy (2010), a geomorphological guide to the Burren (Simms, 2006) and recent books specifically concerning the geology of Ireland by Holland & Sanders (2009) and Meere *et al.* (2013).

1.7 Geological Map of the Region and Stratigraphy

The geology of the Shannon region broadly consists of nearly 3 km of Carboniferous strata that has been folded into a broad syncline that plunges to the west. The youngest strata are found around Spanish Point in the centre of the County Clare coastline (Fig. 1.7.1). Progressively older strata surround this area with the result that Lower Carboniferous limestones are seen both in the northern-most County Clare outcrops of the Burren region, to the east around Ennis, the county town, and to the south on the County Kerry coast at Ballybunnion. However, as noted above, there is progressively more intense Variscan deformation in the southern area; and the southern limb of the synclinal structure consists of secondary folds with wavelengths of hundreds of metres to a few kilometres, whereas the northern limb simply shows a dip of a few degrees to the south. For the most part, these secondary folds are symmetrical.

The oldest rocks in the region belong to the Tournaisian Series of the Early Carboniferous (Mississippian) and consist of fossiliferous, shallow-water

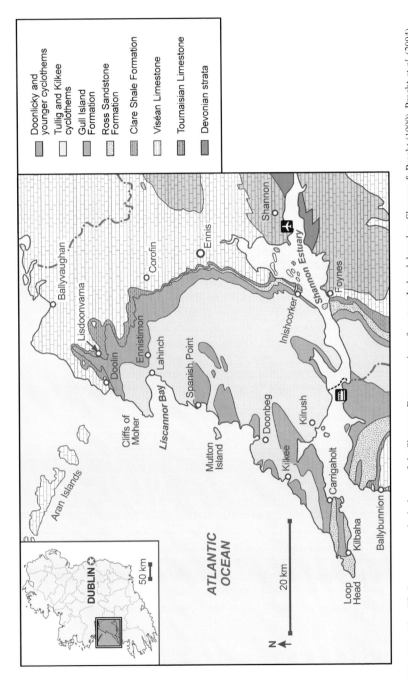

Fig. 1.7.1. Simplified summary geological map of the Shannon Estuary region of western Ireland, based on Sleeman & Pracht (1999), Pracht *et al.* (2004), Gill (1979) and the authors' own data.

limestones and minor shales (Fig. 1.7.2). Outcrops are limited to the banks of the inner Shannon Estuary and are described in Chapter 4. These are succeeded by the Waulsortian Limestones that comprise massive, coalesced mounds of micrite with a large cement component. This enigmatic facies, which was widespread during the middle Dinantian interval, is especially well developed in the Shannon Estuary region from where the thickest developments (~1 km) anywhere in the world are known. The overlying limestones show significant regional variation. In northern County Clare, the thick-bedded limestones contain abundant brachiopods, crinoids and corals, whereas the more southerly correlatives in the Shannon Estuary region are dominated by less fossiliferous, thinner-bedded, more shaly and cherty limestones. Regional thickness trends in these Viséan limestones show the northern County Clare limestones to be slightly thicker than those in the southern area (Fig. 1.7.3) – a thickness trend that is dramatically reversed in the overlying Carboniferous strata.

The transition from the Viséan to Namurian Series (alternatively called the Dinantian-Serpukhovian transition) saw the shutdown of limestone deposition in the Shannon region (and elsewhere, the loss of limestones was a widespread and near-contemporaneous event over a broad area of north-west Europe). This was followed by the deposition of the black shales of the Clare Shale Formation in the Shannon area followed by turbiditic sandstones of the Ross Sandstone Formation and then siltstones of the Gull Island Formation (Fig. 1.7.2). These strata belong to the nearly 1 km thick Shannon Group. Further north in County Clare, the Shannon Group is much thinner (<150 m) and only began to accumulate after a prolonged hiatus of nearly 10 Myr that spanned the entire Serpukhovian Stage. The overlying Central Clare Group consists of thick coarsening-upward packages, known as cyclothems. The lowest two examples, the Tullig and Kilkee cyclothems, outcrop over much of the length of the County Clare coastline and exhibit much less lateral thickness variation, although the major sandbodies that developed within the upper parts of the cyclothems (e.g. the Tullig Sandstone) do show major lateral changes (Fig. 1.7.3). The youngest cyclothems, numbered IV and V, are only seen around Spanish Point in the centre of the regional syncline and so their lateral variation is unknown.

1.8 Organisation of this Field Guide, Areas Covered and Suggested Itineraries

This field guide is broadly arranged in stratigraphic order. The guide commences with a broad analysis of the structural setting and evolution of the Shannon Basin (Chapter 2) and then a chapter summarizing the various

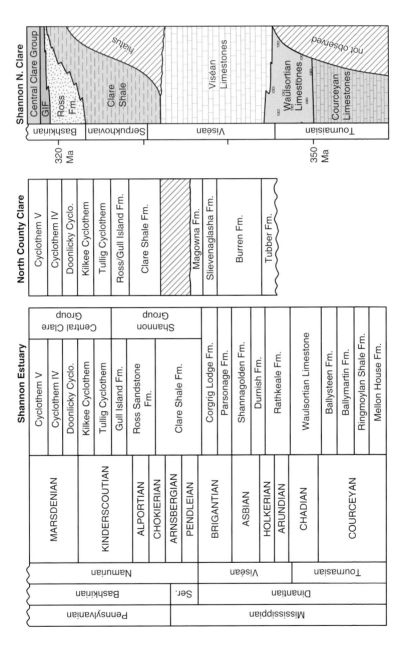

Fig. 1.7.2. Summary of the Carboniferous lithostratigraphy in the Shannon Estuary and northern County Clare regions. The right-hand column shows a chronostratigraphic correlation between the southern Shannon sections and those in northern County Clare, showing the substantial hiatus developed between the Viséan limestones and Clare Shale. Ser. is Serpukhovian.

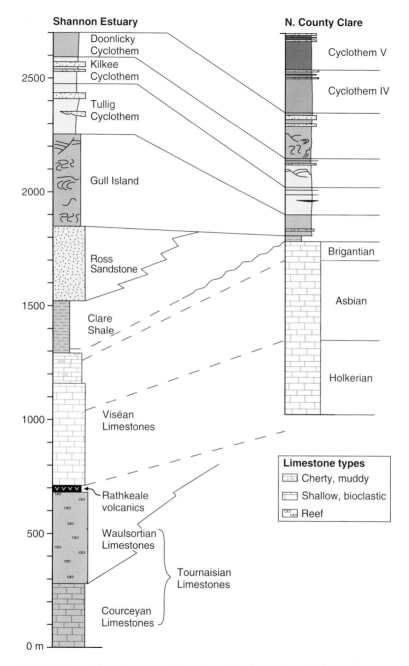

Fig. 1.7.3. Correlation of summary lithostratigraphic columns from the Shannon Estuary region and northern County Clare sections ~50 km to the north. This diagram shows the substantial lateral thickness variations at the level of the Clare Shale, Ross Sandstone and Gull Island formations and much less lateral variation at other levels.

models of sedimentary fill that have been proposed for the Shannon Basin (Chapter 3). The intention of these two chapters is to set the scene for readers, as they use the field guide to examine the stratigraphy and sedimentology described in the following chapters. After these two review chapters, the sediments are subsequently detailed in eight chapters that progressively work up-stratigraphy. Each chapter presents details on a range of localities, with a summary of these locations being shown in Fig. 1.8.1. It is worthwhile to examine this map when planning an itinerary to the region, as some parts of the stratigraphy detailed in separate chapters may lie close to each other geographically in the field. Fig. 1.8.1 shows the

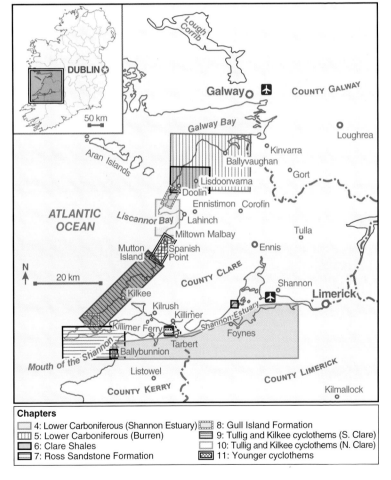

Chapters

- 4: Lower Carboniferous (Shannon Estuary)
- 5: Lower Carboniferous (Burren)
- 6: Clare Shales
- 7: Ross Sandstone Formation
- 8: Gull Island Formation
- 9: Tullig and Kilkee cyclothems (S. Clare)
- 10: Tullig and Kilkee cyclothems (N. Clare)
- 11: Younger cyclothems

Fig. 1.8.1. Overview map of the field localities detailed in this field guide.

regions of occurrence of the localities detailed in each chapter and allows easy visual recognition of the overlap of the differing localities detailed herein.

In addition to the figures printed in the hard copy of the field guide, there are a range of online resources available on the linked field guide website (www.wiley.com/go/best/shannonbasin). These include:

1 Google Earth kmz, Excel and text files of all outcrop locations (in UTM and Irish Grid co-ordinates), subdivided according to Chapter. These allow easy location of all the field stops detailed herein, as well as some key access points to certain localities.

2 High-resolution copies of all figures in the guidebook as PowerPoint slides. These will allow users to utilise these diagrams in their own work in the field, in teaching applications and also allow closer inspection of detail.

3 High-resolution jpeg GigaPan photo-montages for several key localities that can be viewed to better examine the detail within these outcrops. A listing of these GigaPan images, with cross-referencing to the appropriate chapters, is also given in Appendix 1.

4 A list of addresses for useful websites, including the Ordnance Survey of Ireland, Geological Survey of Ireland and various useful field contacts.

Chapter 2
The Shannon Basin: Structural Setting and Evolution

JOHN GRAHAM

2.1 Introduction

During the mid-Carboniferous, around 330 million years ago, the area around the modern Shannon Estuary of western Ireland provided significant accommodation for sediments that are now visible in spectacular coastal exposures. This depositional basin has been referred to as the Shannon Trough, the Shannon Basin, the Clare Basin and the Western Irish Namurian Basin. Despite the considerable interest in, and study of, the coastal exposures, the geographical limits of this basin are very poorly constrained. To the west, the margins are under the Atlantic and geophysical data on their location is equivocal. To the south and the south-east, information is present but it has not been interrogated in detail, partly due to limited exposure of the basin fill. Whilst this does not affect many of the detailed studies that have occurred, it does severely limit those that attempt to make larger scale reconstructions.

In gross terms, the succession represents progressive deepening through a lower carbonate portion, a middle part dominated by deep water muds and then progressive shallowing through turbiditic sands and muddy slope sediments to deltas that have been proposed to bear close comparison to the modern Mississippi system. The history contained in these rocks is, of course, more complex and more informative than this as the subsequent chapters will demonstrate. Since the pioneering work of Dan Gill in the 1950s and 1960s, these rocks have provided useful analogues for petroleum geologists and, as such, that they have become some of the most visited sections in the world. The succeeding chapters discuss many aspects of this succession as well as guiding the reader to where the evidence can be seen.

A Field Guide to the Carboniferous Sediments of the Shannon Basin, Western Ireland, First Edition. Edited by James L. Best and Paul B. Wignall.
© 2016 International Association of Sedimentologists.
Published 2016 by John Wiley & Sons, Ltd.
Companion website: www.wiley.com/go/best/shannonbasin

The purpose of this introduction is twofold; firstly to consider briefly the pre-Carboniferous history of the area and how it might have influenced the siting of this depositional basin, and secondly to briefly review the subsequent deformational and metamorphic history that affects what is seen in these world-famous exposures.

2.2 Pre-Carboniferous History

During the Caledonian orogenic cycle, terminating around 400 million years ago, the north-western and south-eastern parts of Ireland developed on opposite sides of a large ocean, the Iapetus Ocean. As this ocean started to close, a major but short-lived orogenic event affected the north-western margin during the mid-Ordovician (~465 Ma). This is thought to have been caused by collision of a volcanic arc with the Laurentian continent and is termed the Grampian Orogeny (see Chew (2009) and Dewey (2005) for recent summaries). The Iapetus Ocean continued to narrow throughout the Ordovician and early Silurian (485 to 420 Ma) until the south-eastern continent of Avalonia collided with the north-western continent of Laurentia and its accreted arcs.

This continental collision was strongly oblique (Dewey & Strachan, 2003) and shows abundant evidence for sinistral transpression and no substantial crustal thickening. This collisional event is generally referred to as the Acadian Orogeny, with deformation in Ireland being constrained by means of unconformities and strain discontinuities, such as that occurring between the late Wenlock and early Devonian (Chew & Stillman, 2009).

The junction between the Laurentian continent, its accreted arcs and Avalonia is referred to as the Iapetus Suture. Rocks of the Southern Uplands – Longford Down terrane represent deep marine muds and turbidites deposited on the north-western side of the ocean. Many workers have viewed this as an accretionary prism, although alternative views (back arc basin, successor basins) have been expressed (Graham, 2009a; Chew, 2009). Successions developed in south-east Leinster are definitively Avalonian but there is more uncertainty for the limited Lower Palaeozoic exposures north of this area, although south of the Longford-Down (Graham & Stillman, 2009). In eastern Ireland, assignment of blocks to particular margins relies largely on faunal provinciality. This works reasonably well in the Ordovician but less well in the Silurian, as the remnant Iapetus ceased to be a migrational barrier and the proximity of the margins provided little biogeographical distinction of the faunas (Graham & Stillman, 2009; Todd *et al.*, 1991).

Combination of faunal data with structural data suggests that in eastern Ireland the Tinure Fault represents the surface trace of the Iapetus Suture

Zone (Chew & Stillman, 2009; Vaughan & Johnston, 1992), although these data do not define its position at depth. Using similar faunal and structural data, the Iapetus Suture Zone can be projected westwards but the greater degree of Upper Palaeozoic cover makes this projection less well-constrained. However, most workers draw the surface trace of the Iapetus Suture through the Shannon Estuary (Fig. 2.2.1). If this is correct, the presence of such a major crustal discontinuity could well be expected to influence later extensional events, such as the generation of the Shannon Trough, as well as fluid migration pathways.

The nature of the subsurface structure can be indicated to some degree by seismic experiments, by examination of xenoliths carried upwards by later intrusions and by isotopic constraints. However, the lack of granites near the suture zone makes isotopic methods less applicable. The most important seismic traverses with respect to the Shannon Trough are the WIRE1 and WIRE1B (West IRElind geophysical survey) lines shot just off the west coast and the VARNET-96 (VARiscan NETwork 1996 project) lines (A and B) shot on land (Fig. 2.2.2A). An interpretation of the VARNET-A profile is shown in Fig. 2.2.2B and one of the WIRE profiles (1 and 1B) is shown in Fig. 2.2.3. Assigning ages to specific reflectors is, of necessity, interpretative and it is quite possible that some reflectors represent development during more than one orogenic event (Ford *et al.*, 1992). The current interpretation suggests that the Iapetus Suture Zone is about 50 km wide at depths of 5 to 35 km and is characterised by reflectors dipping *c.* 30° N. This is taken to represent the northward subduction of the Leinster Terrane beneath the Longford-Down and the zone coincides geographically with the Shannon Estuary. This interpretation is supported by evidence from xenoliths in lamprophyre dykes in the Longford-Down which have been used to suggest under-thrusting by Leinster type crust (Anderson & Oliver, 1996). Xenoliths from Lower Carboniferous volcanic pipes that occur above the Iapetus Suture Zone have also been compared to Avalonian basement (Van den Berg, 2005).

The data from the VARNET profiles have been used to construct 2D p-wave velocity models (Landes *et al.*, 2000; 2005; Vermeulen *et al.*, 1999). Whilst changes such as the broad dome of the Moho and lateral velocity changes in the mid-crust have been used to delimit the Iapetus Suture Zone (Fig. 2.2.2B), there are major problems with these models that, at shallow levels, are largely due to the limited velocity contrast between the different units of the Palaeozoic geology. Thus the modelled velocity profile across the Shannon Estuary shown in Fig. 2.2.2B is difficult to reconcile with what is known from the surface geology. Moreover, interpretations based on seismic shear wave models offer conflicting views of the structure at depth (Readman *et al.*, 2009; O'Reilly *et al.*, 2010). A geophysical-based model of the Iapetus Suture Zone by O'Reilly *et al.*

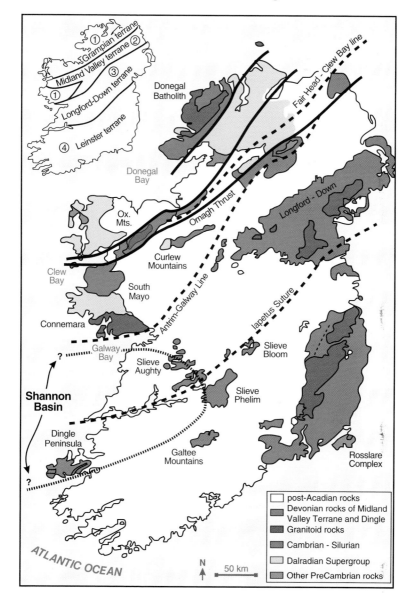

Fig. 2.2.1. Simplified geology of Ireland showing the Iapetus Suture Zone and possible limits of the Shannon Basin (after Chew & Stillman, 2009).

(A)

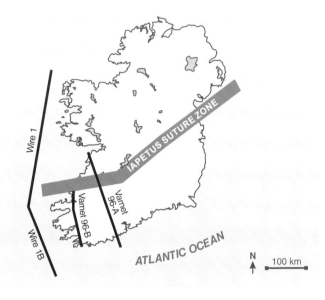

(B)

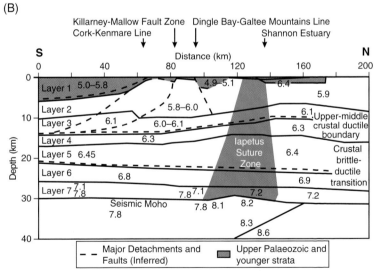

Fig. 2.2.2. A) Location of WIRE and VARNET geophysical profiles. B) Interpretation of data from the VARNET-A profile (after Vermeulen *et al.*, 1999).

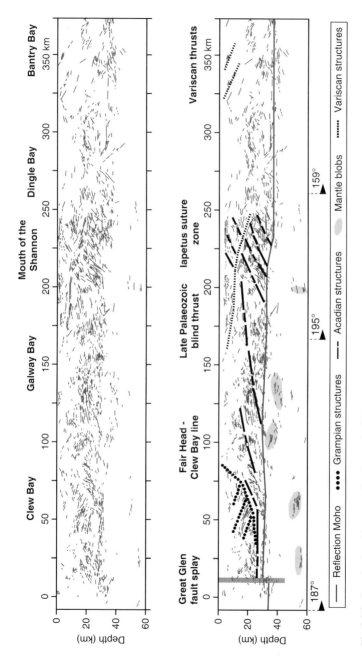

Fig. 2.2.3. Data (top) and interpretation (bottom) of WIRE1 and WIRE1B geophysical profiles (after Klemperer *et al.*, 1991). Reproduced with permission from the Geological Society of London.

(2010) suggests that it may be underlain by numerous lenses of mafic magma consistent with earlier suggestions (Ford *et al.*, 1991) of mafic material at depths of 4 to 5 km.

Some further information concerning the pre-Carboniferous nature of the Iapetus Suture Zone can be derived from a study of Devonian rocks preserved in the northern part of the Dingle Peninsula. The northerly-derived Smerwick Group, which is confined to the north-west of the Dingle Peninsula, is bracketed in age as being younger than the Dingle Group and older than the unconformably overlying Pointagare Group (Graham, 2009b). The most probable age is late Early Devonian (~395 Ma). This group contains numerous coarse conglomeratic levels and the source area is thought to have been controlled by a major fault line, the North Kerry Lineament (Todd, 2000) that is approximately coincident with the projected trace of the Iapetus Suture Zone. This suggests that the Acadian deformation had produced relief associated with some of the structures parallel to the suture as a result of the transpressive deformation. Any positive relief had clearly disappeared by Carboniferous times but it is very possible that the line of this North Kerry Lineament was re-exploited during basin development, as was that of the Dingle Bay Fault Zone further south (Graham, 2009b).

It is clear from the above that the Shannon Trough has a close spatial relationship with the Late Caledonian Iapetus Suture Zone. This zone has a roughly ENE-WSW trend and discontinuities in this zone are likely to have been exploited during subsequent stretching of this area during the formation of the Shannon Trough. The development of the Shannon Trough is discussed more fully by Wignall & Best in the following chapter.

2.3 Variscan Deformation and Metamorphism

Ireland lies near the northern limits of the Variscan orogenic belt but all rocks in the Shannon Trough have been affected by Variscan deformation. Generally, the amount of Variscan deformation decreases northwards but the development of structures depends also on the thickness and nature of the Upper Palaeozoic succession (Graham, 2009c). There is considerable debate regarding the level of detachment of structures developed in Upper Palaeozoic rocks from their basement in the southern part of Ireland, but the consensus view is that such detachment exists, albeit probably in the Lower Palaeozoic basement and that there is a significant northward translation of Upper Palaeozoic successions. The location where the basal detachment becomes zero is predicted to lie in North County Clare (Bresser, 2000), which can thus be regarded as the northern limit of the Variscan fold-thrust belt.

In southern County Clare, there are regional scale box folds indicating 10 to 20% shortening and suggesting a detachment at a depth of about 2 km (Dolan, 1984; Ford *et al.*, 1992; Bresser & Walter, 1999). However, north County Clare shows negligible Variscan shortening and such weak deformation argues against any well-developed deep Variscan detachment in this region. Fig. 2.3.1 shows a simplified structural profile across southern Ireland, including the Shannon Trough, and Fig. 2.3.2. illustrates some of the lateral variation in structure. In much of central County Clare, there is insufficient exposure, or there has been insufficient work, to prove the fold geometries. Where there is reasonable control on the large-scale map pattern of folds, these are seen to be periclinal (Dolan, 1984; Gill, 1979). There is also a marked lithological control on fold geometries, with open, long, wavelength structures in the early Mississippian carbonates (Fig. 2.3.3A) and shorter wavelength folds with numerous monoclines in the overlying clastic successions (Fig. 2.3.3B, C, D and E). This disharmony of structure implies some accommodation of strain within the muddy Clare Shales.

A detailed, balanced cross-section of west County Clare has been produced by Tanner *et al.* (2011; Fig. 2.3.4). Tanner *et al.* (2011) noted that the folds had statistically-horizontal plunges and that the fold geometry was dependent on lithology. The thick sandstone beds of the Ross Sandstone Formation are characterised by box folds, whereas higher stratigraphic units are characterised by parallel folds, with the geometrical differences accommodated by the muddy Gull Island Formation. Tanner *et al.* (2011) used commercial software (MOVE 2008.1) containing an algorithm that enabled de-compaction of the section and substituted porosity data from comparative North Sea sediments, in an attempt to allow for over-pressuring. Application of their retro-deformation technique suggested that the Kilkee and younger cyclothems varied in thickness above pre-existing folds in the lower part of the section, thus indicating syntectonic deformation. Tanner *et al.* (2011) suggested an increase in shortening with increasing age of the rocks (from 1.1% to 7.4%) and also proposed that the most marked shortening occurred between 318.25 and 317.75 Ma at *c.* 30 mm a^{-1}. The numerical ages were based on correlation of the goniatite biostratigraphy with the radiometric time scale of Menning *et al.* (2000). This overall estimate of shortening is consistent with that of Dolan (1984) and Bresser & Walter (1999) of 11.5%, as this included more strongly-deformed areas further south, but it is less than the 15% estimated by Le Gall (1991) for the Shannon Basin. Estimates for the base of the Tullig Cyclothem (Tanner *et al.*, 2011) suggest 2.64% of shortening was due to folds and 4.8% was due to thrusts. A general lateral change from kink geometries in the south to concentric geometries in the north was interpreted as due largely to lithological control.

Cleavage is present throughout much of the Shannon Trough, although it is only weakly developed or even absent on the undeformed 'flats' of the

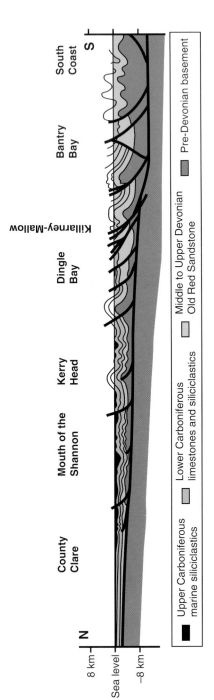

Fig. 2.3.1. Simplified structural profile of southern Ireland (after Bresser & Walter, 1999).

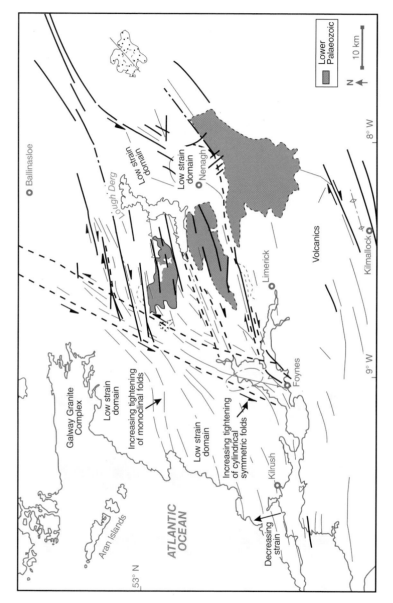

Fig. 2.3.2. Simplified structural map of County Clare (after Coller, 1984). Reproduced with permission from the Geological Society of London.

(A)

(B)

Fig. 2.3.3. A) Typical open fold in Mississippian carbonates of the Burren, Mullaghmore (Photo taken from UTM 498989 m E 5872143 m N; Map 51, 131898 m E, 194683 m N). B) Anticline in Ross Sandstone Formation at Bridges of Ross (UTM 440853 m E 5827157 m N; Map 63, 073110 m E, 150486 m N).

box folds. Cleavage also gradually decreases in intensity and occurrence northwards. The cleavage is a spaced, pressure solution cleavage (Fig. 2.3.3F, G and H) that is generally at a high angle to bedding in the competent limestone and sandstone lithologies with marked fanning

(C)

(D)

Fig. 2.3.3. (*Cont'd*) C) Fanning of cleavage around a fold in Ross Sandstone Formation near Dunmore Head (UTM 439045 m E 5823411 m N; Map 63, 071249 m E, 146763 m N). D) Typical monocline in Ross Sandstone Formation east of Ross (UTM 441899 m E 5827504 m N; Map 63, 074161 m E, 150818 m N).

(E)

(F)

Fig. 2.3.3. (*Cont'd*) E) Small scale parasitic fold on the larger fold shown in D). These small scale folds are relatively uncommon except in the multilayer packages such as shown here. (UTM 442041 m E 5827468 m N; Map 63, 074303 m E, 150780 m N). F) Spaced pressure solution cleavage in the Ross Sandstone Formation seen in both plan and profile, Kilbaha (UTM 442085 m E 5824880 m N; Map 63, 074311 m E, 148190 m N).

(G)

(H)

Fig. 2.3.3. (*Cont'd*) G) Pressure solution cleavage seen as an anastamosing pattern on bedding plane surface, Ross Sandstone Formation, Kilbaha (UTM 442085 m E 5824881 m N; Map 63, 074311 m E, 148191 m N). The pressure solution seams of insoluble residue typically weather preferentially to produce a series of anastamosing cracks in plan view on many exposed bedding surfaces throughout the basin. H) Anastamosing pressure solution seams on bedding plane of Ross Sandstone, Bridges of Ross (UTM 441574 m E 5827293 m N; Map 63, 073833 m E, 150612 m N). The cleavage is slightly oblique to the fold hinge.

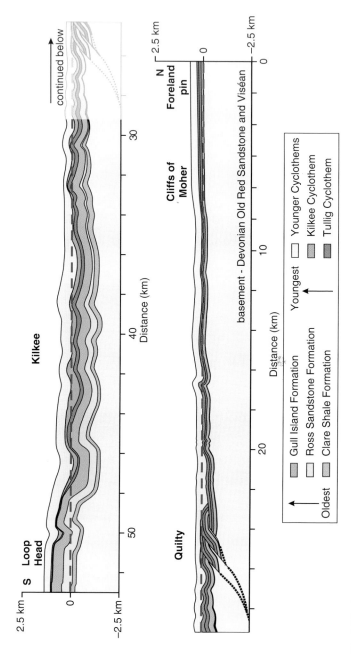

Fig. 2.3.4. Structural profile of the well-exposed west coast sections (from Tanner *et al.*, 2011). Reproduced with permission from the Geological Society of London.

around the folds (Fig. 2.3.3C). The cleavage is locally a penetrative fabric in mudrocks, particularly in fold hinges in the southern part of the basin. The pressure solution cleavage is partly a temperature-controlled fabric that is typical of rocks that have been moderately hot at temperatures <350 to 400° C, at which temperatures crystal plastic processes start to dominate (Rutter, 1976, 1978). Such fabrics can develop in areas of relatively low strain, such as County Clare, if palaeotemperatures were sufficiently high. The pressure solution seams are loci of quartz dissolution with consequent formation of a clay-rich residue (Nenna & Aydin, 2011a).

These pressure solution seams were referred to as 'closing mode structures' by Nenna & Aydin (2011b), who showed these seams are genetically and temporally related to a series of joints and veins ('opening mode structures') that occur almost orthogonal to the strike. In an area of detailed investigation on the Loop Head Peninsula, the pressure solution seams have an average orientation of 075°, whereas the joints and veins have a mean orientation of 170°. These veins form dextral vein arrays trending NNW that are themselves en echelon and define larger-scale ENE trending dextral shears along anticlinal fold hinges, a pattern consistent with overall dextral transpression. The arrays exposed on bedding planes are pinnate fractures (edge effects) at terminations of large individual NNW trending veins. Some of these NNW trending dextral vein arrays have recently been described in detail by Nenna & Aydin (2011b). These NNW trending veins cross-cut earlier bed-parallel veins formed by flexural slip. The veins can be interpreted as synchronous with the late folding history because axial planar cleavage intensifies into the vein arrays and the veins themselves are deformed by the cleavage. Splays of both the joint-vein sets and the pressure solution seams are consistently 35 to 40° clockwise of the main structures and were interpreted as suggesting a clockwise rotation of the principal stresses during deformation (Nenna & Aydin, 2011b). This is entirely consistent with Variscan structures further south (see Graham (2009c) for a summary). The veins contain fibrous quartz crystals that are thin near vein walls and thicken towards the vein centre. The fibres are commonly straight but oblique to the vein walls, suggesting precipitation contemporaneous with strike-slip motion. The geometry of the various structures (folds, cleavage, faults, veins, joints) indicates that they are all related to a single stress system and are probably roughly contemporaneous.

The thermal history of the Shannon Trough has been investigated by examining the maturation of organic matter, conodont alteration indices and the crystallinity of white mica and chlorite. All of these measures are known to be both time and temperature dependent, but it is generally thought that organic maturation can provide useful estimates of peak burial temperatures. The maturation map of Ireland (Clayton *et al.*, 1989) plots calibrated values from vitrinite and conodont data for rocks at the surface (Fig. 2.3.5). It can be seen that much of south and west Ireland

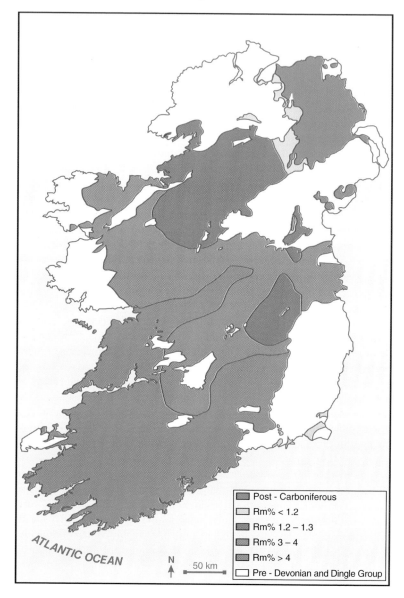

Fig. 2.3.5. Thermal maturation levels in surface exposures of the Devonian and Carboniferous rocks of Ireland (after Clayton *et al.*, 1989).

shows high maturation values that appear to be independent of stratigraphic position and they also show little correlation to the amount of strain. These levels are much greater than might be anticipated from simple overburden estimates (Fitzgerald *et al.*, 1994). In western County Clare, siltstones and coals of Namurian age (310 Ma) yield vitrinite with maximum reflectance (R_{max}) of approximately 6 to 7.5% and conodonts from limestone concretions have conodont alteration indices (CAI) of 5. These data suggest peak burial temperatures of 340 to 370°C. Such temperatures would be consistent with the development of the pressure solution cleavage.

Clayton & Baily (1999) examined oriented coal blocks from the limbs of folds at Croan Rocks and Killard. They showed that the optic fabric of the coals ('vitrinite reflectance indicatrix' *sensu* Levine & Davis, 1984; 1989) is uniaxial negative, with R_{max} developed on planes parallel to bedding and R_{min} consistently perpendicular to bedding. This type of fabric is considered to typify pre-tectonic coalification, in contrast to the strongly biaxial fabrics of syntectonic coalification. This suggests that the folds post-date maximum heating and is difficult to reconcile with the interpretation of synsedimentary folding suggested by Tanner *et al.* (2011).

Where there is evidence of vertical variation of maturation from boreholes, the observed pattern is not that which would be expected from simple overburden-induced heating. Data from the 3 km deep Doonbeg No. 1 Well do not show the expected increase in maturation with depth but rather a more complex profile with an apparent reverse gradient in the upper part of the Dinantian carbonates (Fig. 2.3.6; Fitzgerald *et al.*, 1994; Clayton & Baily, 1999; Goodhue & Clayton, 1999). A similar reverse gradient is seen in the Slievecallan borehole, although here it occurs near the base of the Clare Shale (Goodhue & Clayton, 1999). Simple burial metamorphism seems inadequate to explain these data, and fluid advective heating has been suggested as the most likely mechanism (Fitzgerald *et al.*, 1994; Clayton & Baily, 1999).

Fitzgerald *et al.* (1994) present arguments to suggest that fluids in veins would have been in thermal equilibrium with the enclosing rocks. Analysis of fluid inclusions from these quartz veins permits some estimation of burial temperatures, although these estimates do make several assumptions. The maximum entrapment temperatures (240 to 400°C) overlap the calculated vitrinite reflectance-based burial temperature estimates (340 to 370°C). The modal estimate of entrapment temperatures for the fluid inclusions (220 to 250°C) is significantly lower but is consistent with maximum burial temperatures being achieved before folding (Clayton & Baily, 1999), followed by cooling during folding, uplift and vein formation.

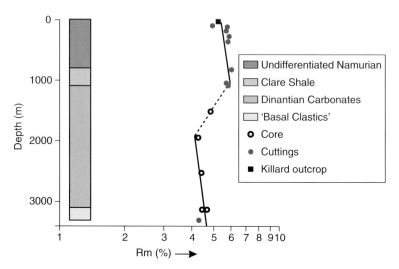

Fig. 2.3.6. Distribution of vitrinite reflectance (R_m) in the Doonbeg No. 1 Well (after Clayton & Baily, 1999).

Thus, the rocks of the Shannon Trough indicate relatively low levels of tectonic deformation but higher than expected burial temperatures, and there is some evidence that the peak temperatures preceded fold development. The overall stress regime was one of dextral transpression. It is difficult to account for the acquired burial temperatures by tectonic overburden and it thus seems probable that some form of advective heating was involved.

Chapter 3
Basin Models

PAUL B. WIGNALL & JIM BEST

3.1 Introduction

The outcrops of the Shannon region provide some of the best Carboniferous sections to be seen in north-west Europe. They are also amongst the most debated. Basin style is an especially contentious issue and several conflicting, and to some extent irreconcilable, models have been proposed. Each model differs in the implications it has for depositional conditions, the orientation of depositional systems and sediment provenance. All of these debates can be pondered when examining the outcrops detailed in this field guide, which provides a chance to see some of the key evidence and, we hope, allow the visitor to come to their own conclusions based on what they have seen. This chapter highlights the key attributes of the various competing depositional models that have been proposed for the Shannon Basin.

The Carboniferous basin has acquired several names in various studies over the years: the Shannon Trough (Sevastopulo, 1981a; Sleeman & Pracht, 1999; Tanner et al., 2011), the Shannon Basin (Strogen, 1988; Sevastopulo, 2001; Martinsen et al., 2003; Pyles, 2007, 2008), the Clare Basin (Deeny, 1982; Croker, 1995), the West Clare Basin (Gill, 1979) and the Western Irish Namurian Basin, usually abbreviated to WINB (Collinson et al., 1991). All are valid (and the Shannon Basin is used here), although Martinsen (et al., 2003) recommended the WINB name be discarded because basin development began prior to the Namurian.

The Shannon Basin is generally recognized to have had an elongate east-north-east to west-south-west orientation aligned along the trend of the modern Shannon Estuary. The basin stretches from central Ireland and

A Field Guide to the Carboniferous Sediments of the Shannon Basin, Western Ireland, First Edition. Edited by James L. Best and Paul B. Wignall.
© 2016 International Association of Sedimentologists.
Published 2016 by John Wiley & Sons, Ltd.
Companion website: www.wiley.com/go/best/shannonbasin

passes westward into the Atlantic Ocean where it is sharply truncated by the north-south trending Porcupine Basin; a much younger structure developed around 150 km west of the County Clare coastline (Strogen, 1988). Overall, the Shannon Basin was approximately 250 to 300 km wide along its roughly east-west axis and at least half that value in a north-south orientation. However, neither its northern or southern terminations are well defined due to a combination of erosion and/or poor inland outcrops. For example, the Namurian stratigraphy of north-west County Cork is nearly identical to that seen in County Clare (Morton, 1965), suggesting that the County Cork strata should be included within the Basin even though this area is usually considered to be south of it. Despite its considerable extent, most studies (and this field guide) concerning the Shannon Basin have focused on the best-exposed sections around the Shannon Estuary and the County Clare coastline, an area that only measures 80 km by 80 km – a small part of the whole Basin.

3.2 Dinantian Basin History

Broad regional subsidence began at the start of the Early Carboniferous (Tournaisian Series; see Figure. 1.7.2) over a broad area, with the result that shallow-water heterolithic deposition became widespread (Strogen, 1988; Sevastopulo & Wyse Jackson, 2009). At this point, the Shannon Basin was not a clearly distinguishable topographic entity but instead was merely part of an extensive shelf sea extending across central southern Ireland. These facies pass southwards into a thick succession of clastic sediments (the Culm Facies) that accumulated in the South Munster Basin, a north-west to south-east trending trough (Price & Todd, 1988).

A topographically distinct Shannon Basin became established in the later Tournaisian as subsidence rates accelerated along the axis of the Shannon Estuary. This coincided, presumably coincidentally, with the onset of highly distinctive, but rather enigmatic, Waulsortian mudbank carbonate deposition. Sometimes forming coalescing, reef-like mounds, these Waulsortian carbonates lack constructing organisms like normal reefs and instead have a high proportion of cement that infills large-stromatactis-like cavities and fissures (Lees, 1961; Lees & Miller, 1995). The Waulsortian facies of southern Ireland are the most extensive strata of this unusual limestone, and the huge thickness in the Shannon Basin, where it approaches 1200 m thick, is by far the thickest development in the world (Shephard-Thorn, 1963; Sleeman & Pracht, 1999). The Waulsortian limestones can be seen in the excursions described by Somerville in Chapter 4.

The Shannon Basin Waulsortian limestones accumulated in the final ~6 Myr of the Tournaisian (Fig. 1.7.2), indicating that accumulation rates were of the order of $200\,mm\,kyr^{-1}$. However, the absence within these limestones of any wave reworking or organisms that lived in the photic zone (e.g. calcareous algae) suggests that water depths were more than 200 m deep, despite the impressive aggradation rates (Lees & Miller, 1995). Clearly subsidence was extremely high in the Shannon Estuary area at this time.

The rapid formation of the Shannon Trough was also associated with volcanic activity. Around Limerick, on the eastern edge of the super-thick Waulsortian mud mound development, the limestones are overlain by a thick succession of earliest Viséan (Chadian Stage) alkali basalts and trachytes with tuffs. Further volcanics were also developed south of Limerick (Shelford, 1967; Strogen, 1988). The later Viséan strata in the Shannon Estuary area record an overall deepening-upwards trend, with the development of fine-grained, thin-bedded, limestones of the Parsonage Formation. The presence of some rudaceous limestones (the Inishtubrid Beds) found in the Corgrig Lodge Formation suggests steep slopes were developed, at least locally, probably associated with small reefs (Sleeman & Pracht, 1999). However, overall a ramp was developed in the Shannon Trough, deepening to the west and north-west from Limerick (Somerville & Strogen, 1992; Sleeman & Pracht, 1999). Ultimately, by the Brigantian Stage, calciturbidites developed at Ballybunnion in the western-most outcrops of the Shannon Basin. In contrast, contemporaneous limestones in northern County Clare remained at shallow depths throughout the Viséan and several emergence horizons with palaeosols are seen (Sleeman & Pracht, 1999; Somerville, Chapter 5). Thus, a northern carbonate platform region appears to have developed adjacent to a deep Shannon trough. After the Waulsortian Limestone development and its exceptional local thickness development in the Shannon Estuary area, lateral variations in thickness of Viséan strata become much more subdued (Fig. 1.7.3).

So, what type of basin was the Shannon Basin in the Dinantian? Several ideas have been suggested. Most agree that the site of greatest subsidence lay along the Shannon Estuary and probably reflects the presence of an underlying Iapetus Suture (e.g. Sevastopulo & Wyse Jackson, 2009; Warr, 2012; Graham, Chapter 2). However, despite this consensus, there is disagreement concerning basin style. Several studies have suggested that the Shannon Basin was the product of extensional rifting in a similar manner to that seen in contemporaneous basins in northern Britain (Deeny, 1982; Haszeldine, 1984; Collinson *et al.*, 1991; Martinsen *et al.*, 2003). However, the sudden but short-lived accelerated pulse of subsidence in the late Tournaisian and the development of volcanoes on the eastern margin of the Shannon Basin are both features atypical of extensional basins. Volcanism produced by extension and decompression melting should see eruptions occurring in the centre of a basin. Instead, these observations better fit a transtensional origin for the Shannon Basin (Haszeldine, 1984; Strogen, 1988; Warr, 2012). A third alternative is that the onset of foreland flexure, as the Gondwanan continent approached the Laurasian continent, may have caused subsidence along the Iapetus Suture (Tanner *et al.*, 2011). In this scenario, the South Munster Basin to the south of the Shannon Estuary became an under-filled foreland basin in the Dinantian, whilst the Shannon Basin was the product of block faulting in the fore-bulge region (Higgs, 2004). A problem for this third model is, once again, the volcanism present at Limerick, which is not anticipated in foreland settings.

3.3 Namurian Basin History

The Late Carboniferous witnessed a fundamental and dramatic change in depositional style. Limestones disappeared and anoxic black shale facies of the Clare Shales Formation developed in the Shannon region in the early Namurian (Hodson, 1954a; Braithwaite, 1993). Identical black shale facies developed over much of Ireland and the British Isles at the same time (Davies *et al.*, 2012). Despite the carbonate-to-clastic transition, the same north-south thickness trends persisted from the Dinantian through to the Namurian. Thus, the Clare Shales are extremely thick, peaking at 232 m at Inishcorker in the centre of Shannon Estuary region (this is the same area of peak thickness of the Waulsortian limestones), and they are much thinner (<20 m) in northern County Clare. The succeeding deep-water clastic sediments of the Shannon Group also show the same thickness trends, although the thickness variation is much reduced in the overlying Central Clare Group.

The similar thickness trends between the Dinantian and Namurian sediments have led some to suggest that the Namurian Shannon Basin inherited the same bathymetry and subsidence regime as seen in the Dinantian (Haszeldine, 1984; Collinson *et al.*, 1991; Sevastopulo & Wyse Jackson, 2009). However, continuous deposition between Viséan and Namurian strata is only seen within the Shannon Estuary area. Elsewhere in the Shannon Basin, and across much of central and southern Ireland generally, there is a major hiatus at the boundary during which uplift, folding and erosion occurred (Hodson, 1959). Thus, in the Doonbeg No. 1 borehole in central County Clare, the Dinantian limestones dip at a different angle to the unconformably overlying Namurian clastics (Croker, 1995). The significance attributed to this tectonic episode varies considerably from model to model but, before discussing the merits of the various alternatives, we first need to present them.

3.4 The CoMa Model

In a highly influential paper, Collinson *et al.* (1991) presented an inclusive model for the Shannon Basin that built on the earlier observations of Hodson & Lewarne (1961). The model was subsequently challenged (see below) and vigorously defended (e.g. Martinsen & Collinson, 2002; Martinsen *et al.*, 2003). Here we term it the CoMa model after its two principal proponents: Collinson and Martinsen.

In essence, the CoMa model envisages business as usual between the Dinantian and Namurian phases of basin development. Thus, continued extensional rifting along the Iapetus Suture/Shannon Estuary trend caused a narrow, confined basin "elongated along an ENE-WSW line" (Martinsen *et al.*, 2003, p. 791) to persist, with shallower conditions lying to the north-east in northern County Clare. After the initial phase of fill by thick, black shales, the sandstones of the Ross Sandstone Formation turbidite system arrived from a source area to the north-west and, having been deflected along the axis of the basin, aligned along the Shannon Estuary, then prograded along the basin floor. The majority of palaeocurrents in the Ross Sandstone Formation consequently record "flow to the north-east" (Fig. 3.4.1A; Collinson *et al.*, 1991, p. 235).

The biostratigraphic framework established by Hodson & Lewarne (1961) shows that the Ross Sandstone Formation passes north-eastward into a thin development of the Clare Shales. The black shales in this region are considered to have accumulated on a basin margin in a "shallow, marginal area [where] agitation kept fines in suspension" (Collinson *et al.*, 1991, p. 237). Slumps are abundant in the upper Ross Sandstone Formation and they become an even more common style of soft-sediment deformation in the overlying siltstone-dominated Gull Island Formation (Strachan & Pyles, Chapter 8). The down-slope movement direction of the majority of the slumps is interpreted to be to the south-east (Martinsen & Bakken, 1990) and thus is at 90° degrees to the progradation direction of the turbidites. In the CoMa model, this is interpreted to represent collapse of a slope prograding to the south-east whilst the turbidites are flowing along the foot of the slope to the north-east (Fig. 3.4.1B). The slope system of the Gull Island Formation was fed by shelf-edge deltas and this unit is thus overlain by a series of deltaic units, beginning with the Tullig Cyclothem (Fig. 3.4.1C). Like the Gull Island Formation, the down-slope direction, in this case palaeocurrent indicators in channel sandstones, are said to suggest "flow dominantly to the south-east" (Collinson *et al.*, 1991, p. 236), in accord with Pulham's (1989) study of the cyclothem sediments of the Central Clare Group.

In the original manifestation of the CoMa model, there was some discrepancy in the suggested orientation of the deep-water trough in the

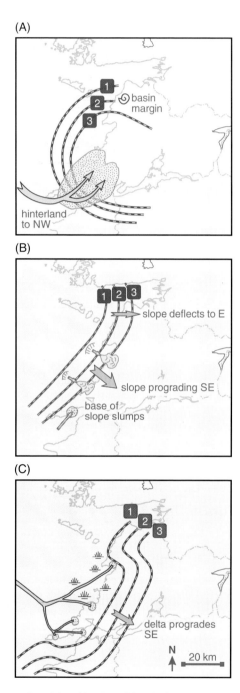

Fig. 3.4.1. Reconstruction of depositional conditions during Namurian infill of the Shannon Basin, based on the CoMa model (after Collinson *et al.*, 1991). Contours, labelled 1 to 3 from shallow to deep, show slope orientations, whilst the arrows show dominant sediment dispersal directions. A) the Ross Sandstone Formation turbidite system; B) the Gull Island Formation slope system, with slumps and slides; C) the Tullig Cyclothem deltaic system.

Shannon Basin (Wignall & Best, 2000). Although Collinson and colleagues considered the basin to be developed above the Iapetus Suture, which runs east to west along the Shannon Estuary, the Ross Sandstone Formation turbidites were stated to flow along this axis to the north-east in the direction of the northern basin margin in the CoMa model.

In a re-validation of the CoMa model, Martinsen *et al.* (2003) modified it somewhat. They suggested that the early phase of fill, up to a level in the mid Gull Island Formation, was primarily an aggradational infill of a deep, narrow, trough centred at Loop Head in south-western most County Clare. At this point, the topographic low had been infilled, thus allowing the turbidity currents in the upper Gull Island Formation to spill-over and prograde northwards. Thus, the "basin high [in northern County Clare] received turbiditic sediments once the central depression of the basin had been filled and margins onlapped" (Martinsen *et al.*, 2003, p. 802). Nonetheless, the principal progradation of the Gull Island Formation slope remained to the south-east in this modified version of the CoMa model. Intriguingly, this modification has turbidity currents flowing along a slope up onto a basin high. The healed slope accommodation is then downlapped by the delta slope of the Tullig depositional system prograding south-east (Martinsen *et al.*, 2003).

3.5 The Clockwise Minibasin Model

Pyles (2007, 2008) has proposed a similar basinal model to the modified CoMa model in which the Ross Sandstone Formation is once again envisaged to be a ponded submarine fan (Fig. 3.5.1; Pyles, 2007, 2008; Pyles & Jeanette, 2009). However, the model of Pyles differs in significant details. The key difference is that the Shannon Basin is no longer considered to be an elongate trough but rather a minibasin of circular outline analogous to "structurally confined, salt-withdrawal basins [such as those found today] on the northern Gulf of Mexico continental slope" (Pyles, 2008, p. 568). Palaeocurrent data from the Ross Sandstone Formation show a dominant north-eastwards flow but with considerable spread from north-west to south-east (Pyles, 2007; Pyles & Jeanette, 2009). This is interpreted to record a radially-dispersive, lobe-dominated, aggradational turbidite system (Fig. 3.5.1A). The Minibasin model also differs in its interpretation of the Clare Shales/Ross Sandstone Formation contact in the Shannon Estuary. In the CoMa model this is seen to be a downlap relationship, whereas in the Minibasin model the Ross Sandstone Formation turbidites onlap a lateral slope of Clare Shale (Pyles & Jeanette, 2009). It should be noted though, that this purported slope of organic-rich muds lacks any evidence for slope failure.

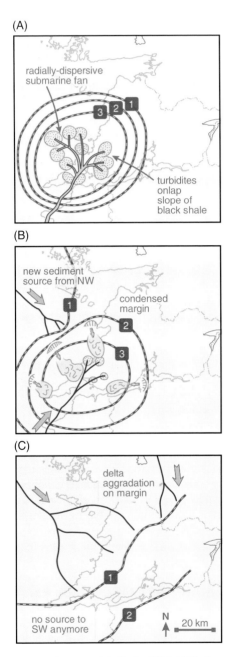

Fig. 3.5.1. The Clockwise Minibasin Model of Pyles (2007, 2008). Contours, labelled 1 to 3 from shallow to deep, show slope orientations, whilst the large grey arrows show dominant sediment dispersal directions. A) a ponded submarine fan model for the Ross Sandstone Formation; B) the Gull Island Formation slope system, with slumps and slides. C) the Tullig Cyclothem deltaic system.

The Minibasin model also differs from the CoMa model in the inferred provenance of the sediment supply. The CoMa model considers the hinterland to be a Caledonian Province to the north-west of the Shannon Basin whereas Pyles (2008) suggests initial sediment supply was from the south-south-west. This source was then joined by westerly-supplied sediment during deposition of the upper Ross Sandstone Formation. Later, in the Gull Island Formation, only the western source was active and finally the sediment source was proposed to change again in the Central Clare Group to a north to north-west provenance. Thus, there is "a clockwise change in the sediment transport direction through time associated with the filling of the basin" (Pyles & Jeanette, 2009, p. 1975–6; Pyles & Strachan, Chapter 7) – hence a clockwise minibasin model (Fig. 3.5.1).

3.6 The WiBe Model

A fundamentally different basin model was proposed by Wignall & Best (2000, 2002, 2004) for the Namurian infill of the Shannon Basin, here abbreviated to the WiBe model, in which basin style is envisaged to change fundamentally between the Dinantian and the Namurian. In this alternative model, Namurian clastic sediment was proposed to be derived from the south and south-west with the basin being infilled by depositional systems that were building, down-dip, to the north-east (Fig. 3.6.1). The Ross Sandstone Formation turbidite system is viewed as unconfined and showing north-eastward downlap onto the deeper water black shales of the Clare Shale Formation (Fig. 3.6.1A). A plot of the extent of turbidite deposition accords with this north-eastward progradation (Wignall & Best, 2000, fig. 7; Pyles & Strachan, Chapter 7). In contrast, the turbidites onlap a distal basin margin in the modified CoMa model (Martinsen *et al.*, 2003). The Gull Island Formation remains a slope system in the WiBe model but with the primary down-dip directions, based on evidence from growth faults and slumps, to the north-east (Fig. 3.6.1B). The deltaic systems of the Tullig Cyclothem continue the theme of north-easterly progradation (Fig. 3.6.1C).

This WiBe model interpretation of Tullig Cyclothem migration is in accord with the earlier work of Rider (1974) who also suggested that the Tullig deltas show progradation to the north-east. This is at variance to the proposed south-eastward direction proposed by Pulham (1989). However, Pulham's own data clearly shows that nearly all the palaeocurrents in the Tullig Cyclothem record flow to the north-east (Pulham, 1989, fig. 14), but he suggested that this was because the measurements come from a distributary channel developed at a high angle (~90°) to the main progradation direction (Pulham, 1989, fig. 19).

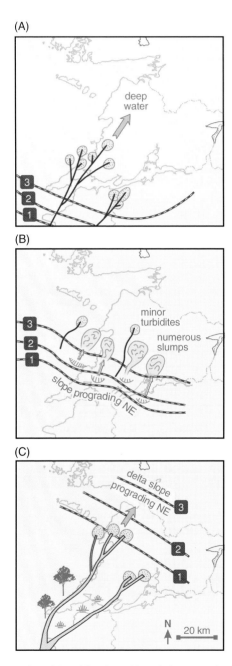

Fig. 3.6.1. Reconstruction of depositional conditions during Namurian infill of the Shannon Basin, based on the WiBe model (after Wignall & Best, 2000, 2002). Contours, labelled 1 to 3 from shallow to deep, show slope orientations, whilst the large grey arrows show dominant sediment dispersal directions. A) the Ross Sandstone Formation turbidite system; B) the Gull Island Formation slope system, with slumps and slides; C) the Tullig Cyclothem deltaic system.

The CoMa and Clockwise Minibasin models both consider the area of thickest Shannon Group sediments to record the deepest water parts of an extensional basin where infill was predominantly aggradational. In contrast, the WiBe model views the infill as essentially progradational, albeit with thickening in the Shannon Estuary region due to rapid subsidence along the Iapetus Suture. This has been attributed to flexural loading (Higgs, 2004) and suggests that the Namurian phase of the Shannon Basin saw a foreland basin develop. Thus the change in structural style between the Viséan and Namurian coincides with a regional phase of uplift and moderate folding, probably of a transient fore-bulge, especially to the south of the Shannon Basin (Strogen, 1988). The tectonic readjustment from a transtensional to compressional regime explains the change of basin configuration from one that deepened from east to west in the Viséan (Sleeman & Pracht, 1999) to one that deepened from south-west to north-east in the Namurian.

The development of a foreland basin in the Namurian also provides an explanation for the shallow- to deep- water transition seen in northern County Clare in the WiBe model (Wignall & Best, 2002). In this region, the Brigantian platform carbonates are succeeded by Clare Shales but are separated from them by a thin development of phosphatic pebbles (the St. Brendan's Well Phosphate Bed – see Chapter 6; Braithwaite, 1993). These phosphates rest on a truncation surface that records a hiatus spanning the late Brigantian to Chokerian stages – an interval of ~10 Myr (Wignall & Best, 2000; Barham *et al.*, 2014). The region was probably emergent during some of this time, as evidenced by rare, reworked limestone pebbles that occur in the Phosphate Bed; but the renewed onset of subsidence was not compensated by sediment infill until the onset of black shale accumulation. Even if subsidence only spanned half the duration of the hiatus and with modest rates of $0.5\,mm\,kyr^{-1}$, water depths of ~200 m could have been achieved in northern County Clare prior to black shale deposition. Subsidence rates were undoubtedly higher in southern County Clare, but here subsidence was compensated for by the accumulation of hundreds of metres of black shales and turbidites.

Support for a foreland basin model has also come from the kinematic retro-modelling of fold and thrust structures seen in the Namurian sections of the County Clare coastline. Previously considered to be a product of Variscan compression in the latest Carboniferous, Tanner *et al.* (2011) have shown syn-sedimentary compression and folding had begun by at least the base of the Tullig Cyclothem. Tanner *et al.* (2011) therefore favour a foreland basin model, but they note that subsidence rates in the Shannon Estuary region were exceptionally high for such a basin style. This was presumably due to the ready subsidence on the weak Iapetus Suture.

3.7 Debate on Basin Development

All three alternative models outlined above raise key questions that can be pondered whilst visiting the Namurian sections detailed in this field guide. Pertinent areas for debate include:

1 What was the nature of the northern County Clare sections? In particular, are the Namurian strata in this region deeper or shallower water than the equivalent strata exposed around the Shannon Estuary? The interpretation of the Clare Shale in northern County Clare is especially contentious. In the WiBe model, the black shales are deep water, low energy deposits accumulated in a distal basinal setting, whilst the other models suggest high energy, winnowed accumulation on a basin margin.

2 What direction was downslope? There are an unrivalled variety and abundance of slope collapse indicators to be seen in the Shannon Basin. They include spectacular slumps, debris flows and numerous growth faults that provide the opportunity to assess their movement direction (Rider, 1974; Gill, 1979; Martinsen *et al.*, 2003; Wignall & Best, 2000, 2002, 2004). The CoMa model envisages slope progradation to the south-east and its proposers have supported this assertion with several studies of large slumps in the Gull Island Formation (Martinsen & Bakken, 1990). However, detailed re-evaluation of several large examples has re-interpreted movement directions to show a narrow range of movement directions spanning east-north-east to north-east (Strachan & Alsop, 2006), in accord with the WiBe model. Whilst difficult to interpret, some major slope collapse features, such as the slumps and growth faults seen at the Point of Relief (see Strachan & Pyles, Chapter 7), unequivocally record down slope collapse to the west – a direction that is impossible to reconcile with the CoMa model.

3 What was the sediment provenance? Whilst difficult to judge in the field, when visiting the Namurian outcrops in County Clare it is worth considering whether there is any lithological change to be seen amongst the various sandstone units. The Clockwise Minibasin model in particular, invokes highly variable provenance at different horizons, whereas the WiBe model proposes a consistent source to the south-west and the CoMa model has sediments with a north-westerly provenance.

In the only provenance study on the basin, Pointon *et al.* (2012) have dated detrital zircons from throughout the Namurian infill and found a range of age groupings. Of these, the most diagnostic are a 500 to 700 Ma group that indicates a source area that was sampling "Cadomian-Avalonian orogenic activity within terranes to the south" (Pointon *et al.*, 2012, p. 77). As there is no available source for these zircons north of the Shannon Basin, their presence unequivocally indicates that Gondwanan sediment was reaching County Clare with the Ross Sandstone Formation turbidity currents – clear and unequivocal support for foreland basin development early in the Namurian.

4 What was the predominant progradation direction of the deltaic systems of the Central Clare Group? There is a clear difference between the CoMa and WiBe models, a conflict that is also mirrored in the different views of Pulham (1989) and Rider (1974). Abundant palaeocurrent evidence is available to test these two models and fit these in with reconstructions of the depositional systems, including the changing size of fluvial-deltaic channels in a proximal-distal traverse and the orientation of any palaeo-valleys within these sediments. Furthermore, there is good evidence that the progradation direction of the deltaic systems changed between the Tullig Cyclothem and the overlying Kilkee and Doonlicky Cyclothems (Rider, 1974; Gill, 1979; Pulham, 1989). Thus, the Tullig palaeocurrents generally record flow to the north and north east, whereas flow directions swing clockwise in the younger cyclothems and the Doonlicky flow directions are to the south east. Despite this trend, provenance data suggest there is little change. The study of Pointon *et al.* (2012) showed that all cyclothem sediments possess a major component of southerly-derived sediment. It is worthy of consideration as to why the predominant flow direction may have changed, how this influenced the resulting depositional geometry and if the source of sediment also changed or whether sediment reworking was a major contributor to the upper cyclothems.

5 What are the various scales of autocyclic and allocyclic control on sedimentation within these Namurian sediments and how are base-level changes recorded in the rock record within these sediments? As one example, although the distinct marine bands show periods of marine incursion, other evidence of marine influence, in the form of fully bioturbated sediments, is highly variable especially in the Central Clare Group. How are base-level changes manifested within both the deeper water sediments of the Clare Shales, Ross Sandstone Formation and Gull Island Formation and what role do autocyclic and allocyclic scour play in the marked erosion surfaces that can be seen within the cyclothems? For instance, the distinct erosion surfaces that are present at the base of some of the fluvial channel sandbodies have been proposed to represent type-1 sequence boundaries (Davies & Elliott, 1996; Hampson *et al.*, 1997) as well as showing the influence of autocyclic scour (Wignall & Best, 2000). It is clear that the nature of the facies above and below these boundaries, and their lateral variation, hold the key to this debate.

Chapter 4
Lower Carboniferous of the Shannon Basin Region

IAN D. SOMERVILLE

4.1 Introduction

The Shannon Estuary (Fig. 4.1.1), separating County Clare to the north from County Limerick to the south, exposes an almost complete Lower Carboniferous (Mississippian) succession of mostly limestones (Chapter 4.2) ranging in age from lower Tournaisian to upper Viséan, succeeded by late Mississippian (Serpukhovian) shales, followed by Upper Carboniferous (Pennsylvanian) shales and sandstones. Lower Carboniferous rocks are also exposed in the Mouth of the Shannon on the north Kerry coast at Ballybunnion (Chapter 4.3).

The lithostratigraphy of the Tournaisian and Viséan rocks of the Shannon Estuary was first described by Shephard-Thorn (1963), based on field outcrops, and subsequently modified using borehole data (Somerville & Jones, 1985; Somerville et al., 1992; Sleeman & Pracht, 1999) (Fig. 4.1.2). Douglas (1909) first applied the Vaughan (1905) divisions of the Lower Carboniferous of the Bristol area, SW England, to the shallow-water platform limestone sequence in County Clare, north of the Shannon Estuary (see Chapter 5.2). These zones were then later applied by Shephard-Thorn (1963) to a similar but much thicker (up to 2 km) succession on the south side of the Shannon Estuary, comprising both shallow-water and deep-water limestone facies (Fig. 4.1.2). Douglas (1909) was also the first geologist to apply the term 'Waulsortian' to the exposures of massive, pale grey, fine-grained limestones of mud-mound facies in the Clare-Limerick region of Ireland, as used in Belgium.

A Field Guide to the Carboniferous Sediments of the Shannon Basin, Western Ireland, First Edition. Edited by James L. Best and Paul B. Wignall.
© 2016 International Association of Sedimentologists.
Published 2016 by John Wiley & Sons, Ltd.
Companion website: www.wiley.com/go/best/shannonbasin

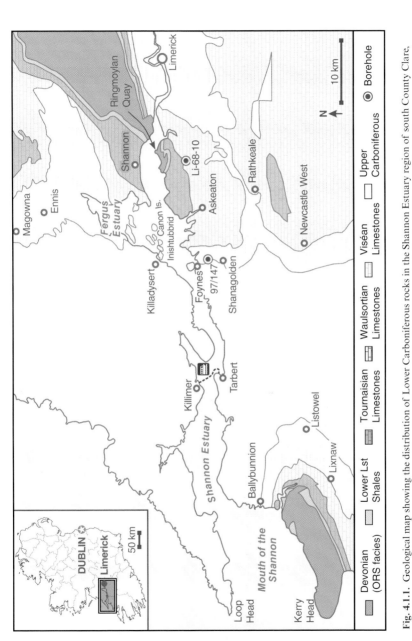

Fig. 4.1.1. Geological map showing the distribution of Lower Carboniferous rocks in the Shannon Estuary region of south County Clare, west County Limerick and north County Clare (from Sleeman & Pracht, 1999; copyright Geological Survey of Ireland).

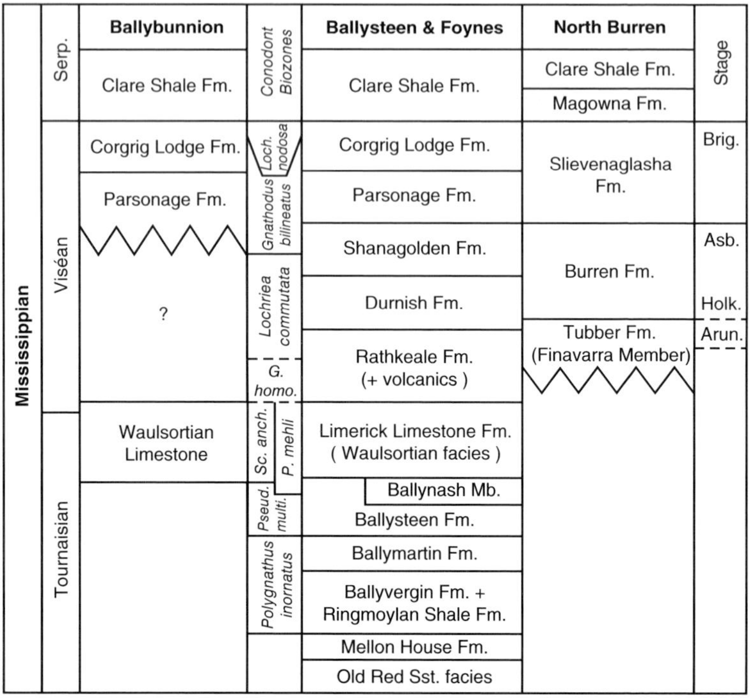

Stage	Ballybunnion	Conodont Biozones	Ballysteen & Foynes	North Burren	Stage
Mississippian / Serp.	Clare Shale Fm.	*Gnathodus bilineatus / Loch. nodosa*	Clare Shale Fm.	Clare Shale Fm.	
				Magowna Fm.	
Viséan	Corgrig Lodge Fm.		Corgrig Lodge Fm.	Slievenaglasha Fm.	Brig.
	Parsonage Fm.		Parsonage Fm.		
	?	*Lochriea commutata*	Shanagolden Fm.	Burren Fm.	Asb.
			Durnish Fm.		Holk.
		G. homo.	Rathkeale Fm. (+ volcanics)	Tubber Fm. (Finavarra Member)	Arun.
Tournaisian	Waulsortian Limestone	*Sc. anch.* / *P. mehli*	Limerick Limestone Fm. (Waulsortian facies)		
		Pseud. multi.	Ballynash Mb.		
			Ballysteen Fm.		
		Polygnathus inornatus	Ballymartin Fm.		
			Ballyvergin Fm. + Ringmoylan Shale Fm.		
			Mellon House Fm.		
			Old Red Sst. facies		

Fig. 4.1.2. A summary of the Lower Carboniferous (Mississippian) stratigraphy of the region (adapted from Shephard-Thorn (1963), Somerville & Jones (1985), Sleeman & Pracht (1999; copyright Geological Society of Ireland) and Gallagher *et al.* (2006)). Arun. = Arundian; Holk. = Holkerian; Asb. = Asbian; Brig. = Brigantian; Serp. = Serpukhovian; *G. = Gnathodus*; *P. = Polygnathus*; *Pseud. multi. = Pseudopolygnathus mulitistiatus*; *Loch. = Lochriea*; *Sc. anch. = Scaliognathus anchoralis*.

4.2 Tournaisian and Viséan Limestones from the Shannon Estuary, Co. Limerick

This one-day excursion provides an opportunity to traverse the entire lower Tournaisian to Upper Viséan succession, from the basal siliciclastics of the 'Old Red Sandstone facies' in the Mellon Point Inlier to the Namurian (Serpukhovian and Pennsylvanian) siliciclastics of the Clare Shale Formation at Foynes (Fig. 4.1.2). The excursion will examine the succession in stratigraphic order, commencing with the oldest rocks of the Mellon House Formation exposed west of Ringmoylan Quay (Stop 1). After this the suggested route travels west down the Shannon Estuary, thereby passing stratigraphically upwards through the Tournaisian and Viséan limestones (Stops 2 to 6).

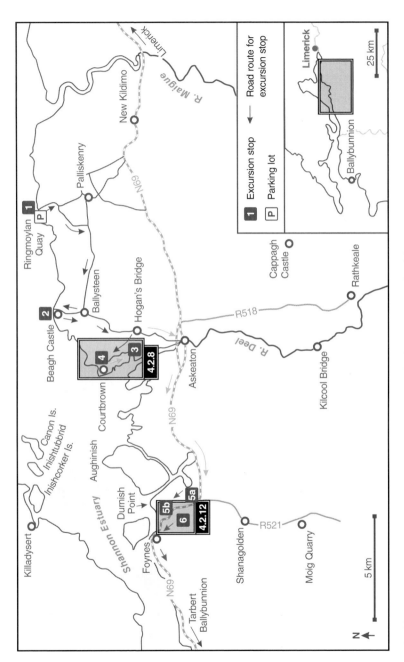

Fig. 4.2.1. Location of excursion stops in the Shannon estuary region in west County Limerick.

Warning: it is recommended that the excursion be undertaken on a falling tide as many of the sections lie below the high water mark. Also, be careful of the sticky estuarine muds of the Shannon Estuary when traversing the foreshore.

Directions to Stop 1

The trip begins at Ringmoylan Quay on the Shannon Estuary (Fig. 4.2.1 and 4.2.2), opposite Shannon airport. Access to Ringmoylan Quay from Limerick city to the east is via Mungret, Clarina, Ferrybridge and New Kildimo (N69). Continue westwards on the N69 for 2 km, then take the minor road on the right and head north for 5 km, passing through Pallaskenry village down to the coast, where car parking is possible at the quay.

Stop 1: Ringmoylan Quay (UTM 508239 m E, 5835456 m N; Map 65, 140641 m E, 157853 m N)

Proceed west from Ringmoylan Quay, built mainly of local Devonian 'Old Red Sandstone', for over 1 km around the first headland and bay, to the west side of the second main headland (UTM 507508 m E, 5834793 m N; Map 65, 139900 m E, 157200 m N), passing cliffs made of boulder clay to the first rock outcrop. The oldest beds exposed here form a small, one-metre high scarp, and dip gently to the south-west (Fig. 4.2.3). They are grey, finely laminated, medium-grained sandstones that belong to the upper part of the Mellon House Formation, which is of lower to middle Tournaisian age (Somerville & Jones, 1985; Higgs *et al.*, 1988). The contact with the underlying 'Old Red Sandstone' interval is not exposed but red sandstones in the core of the anticlinal inlier are occasionally exposed around the headland at Mellon Point to the east (Fig. 4.2.2). The Mellon House Formation beds are mostly siliciclastic and comprise laminated siltstones, fine-grained grey sandstones, dark grey calcareous shales, green silty mudstones and rare thin silty limestones. A complete section through the formation (37.4 m thick) is present in the Pallaskenry Borehole (Li-68-10), drilled 4.5 km to the south (Fig. 4.2.2; Somerville & Jones, 1985). The limestones in the formation contain oncoids of *Ortonella*, conodonts and foraminiferans, indicating a marine influence. In the Mellon House Formation in the Ballyvergin Borehole (BV-14) in County Clare, Higgs *et al.* (1988) recorded spores of the HD and BP palynozones. The top of the latter zone lies within the upper part of the *Polygnathus spicatus* conodont Biozone.

At this first outcrop west of Ringmoylan Quay, the brownish weathering grey-laminated sandstones on the foreshore are slightly calcareous. The finer-grained sandstone and siltstone beds show ripple marks (Fig. 4.2.4) and desiccation cracks (Fig. 4.2.5). In the type section, a quarry 180 m

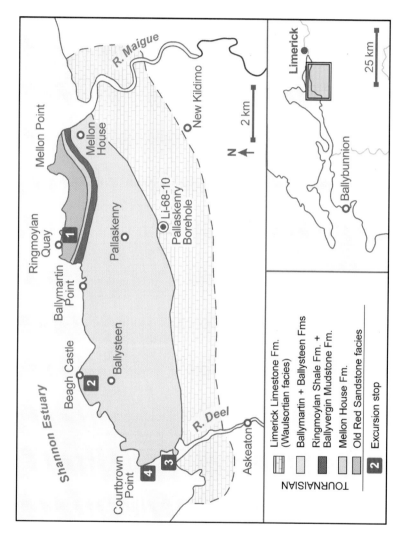

Fig. 4.2.2. Geological map of the Ringmoylan-Ballysteen area of the Shannon Estuary (adapted from Shephard-Thorn (1963) and Somerville & Jones (1985)).

Fig. 4.2.3. Sandstones of the Mellon House Formation, Ringmoylan Quay (Stop 1). Hammer is 40 cm long.

Fig. 4.2.4. Ripples in fine-grained sandstones of the Mellon House Formation, Ringmoylan Quay (Stop 1). Hammer is 40 cm long.

NNW of Mellon House (UTM 512101 m E, 5835156 m N; Map 65, 144500 m E, 157500 m N), Shephard-Thorn (1963) recorded *Chondrites* burrows and bivalves (*Modiola*). The Mellon House Formation represents mostly an intertidal and shallow subtidal, fairly low-energy shelf facies. The top of the formation is marked by the presence of the first shelly

Fig. 4.2.5. Desiccation cracks in grey siltstones of the Mellon House Formation, Ringmoylan Quay (Stop 1). Hammer is 40cm long.

limestone bed, defining the base of the succeeding Ringmoylan Shale Formation. In the Pallaskenry Borehole, this latter formation is 30.8 m thick (Somerville & Jones, 1985) and is composed of dark grey to black calcareous shales with thin bands of bioclastic limestone that are rich in brachiopods, crinoids, bryozoans and solitary zaphrentid corals. In drill-core, some of these grainstone beds are stained pink due to the presence of hematite. The shales and limestones of the Ringmoylan Shale Formation are best exposed at Stop 1 on the foreshore at low tide. The lowest limestone bed, *c*. 10 cm thick, is a dark grey, coarse-grained, crinoidal-brachiopod grainstone. It marks the beginning of a major transgression, which moved from south to north across the Shannon Basin region in the early Carboniferous (Sevastopulo & Wyse Jackson, 2009). Conodonts recovered from this formation belong to the *Polygnathus inornatus* Biozone (Somerville & Jones, 1985) and Austin *et al.* (1970) have recorded shallow-water taxa including *Patrognathus variabilis*.

Walking farther to the south around the bay, which is protected by a wall and high earth bank, is the discontinuous passage into the Ballymartin Formation exposed north of Ballymartin Farm (UTM 507310 m E, 5834590 m N; Map 65, 139699 m E, 157000 m N: Stop 1 Ballymartin Farm) and at Ballymartin Point (UTM 506812 m E, 5834483 m N; Map 65, 139200 m E, 156900 m N), which shows an increase in the proportion of limestone to shale (approximately 50:50). This formation represents the lower, more argillaceous unit of the Ballysteen Group that is succeeded by the limestone-dominated Ballysteen Formation

(Somerville *et al.*, 2011; Fig. 4.1.2). The limestones and shales of the Ballymartin Formation have abundant corals, especially tabulate colonies of *Michelinia* and *Syringopora*, and brachiopods. In the Pallaskenry Borehole, the Ballymartin Formation is 45.6 m thick (Somerville & Jones, 1985) and rests on a distinctive green-grey non-calcareous mudstone layer 5.4 m thick (Ballyvergin Formation), which is not exposed on the coast. However, this mudstone unit is present in most boreholes in the Shannon Basin region, including the type section at Ballyvergin in County Clare (Hudson & Sevastopulo, 1966; Clayton *et al.*, 1980; Philcox, 1984), 30 km to the north, across the Shannon Estuary. The Ballyvergin Formation is unique because of its assemblage of spores and acritarchs of Lower Palaeozoic age, probably derived from reworking of older rocks to the north (Clayton *et al.*, 1980). The mudstone is sparsely fossiliferous with rare bands of brachiopods. It represents a virtual isochronous surface, although it is not volcanically derived (Sevastopulo & Wyse Jackson, 2009).

Transfer to Stop 2

Return to Ringmoylan Quay and follow the minor road south from the coast for 2.5 km to the village of Pallaskenry and then head west for 6 km to Ballysteen village (Fig. 4.2.1). Turn right in the village and proceed north for 1.5 km to the coast at Beagh Castle.

Stop 2: Beagh Castle (UTM 503413 m E, 5834536 m N;
Map 64, 135800 m E, 157000 m N)

The rocks exposed on the coast west of Beagh Castle belong to the Ballysteen Formation and are dominated by argillaceous bioclastic limestones (Fig. 4.2.2). The formation is 190.5 m thick in the Pallaskenry Borehole (Somerville & Jones 1985), with the upper part containing fine-grained limestones and a higher proportion of interbedded shale. Access to the foreshore and cliff section on which Beagh Castle is built is under the arch, immediately west of the quay. Here, the well-bedded and nodular limestones are horizontal (Fig. 4.2.6) and are from the middle part of the formation. The argillaceous skeletal wackestones are interbedded with thin shale seams and are richly fossiliferous with crinoids, bryozoans, brachiopods and solitary zaphrentid and caniniid corals, including large *Caninophyllum*. The most conspicuous fossils though, are the cerioid colonies of the tabulate coral *Michelina favosa*. Many of the colonies are inverted (Fig. 4.2.7), but some are on their side or the right way up, in growth position. The base of the Ballysteen Formation in the Pallaskenry Borehole marks the first appearance of the zonal conodont taxon *Polygnathus multistriatus* (Somerville & Jones, 1985), which has also been recorded from Beagh

Fig. 4.2.6. Horizontally-bedded bioclastic limestones of the Ballysteen Formation, Beagh Castle (Stop 2).

Fig. 4.2.7. Silicified inverted cerioid colony of the tabulate coral *Michelinia favosa* (arrowed) in wavy-bedded argillaceous skeletal wackestone with thin shale seams, Ballysteen Formation, Beagh Castle (Stop 2). Hammer is 40 cm long.

Castle (Austin *et al.*, 1970). The fossils in the limestones often show partial silicification. The Ballysteen Formation limestones formed in a mid-ramp setting and represent part of an upward-deepening trend (Somerville & Jones, 1985; Strogen *et al.*, 1996). This interpretation differs from that of Shephard-Thorn (1963), who considered that the influx of shale in the upper part of the Ballysteen Formation was linked to a shallowing-upward trend. Nevertheless, the presence of overturned colonies implies some degree of bottom current activity.

Transfer to Stop 3

From Beagh Castle, follow the minor road south from the coast for 4.5 km, passing through the village of Ballysteen to Hogan's Bridge (Fig. 4.2.1). Turn right here and proceed along the road westwards for about 1 km to Goleen Bridge (UTM 501239 m E, 5830564 m N; Map 64, 133570 m E, 153057 m N) and park vehicles (Fig. 4.2.8). Pass through the gate and proceed along the track heading north-west to Ballynash Farm (Mantlehill) and then head down to the coast nearby on the eastern bank of the River Deel Estuary (Fig. 4.2.8). From here you can then proceed on foot along the coast north-westward for *c*. 1 km to Courtbrown Point (Stop 4). Alternatively, return to Goleen Bridge and take the road north for 1 km where it ends at Courtbrown Farm. Following this, head across the fields in a NW direction for 500 m to Courtbrown Point at the northern end of the section (Stop 4A), where there are splendid views across the River Shannon to the islands in the Fergus Estuary to the north-west. From here, you can then proceed south along the coast towards Ballynash.

Stop 3: Ballynash (UTM 500601 m E, 5830807 m N; Map 64, 132935 m E, 153308 m N)

The top of the Ballysteen Formation is marked by a change of facies to cherty nodular limestones with shales, known as the Ballynash Member, named after the coastal section at Ballynash (Figs 4.2.2 and 4.2.8). In the Pallaskenry Borehole, this member is 9.4 m thick (Somerville & Jones, 1985). These cherty beds are well-exposed in a quarry at Ballynash, where some 4 m of thinly-bedded silicified nodular limestones rich in crinoids with thin interbedded shales dip gently to the west. The Ballynash Member has yielded the important conodont taxa *Polygnathus mehli*, *Pseudopolygnathus pinnatus* and *Dollymae bouckaerti* indicating the *P. mehli* Biozone (Somerville & Jones, 1985; Jones & Somerville, 1996).

Just around Illaunlea headland to the north (UTM 500381 m E, 5831198 m N; Map 64, 132721 m E, 153703 m N) is an exposed contact with the overlying pale grey, fine-grained limestones of the Waulsortian

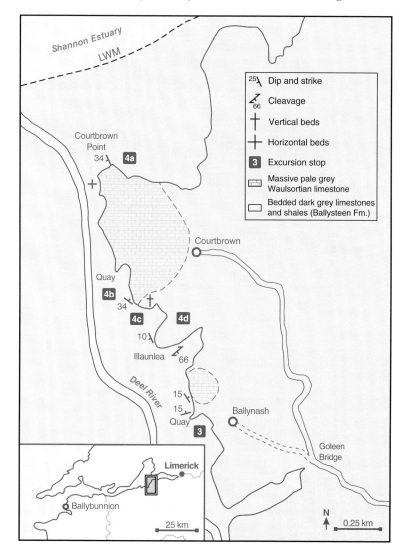

Fig. 4.2.8. Location of excursion stops in the Courtbrown-Ballynash area.

Limerick Limestone Formation (Fig. 4.2.9). The transitional beds are composed of thin-bedded, pale grey, fine-grained nodular crinoidal wacke-stone, very similar in composition to the massive thick beds of limestones of the overlying Waulsortian facies (see Devuyst & Lees, 2001, for a detailed description of this transition).

Fig. 4.2.9. Contact of the thin-bedded, dark grey argillaceous bioclastic limestones of the Ballynash Member (Ballysteen Formation) (at base of hammer) and massive pale grey Waulsortian limestones, Ballynash (Stop 3). Hammer is 40 cm long.

Stop 4: Courtbrown Point (UTM 500051 m E, 5832065 m N; Map 64, 132402 m E, 154575 m N)

At Courtbrown Point (Stop 4A; Fig. 4.2.8), the Waulsortian facies is well-bedded and dips 34° to the SW, towards the alumina plant on Aughinish Island. However, south of the Point, the beds become horizontal. Some 500 m to the south (UTM 500155 m E, 5831592 m N; Map 64, 132500 m E, 154100 m N), the Waulsortian facies contains thin shale seams and is affected by Variscan folding (Stop 4B). The contact of the Waulsortian Limestone with the underlying Ballysteen Formation is very sharply defined at the entrance to the bay north of Illaunlea (Stop 4C), where 3 m of vertically-bedded, dark grey nodular argillaceous limestones and shales containing chert bands strike N-S and pass up abruptly to thick-bedded, pale grey Waulsortian facies. In the vicinity, some of the limestone nodules in the Ballysteen Formation are up to 25 cm in diameter. On Illaunlea headland to the south (UTM 500381 m E, 5831198 m N; Map 64, 132721 m E, 153703 m N; Stop 4D), the Ballysteen limestones dip gently to the SW (Fig. 4.2.10) and are strongly cleaved (048/66° SE). Remarkably, such cleavage, associated with the Variscan deformation, is seldom developed in these rocks, despite the abundance of argillaceous limestones in the succession.

Fig. 4.2.10. Gently dipping nodular argillaceous limestone and shale (Ballynash Member) cut by steeply inclined Variscan cleavage dipping south (white line). Illaunlea headland (Stop 4D). Hammer is 40 cm long.

Waulsortian Limestone: depositional setting, age and thickness

This transition from the cherty nodular limestones of the Ballynash Member into the overlying Waulsortian limestones continues the deepening trend and it is interpreted that the Waulsortian initially formed isolated mud-mounds that quickly coalesced on the distal part of a ramp (Somerville & Jones, 1985; Strogen, 1988; Somerville *et al.*, 1992; Strogen *et al.*, 1996; Sleeman & Pracht, 1999). The Waulsortian represents a very thick succession of mostly massive, pale grey, fine-grained limestones (lime mudstones and wackestones) with local concentrations of brachiopods, bivalves and fenestellid bryozoans. Conodonts are rare in the mud-mounds, but *Polygnathus bischoffi*, *Gnathodus texanus* and *G. typicus* have been recorded from the Pallaskenry Borehole (Somerville & Jones, 1985) and Hill (1971) recorded *Scaliognathus anchoralis* from mud-mounds near Askeaton (Fig. 4.1.1). This establishes a late Tournaisian age for most of the mud-mound development (*Scaliognathus anchoralis* Biozone). However, the top of the Waulsortian in north Kerry and in north Cork is of early Viséan age (see Sevastopulo, 1982; Somerville, 2003; Sevastopulo & Wyse Jackson, 2009; Fig. 3.1.2). There is considerable variation in the thickness of the Waulsortian in the Shannon region. Shephard-Thorn (1963) suggested a thickness of between *c*. 1000 to 1200 m, Strogen (1988) *c*. 1000 m and Lees & Miller (1995) suggested a thickness of *c*. 900 m. These mud-mounds are characterised by the presence of allochems of mostly deep-water phases A and B of Lees & Miller (1985, 1995) and show little interbank facies (Somerville, 2003). The Waulsortian limestones thin considerably to the north in County Clare and in the Gort Borehole (G1) in the Gort Lowlands (*c*. 50 km north of Askeaton) where they are less than 100 m thick (Pracht *et al.*, 2004).

What is particularly interesting is that this vast accumulation of Waulsortian, mud-bank, carbonate sediment (the thickest anywhere in the world during the late Tournaisian (Lees & Miller, 1995)) coincides with an episode of increased subsidence centred on the axis of the Shannon Trough (Strogen, 1988). This depocentre was marked by a period of renewed subsidence in the early Namurian (Serpukhovian) (Sevastopulo, 2009). It has been suggested (Strogen *et al.*, 1996) that this linear depocentre in the Shannon Basin, which is aligned approximately ENE-WSW and centred on the Shannon Estuary, may coincide with the position of the Iapetus Suture (see Phillips, 2001; and further discussions in Chapters 2 and 3).

Transfer to Stop 5

From Courtbrown, proceed south to Askeaton and turn right onto the N69, then travel west for *c*. 8 km towards Foynes. About 2 km SSE of Foynes there is a major bend in the main road at the junction with the R521 road to Shanagolden (Fig. 4.2.1), where several small disused limestone quarries occur north of the road.

Stop 5: Ardaneer, SSE Foynes (UTM 494018 m E, 5827209 m N; Map 64, 126300 m E, 149800 m N)

The underlying Rathkeale Formation is very poorly exposed in the low marshy ground south-east of Durnish Point (Fig. 4.2.11).

South of Ardaneer Cottage, several small quarries north and east of the N69 expose limestone beds of the Durnish Formation (Figs 4.2.1 and 4.2.12). The type section of the Rathkeale Formation is in the River Deel stream section upstream from Kilcool Bridge (UTM 501805 m E, 5820719 m N; Map 64, 134000 m E, 143200 m N), 2.5 km NW of Rathkeale (Fig. 4.2.1). There, a thick cleaved sequence of poorly fossiliferous basinal argillaceous limestones and shales is exposed (*c*. 450 m thick according to Shephard-Thorn, 1963). Austin *et al.* (1970) recovered conodonts from this formation including *Gnathodus homopunctatus*, *G. texanus*, *G. semiglaber* and *G. symmutatus*, which together with foraminiferans recorded by Somerville & Strogen (1992) from Cappagh Castle 4 km NE of Rathkeale (Fig. 4.2.1), suggest a probable Arundian age. The Rathkeale Formation

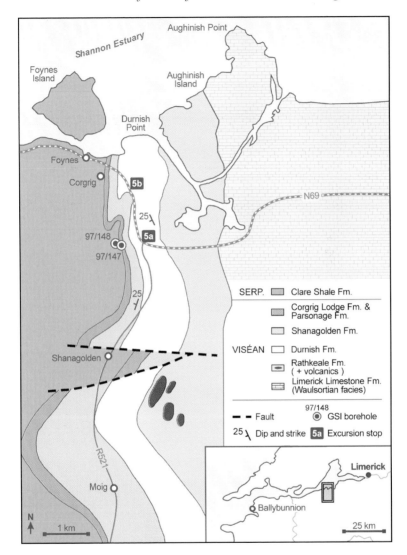

Fig. 4.2.11. A map of the geology of west County Limerick near Foynes (adapted from Shephard-Thorn, 1963). SERP. = Serpukhovian.

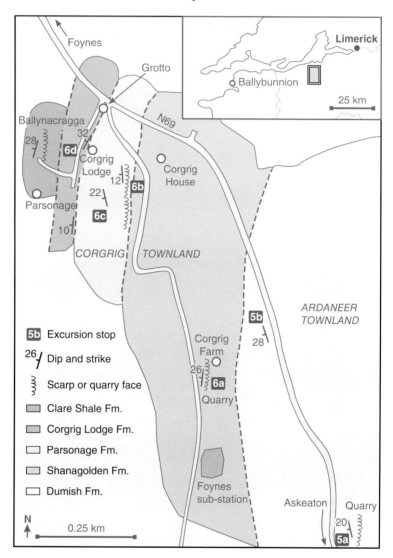

Fig. 4.2.12. A detailed map of the geology south-east of Foynes showing location of sections and excursion stops (from Somerville, 1999).

Fig. 4.2.13. Thick-bedded and thin-bedded limestones of the Durnish Formation in N69 roadside scarp (west side), near Foynes (Stop 5B). Height of road-cut section *c*. 12 m.

is also distinctive for the presence of interbedded volcanic agglomerate (well-sorted vitric lapilli tuffs), which has been recorded as debris in fields 2 km SE of Shanagolden (Fig. 4.2.11; Ashby, 1939; Shephard-Thorn, 1963; Sleeman & Pracht, 1999).

The limestones in the roadside quarries (UTM 494018 m E, 5827209 m N; Map 64, 126300 m E, 149800 m N) dip 25° to the west and comprise medium-grained, dark grey crinoidal wackestones, packstones and grainstones with chert nodules, and represent the upper part of the Durnish Formation. Occasionally, horizons rich in brachiopods and bryozoans can be found. In a new road-cut section closer to Foynes (UTM 493594 m E, 5828173 m N; Map 64, 125889 m E, 150771 m N; Stop 5B), 23 m of the formation is exposed on the west side of the N69 (Figs 4.2.12 and 4.2.13), which yields rare, solitary corals (*Siphonophyllia*) and occasional bands of chert nodules. In the GSI Foynes Borehole (97/147), over 184 m of the formation was intersected but the contact with the Rathkeale Formation was not reached (Sleeman & Pracht, 1999). Shephard-Thorn (1963) estimated that the formation was *c*. 300 m thick and recorded the presence of a rich coral-brachiopod fauna including fasciculate colonial corals (*Solenodendron furcatum* and *Siphonodendron martini*) and solitary rugose corals (*Siphonophyllia benburbensis* = *S. samsonesis* and *Caninophyllum archiaci*) of Asbian age, although the lower part of the formation may be of Holkerian age. This fauna is comparable with that from the lower Burren Formation in

County Clare (Gallagher, 1996; Gallagher *et al.*, 2006; Fig. 4.1.2). Conodonts recovered from the top of the Durnish Formation in Foynes Borehole 97/147 include *Gnathodus girtyi* and *Lochriea commutata* (Somerville, 1999).

Transfer to Stop 6

Continue north-westwards along the N69 road towards the outskirts of Foynes (Fig. 4.2.1). About 1 km before Foynes, take the minor road on the left, immediately before the wayside shrine (grotto) and head SSE towards Corgrig Farm and Foynes power station (Fig. 4.2.12).

Stop 6: Corgrig, SE Foynes (UTM 493211 m E, 5828496 m N; Map 64, 125510 m E, 151099 m N)

At Corgrig Farm, a small quarry (UTM 493468 m E, 5827954 m N; Map 64, 125760 m E, 150553 m N) north of the power station exposes a 60 m long extensive bedding plane in well-bedded, fine-grained, dark grey wackestones of the Shanagolden Formation (Fig. 4.2.12), dipping 26° W (Stop 6A). Close examination of the bedding plane reveals silicified crinoid stems, brachiopods, zaphrentid corals and small colonies of *Michelinia tenuisepta*. The limestones of the Shanagolden Formation are similar to those of the underlying Durnish Formation, except chert is very rare in the former (Shephard-Thorn, 1963). According to Shephard-Thorn (1963), corals are very sparse in the Shanagolden Formation, with mostly solitary caniniids and cyathaxoniids, but the recorded presence of the ammonoid *Beyrichoceras obtusum* confirms an Asbian (B_2) age for the formation. In the Foynes Borehole (97/147), the Late Asbian conodont *Gnathodus bilineatus* first appears in this formation (Sleeman & Pracht, 1999; Somerville, 1999). The Shanagolden Formation is 50 m thick in the Foynes borehole and is named after a small quarry near Moig House (UTM 493090 m E, 5822098 m N; Map 64, 125300 m E, 144700 m N; Fig. 4.2.1), 3 km south of Shanagolden, where 8 m of well-bedded dark grey wackestones are exposed.

The youngest part of the Viséan limestone sequence in the Foynes area is represented by the Parsonage and Corgrig Lodge formations (Shephard-Thorn, 1963), which are combined on the GSI map for Sheet 17 (Sleeman & Pracht, 1999). Both formations are relatively thin and poorly exposed. A 5 m high, N-S trending scarp behind the row of houses (UTM 493211 m E, 5828496 m N; Map 64, 125510 m E, 151099 m N) SW of Corgrig House (Stop 6B, Fig. 4.2.12) and Corgrig Castle ruins belongs to the Parsonage Formation. It comprises thick-bedded, pale grey weathering, dark grey wackestones with thin shaly beds (Fig. 4.2.14) and there are nodular, pseudo-brecciated intervals (limestone 'conglomerates'). Silicified crinoids are conspicuous, but also recorded are rare goniatites and

Fig. 4.2.14. Scarp behind houses at Corgrig (Stop 6B) showing thick beds of banded argillaceous skeletal wackestone separated by thin wavy shale laminae of the Parsonage Formation, which contains silicified crinoids, goniatites and solitary corals. Height of scarp *c*. 5 m.

zaphrentid corals. Across the field to the west, another 5 m high scarp (UTM 493144 m E, 5828433 m N; Map 64, 125442 m E, 151037 m N) reveals limestones of the same formation (Stop 6C), except here they contain two prominent intervals of well-laminated micrites ('striped' beds). The lower interval near the base is 1.5 m thick and the upper interval near the top of the scarp is 40 cm thick. In the 97/147 borehole, the Parsonage Formation is 23 m thick and is composed of fine-grained micritic limestones with a basal 7 m thick interval of laminated striped limestones and nodular conglomeratic beds, with a massive unit 5 m thick of recrystallised mud-mound limestone in the middle of the formation. Similar lithotypes have been encountered in the islands of the Fergus Estuary, south County Clare (Hodson & Lewarne, 1961) and at Ballybunnion (see Excursion 4.3). The distinctive laminated limestones are enigmatic and it has been suggested that they may represent replaced deep-water evaporites (Sevastopulo, 1981a). A Brigantian foraminiferal assemblage with *Howchinia bradyana* and *Asteroarchaediscus* sp. was recorded from thin grainstone beds in the Parsonage Formation (Sleeman & Pracht, 1999; Somerville, 1999).

The Corgrig Lodge Formation is exposed behind the house south of Corgrig Lodge (UTM 493101 m E, 5828495 m N; Map 64, 125400 m E, 151100 m N) and in small scarps in the grounds of the Parsonage (Fig. 4.2.12). Access to these sites is via the road leading south from the N69 towards Corgrig Lodge and the church (Stop 6D). Several metres of thin-bedded argillaceous wackestones interbedded with thin shales are exposed (Hodson & Lewarne, 1961). Unfortunately, the contact with the Clare Shale Formation is not exposed, but must lie below the boggy low ground at the foot of Ballynacragga scarp, 100 m west of Corgrig Lodge, where blue grey shales dip 28° W, identical to that of the limestones. In Foynes Borehole 97/148, the Corgrig Lodge Formation is 24 m thick and the conformable contact with shales of the Serpukhovian Clare Shale Formation is observed (Sleeman & Pracht, 1999). In the borehole, rare thin graded coarse-grained bioclastic limestones occur, with argillaceous lime mudstones and wackestones alternating with shales. The coarser limestones have yielded Brigantian foraminiferans and the wackestones have yielded abundant conodonts including *Mestognathus bipluti* (Austin *et al.*, 1970) and *Vogelgnathus postcampbelli*, as well as the diagnostic taxa *Lochriea nodosa* and *L. mononodosa*, establishing a late Brigantian age for the upper part of the formation (Sleeman & Pracht, 1999; Somerville, 1999). *Lochriea nodosa* and *Cavusgnathus naviculus* have been recorded in the equivalent 'Inishtubbrid beds' on the Fergus Estuary island of Inishtubbrid (Austin *et al.*, 1970; Austin & Husri, 1974; Fig. 4.2.1).

Depositional environments of the Viséan rocks in the Shannon Trough (Basin)

From early in the Viséan (Arundian times), western County Limerick began to accumulate deep-water basinal facies within the Shannon Trough region, in contrast to the shallow-water platform facies in the Limerick Syncline to the east and around Limerick city (Strogen, 1988; Somerville & Strogen, 1992, Somerville *et al.*, 1992; Strogen *et al.*, 1996). In the late Tournaisian to earliest Viséan, the Shannon Trough underwent more pronounced rapid subsidence creating accommodation space for a very thick sequence of Waulsortian facies, *c.* 800 to 1000 m, extending from Askeaton to Ballybunnion (Strogen, 1988; Lees & Miller, 1995; Strogen *et al.*, 1996; Somerville, 2003). In the early-mid Viséan (Arundian-Holkerian) interval, deposition of the Rathkeale Formation was dominated by an influx of mud into the basin and suppression of carbonate sedimentation. The latter was restricted to rare calciturbidite event beds (Somerville & Strogen, 1992). Benthic faunas were very sparse and confined mainly to transported shallow-water assemblages. Volcanism was also manifest at this time with extrusion of pyroclastic material in the Shanagolden area. In contrast, equivalent-aged Arundian rocks in the Limerick Syncline were mostly shallow-water oolitic packstones and grainstones (Herbertstown Limestone Formation) (Somerville *et al.*, 1992). In the Limerick city area, a ramp sequence was developed, with proximal ramp cross-bedded grainstones deposited in the Mungret area passing up into mid-ramp wackestones and mud-mounds (Somerville &

Strogen, 1992). These facies pass westwards into the deeper water distal ramp and basinal facies of the Rathkeale Formation around Rathkeale and Cappagh Castle (Figs 4.1.1 and 4.2.1).

In the upper Viséan (Asbian times), upward-shallowing occurred within the Shannon Trough with the deposition of the carbonate-rich Durnish and Shanagolden formations in probable outer shelf or mid-ramp settings. Benthic faunas were re-established with abundant colonial and solitary rugose corals in the Durnish Formation and brachiopods and crinoids. A similar facies and fauna are recorded north of the Shannon Estuary in County Clare within the lower part of the Asbian Burren Formation (Gallagher, 1996; Gallagher & Somerville, 1997, 2003; Pracht *et al.*, 2004; Gallagher *et al.*, 2006; see Excursion 5.2). In the Limerick Syncline, shal-lower-water platform limestones were deposited (Dromkeen Limestone Formation) with bentonitic clay bands similar to those of the upper Burren Formation (Somerville *et al.*, 1992; Strogen *et al.*, 1996; Gallagher & Somerville, 1997, 2003; Gallagher *et al.*, 2006). Graded grainstone beds in the upper part of the Shanagolden Formation contain abundant foraminif-erans together with red and green calcareous algae (*Koninckopora, Coelosporella, Ungdarella*) typical of the adjacent Late Asbian platform facies in east County Limerick and north County Clare, from which they were probably derived (Somerville *et al.*, 1992; Gallagher & Somerville, 1997, 2003; Somerville, 1999; Gallagher *et al.*, 2006).

A deepening-upward trend is manifest in the Brigantian with a switch to finer-grained carbonates (wackestones and lime mudstones) and an increase in the proportion of mud deposited in the Shannon Trough (Parsonage and Corgrig Lodge formations). Rare thin beds of graded grainstone mark the influx of calciturbidites containing shallow-water-derived benthic microfossil assemblages (foraminiferans and algae). Deep-water mud-mounds are recorded in the Parsonage Formation, particularly in the axial part of the Shannon Trough in the islands of the Fergus River and in north County Kerry at Lixnaw (Hodson & Lewarne, 1961; Tattershall, 1963; Sleeman & Pracht, 1999). Also, deep-water laminated limestones (striped beds), representing possible replaced evaporites, are recorded from these same areas, as well as in the Foynes Borehole (Sleeman & Pracht, 1999). Basin slopes are also evident with the presence at outcrop of slumping at Ballybunnion (see Excursion 4.3) and Canon Island and Inishtubbrid Island in the Fergus River (Hodson & Lewarne, 1961; Tattershall, 1963; Sleeman & Pracht, 1999; Fig. 4.2.1). The occur-rence of breccia beds and limestone conglomerates in the Foynes bore-holes may also be related to instability associated with slumping and reworking of shallow-water carbonate facies.

The Corgrig Lodge Formation differs from the Parsonage Formation in the amount of interbedded shale with the limestones. There is a marked increase in the proportion of mud in the former formation, especially in the upper part, which heralds the passage to the mud-dominant Clare Shale Formation in the Serpukhovian. This marks the final suppression of carbonate production and development of a mud blanket in the Shannon Basin region. Benthic faunas are very sparse in the Corgrig Lodge Formation and restricted to transported assemblages in thin, graded, bioclastic grainstone beds. Nektonic faunas are common with deep-water conodont species of *Lochriea* and *Gnathodus* recorded in some of the fine-grained limestones (Austin *et al.*, 1970; Austin & Husri, 1974; Somerville, 1999). Equivalent beds to the Parsonage and Corgrig Lodge formations north of Ennis in County Clare are in the Brigantian Slievenaglasha Formation, which represents a shallow-water platform sequence (Gallagher *et al.*, 2006; Fig. 4.1.2). Thus, in the Brigantian, there is a lateral passage southwards from thick-bedded, shallow-water limestone with a diverse coral-brachiopod fauna in north County Clare (Burren), to outer platform or ramp mud-mounds interbedded with wackestones and shales showing slumping in south County Clare (Fergus Estuary), to deep-water basinal limestones and shales in the axis of the Shannon Basin in north-west County Limerick and north County Kerry. This continued into the Serpukhovian, with establishment of deep-water basinal mudstones of the Clare Shale Formation.

4.3 Viséan Limestones of the Ballybunnion Region, Co. Kerry

The aim of this half-day excursion is to examine deep-water carbonate palaeoenvironments in a basinal setting. The Mississippian limestones are mostly of Viséan age and represent a westward extension of the succession exposed in the Foynes area, 40 km to the east in west County Limerick (see Excursion 4.2). They include the Waulsortian mud-mound limestones, followed by a calciturbidite limestone and shale sequence (Parsonage and Corgrig Lodge formations) of Upper Viséan age (see Fig. 4.1.2). Beds of these formations are much better exposed here than in the type area near Foynes, with a wealth of sedimentary structures and textures visible in water-worn coastal rocks.

Directions to Stop 1

From Foynes, travel west along the N69 for 19 km to Tarbert, County Kerry, and then continue west along the R551 for 23 km to the town of Ballybunnion, North Kerry. (N.B. from Tarbert you can travel to the sections in west County Clare via the Tarbert to Killimer ferry that regularly crosses the Shannon Estuary).

Stop 1: Ballybunnion coast section below the golf course (UTM 453800 m E, 5817895 m N; Map 63, 085935 m E, 141040 m N)

The Carboniferous succession at Ballybunnion (also spelt Ballybunion) dips uniformly north *c.* 45° and the two sandy beaches (Ballybunnion Bay North and South) are partly separated by a headland (Fig. 4.3.1) on which Ballybunnion Castle is built (Fig. 4.3.2). The cliffs are quite high but there are many paths and roads down onto the foreshore. The description of the

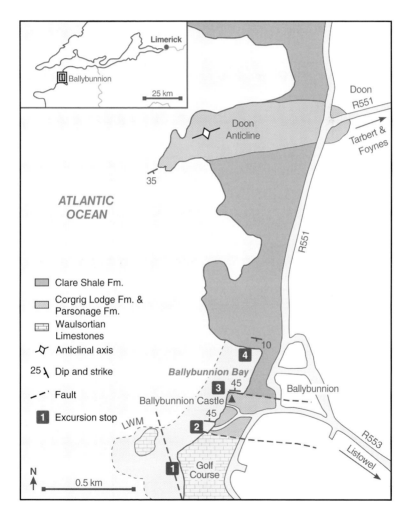

Fig. 4.3.1. Geological map of Ballybunnion showing location of excursion stops (modified from Sleeman & Pracht, 1999; copyright Geological Survey of Ireland).

Fig. 4.3.2. Ballybunnion Castle built on Upper Viséan limestones of the Parsonage and Corgrig Lodge Formations dipping steeply to the north. In the background are cliffs of the Namurian Clare Shale Formation dipping gently to the south.

section begins at the southern end, below cliffs of glacial diamictite. **Note that most of the foreshore section is covered at high tide and is best observed on a falling tide.**

The oldest Carboniferous rocks are exposed below the high-water mark on the wave-cut platform SW of the town, below the golf course (Stop 1; Fig. 4.3.1), although the amount of beach sand covering these outcrops varies considerably from year to year. The outcrops comprise massive, water-worn, pale grey Waulsortian limestones with superb large stromatactis cavities filled with orange and white sparry calcite. The limestones are also locally very fossiliferous with crinoids, gastropods and fenestellid bryozoans, with occasional orthocone nautiloids, goniatites and *Amplexus* solitary corals (Fig. 4.3.3). Dolan (1984) estimated that the Waulsortian is >200 m thick, although borehole data in the area suggests that it could be between 800 to 1000 m thick (Sleeman & Pracht, 1999; Sevastopulo & Wyse Jackson, 2009). The age of the Waulsortian limestones are early Viséan (Sleeman & Pracht, 1999). The top of the Waulsortian is succeeded by 3 m of dark, cherty bioclastic limestones of Arundian age rich in sponge spicules and calcispheres (Sleeman & Pracht, 1999; Sevastopulo & Wyse Jackson, 2009). These cherty limestones may be equivalent to the 'reef cover cherts' of east County Limerick (Lough Gur Formation of Strogen, 1988; Somerville *et al.*, 1992; Strogen *et al.*, 1996).

Fig. 4.3.3. Orange and white spar-filled stromatactis cavities and goniatites in the massive fine-grained Waulsortian limestones, Ballybunnion (Stop 1). Coin is 2.5 cm in diameter.

Stop 2: Ballybunnion coast section (UTM 454061 m E, 5818159 m N; Map 63, 086200 m E, 141300 m N)

In Ballybunnion Bay and east of the N-S trending fault which cuts out most of the Arundian-Asbian (Viséan) succession (Sevastopulo & Wyse Jackson, 2009) is a sequence of bedded, dark grey to black limestones (c. 70 m thick) that have been assigned to the Parsonage Formation (Sleeman & Pracht, 1999). They correspond to the 'Breccia Beds' of Kelk (1960) and unit 1 of the 'supra-reef limestones' of Dolan (1984). The lowest exposed beds beside the wall of the road leading down on to the sandy beach (UTM 453894 m E, 5818033 m N; Map 63, 086031 m E, 141176 m N; Stop 2A) comprise very thin (2 to 3 cm), coarse-grained, graded laminated limestone beds with concentrations of crinoids and brachiopods, interbedded with black shales and nodular crinoidal argillaceous wackestones (Fig. 4.3.4). The graded beds are erosive at their base and are interpreted as calciturbidites. The overlying limestones (c. 20 m thick) amid the sands are mostly fine grained, with a prominent 7 m thick finely-laminated interval near the base (Fig. 4.3.5). The latter are similar to the 'striped limestones' of Hodson & Lewarne (1961), recorded as the Parsonage Formation in the Foynes Borehole (Sleeman & Pracht, 1999) and at Corgrig (4.2; Stop 6C). In addition, there are lenses of coarse-grained bioclastic limestone and breccia (>2 m thick) (Fig. 4.3.5) with evidence of slumping (Fig. 4.3.6), especially near the top of the formation, close to the steps leading down from the road onto the beach beneath the castle.

Fig. 4.3.4. Two thin, graded, skeletal limestone beds, with marked erosive bases (arrowed), above black shale, Parsonage Formation, Ballybunnion (Stop 2). Coin is 2.5 cm in diameter.

Fig. 4.3.5. Finely laminated (pinstripe) micrites ('striped beds') overlain by a limestone breccia bed, Parsonage Formation, Ballybunnion (Stop 2). Coin is 2.5 cm in diameter.

Fig. 4.3.6. Slumping in limestones and shales of the Parsonage Formation, with large block of limestone breccia, Ballybunnion Castle (Stop 3). Coin is 2.5 cm in diameter.

Stop 3: Ballybunnion Castle section (UTM 454022 m E, 5818143 m N; Map 63, 086160 m E, 141285 m N)

At the base of the cliff below Ballybunnion Castle is a >44 m thick sequence of continuously exposed beds (Fig. 4.3.2) dipping 45° N (Stop 3). They comprise interbedded, fine-grained, dark grey argillaceous limestones and shales, with rare, thin, coarser-grained crinoidal limestone beds that have been assigned to the Corgrig Lodge Formation (Sleeman & Pracht, 1999). They correspond to the 'Ballybunnion Castle Beds' of Kelk (1960) and unit 2 of the 'supra-reef limestones' of Dolan (1984). The fine-grained wackestones and lime mudstones are well-bedded (typically 20 to 60 cm thick, but can be up to a metre in thickness), mostly unfossiliferous and frequently contain chert bands and nodules. Some of the thicker fine-grained limestone beds show channelling (Fig. 4.3.7), whereas others are mostly massive, but with laminations developed in the upper part of the bed (Fig. 4.3.8). These are interpreted as distal or dilute calciturbidites. The rare coarser-grained limestone beds (10 to 50 cm thick) are pale grey, laminated, graded and have erosive bases (Fig. 4.3.9). They also contain detrital silt-size quartz, feldspar grains and oolitic clasts. The interbedded shales are up to 25 cm thick and contain pyrite and burrows (Dolan, 1984). One prominent grainstone bed, *c*. 12 cm thick and some 21 m above the base of the section, has a concentration of shells that are mostly concave up, and in the overlying shale bed vertical and horizontal burrows can be observed filled with coarse skeletal material

Fig. 4.3.7. Channel-fill of fine-grained limestone cut into shales, Corgrig Lodge Formation, Ballybunnion Castle (Stop 3). Pen is 14 cm long.

Fig. 4.3.8. Thick, sharp-based, graded fine-grained calciturbidite bed with laminated top (black arrow), Corgrig Lodge Formation, Ballybunnion Castle (Stop 3). Pen is 14 cm long.

Fig. 4.3.9. Thin, graded, coarse-grained, pale grey skeletal grainstone with irregular scoured base interbedded with dark grey mudstones and laminated wackestones, Corgrig Lodge Formation, Ballybunnion Castle (Stop 3). Pen is 14 cm long.

Fig. 4.3.10. Horizontal and vertical burrows (arrowed) in laminated argillaceous wackestones and shales. The pale grey burrow fill is derived from the overlying skeletal packstone bed, Corgrig Lodge Formation, Ballybunnion Castle (Stop 3). Hand lens is 3 cm long.

from a succeeding, thin grainstone bed (Fig. 4.3.10). The proportion of shale increases up through the formation, although the top contact is obscured beneath the sand. The Corgrig Lodge Formation is interpreted as a basinal calciturbidite sequence (Sevastopulo, 1981a; Sleeman & Pracht, 1999) and may correspond to the distal western part of the Limerick Ramp (Somerville & Strogen, 1992).

The age of the Parsonage and Corgrig Lodge formations at Ballybunnion, by comparison with the Foynes Borehole sequence, are of Upper Viséan (Brigantian) age. Ammonoids recovered from near the top of the Corgrig Lodge Formation contain *Hibernicoceras* sp., indicating a P_{1d} (Brigantian) age (Sleeman & Pracht, 1999).

The Clare Shale Formation forms the impressive cliffs on the north side of the north bay at Ballybunnion, a short stroll across the beach from this final Viséan limestone stop (see Chapter 6 for description).

Chapter 5
Viséan Coral Biostromes and Karsts of the Burren

IAN D. SOMERVILLE

5.1 Introduction

The Burren region (Fig. 5.1.1) is formed mostly of Viséan (Mississippian) platform carbonates and represents one of the most extensive and best exposed glaciated limestone karst regions in Europe, covering *c.* 600 km^2 of northern County Clare (Simms, 2006). Limestone plateaus are characteristic (especially in the northern part), rise to over 300 metres and provide some spectacular scenery and extraordinary landscapes (Fig. 5.1.2). The area is bounded to the east by the limestone of the Gort Lowlands (maximum topography 30 m) and to the south by less well exposed, mainly Namurian, sandstone and shale that produces a subdued topography (maximum elevation 100 m) in the southern half of County Clare. The lithostratigraphy of the Viséan rocks (Fig. 5.1.3) has been divided into three formations based on topographic expression, facies and age (see Gallagher, 1996; Pracht *et al.*, 2004; Gallagher *et al.*, 2006).

In the Burren, Douglas (1909) used the Vaughan (1905) divisions of the English Lower Carboniferous, established in the Bristol area, and subsequently modified by Sibly (1908) from his work in Derbyshire. Douglas (1909) identified a thick succession of limestone ranging in age from the S_1 to D_3 zones (Arundian to Brigantian) using Vaughan's coral and brachiopod zonal scheme. Douglas (1909) described black, compact crystalline and crinoidal limestone with some chert and oolitic beds, which he assigned to the S_2 (*Seminula*) Zone. These strata, which

A Field Guide to the Carboniferous Sediments of the Shannon Basin,
Western Ireland, First Edition. Edited by James L. Best and Paul B. Wignall.
© 2016 International Association of Sedimentologists.
Published 2016 by John Wiley & Sons, Ltd.
Companion website: www.wiley.com/go/best/shannonbasin

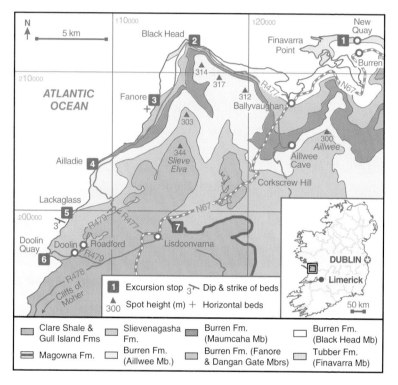

Fig. 5.1.1. Geological map of the Burren region adapted from GSI Sheet 14 (Geology of Galway Bay; Pracht *et al.*, 2004; copyright Geological Survey of Ireland), showing important localities and stops described in the text. Co-ordinates are Irish Grid co-ordinates as on 1:50,000 maps.

form the Tubber Formation (Fig. 5.1.3), are exposed in the Gort Lowlands and along the north Burren coast and are also known from the Gort (G-1) borehole (Pracht *et al.*, 2004). The succeeding *Dibunophyllum* (D) Zone strata that forms most of the Burren region (Burren Formation) is typified in the lower part (D_1 subzone) by dark grey crinoidal limestone, which shows some dolomitization, overlain by pale grey, finely-crystalline, bedded limestone with rare chert. The limestones of the upper D_2 to D_3 subzones are crinoidal and interbedded with black, compact limestone containing chert horizons (Slievenaglasha Formation) (Fig. 5.1.3). The total thickness of the exposed upper Viséan limestone succession in the Burren region is *c.* 500 m (Gallagher *et al.*, 2006; Somerville *et al.*, 2011).

Fig. 5.1.2. View looking north-east towards Aillwee Hill (300 m) from Corkscrew Hill (north Burren region). The lower smooth slope is the massive limestones (M) of the Burren Formation (Maumcaha Member); the main part of the slope is the bedded limestones of the Aillwee Member (A) forming terraces T1 to T9 (see Fig. 5.1.3); the upper part of the hill is represented by the Slievenaglasha Formation (S).

5.2 Upper Viséan Limestones of the Burren Region, Co. Clare

This one-day excursion provides an opportunity to traverse the entire Upper Viséan limestone succession from the Tubber Formation, at the base, to the Slievenaglasha Formation, at the top, and its contact with the Namurian black shales of the overlying Clare Shale Formation. The excursion will examine the succession in stratigraphic order. It commences with the oldest rocks exposed on the north coast of the Burren (Stops 1 and 2) and then climbs up the succession along the western coast road heading south-west towards Doolin (Stops 3 to 6), following the gentle regional dip (3 to 5° SW). The excursion finishes inland at Lisdoonvarna (Stop 7). As most sections (stops) are coastal, it is best to start the excursion on a falling tide. Stop 3 is mostly covered at high tide and Stops 4 and 5 should not be visited at high tide because they can be very dangerous, especially if there are strong westerly winds. Please note that in summer the Burren is a very popular tourist area with lots of vehicles, especially coaches, on the roads

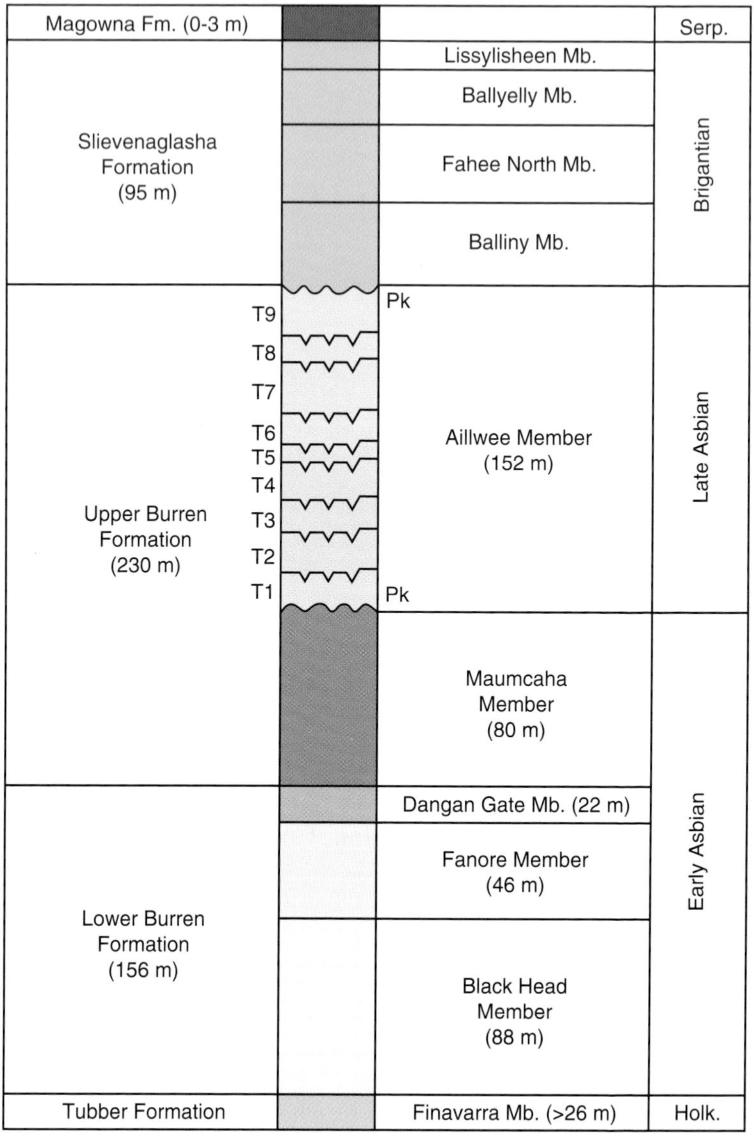

Magowna Fm. (0-3 m)			Serp.
Slievenaglasha Formation (95 m)		Lissylisheen Mb.	Brigantian
		Ballyelly Mb.	
		Fahee North Mb.	
		Balliny Mb.	
Upper Burren Formation (230 m)	T9	Pk	Late Asbian
	T8		
	T7		
	T6	Aillwee Member (152 m)	
	T5		
	T4		
	T3		
	T2		
	T1	Pk	
		Maumcaha Member (80 m)	
Lower Burren Formation (156 m)		Dangan Gate Mb. (22 m)	Early Asbian
		Fanore Member (46 m)	
		Black Head Member (88 m)	
Tubber Formation		Finavarra Mb. (>26 m)	Holk.

Fig. 5.1.3. A summary of the stratigraphy of the upper Viséan carbonates of the Burren. T1 to T9 are numbered terraces in the Aillwee Member (from Gallagher *et al.*, 2006). Pk denotes palaeokarst. Serp. = Serpukhovian, Holk. = Holkerian.

and parking may be difficult at some stops, particularly Stops 2 and 5. **Please do not block the roads**.

The trip begins at Finavarra (also spelt Finvarra) headland in the north of the Burren (Fig. 5.1.1). Access to Finavarra is from the Galway and Dublin roads to the east. From Kinvarra, head west along the N67 towards Ballyvaughan (also spelt Ballyvaghan). Nine kilometres from Kinvarra, at the village of Burren, take the minor road on the right heading north to New Quay. At New Quay, turn left and then follow the coast road west for 1.5 km to arrive on the Finavarra headland. From here, there is a spectacular view across Galway Bay to the remarkably straight northern shoreline formed of Galway granite, with the Connemara peaks beyond. Erratics of Galway granite can be found resting on the Burren limestone pavements and these demonstrate a north to south flow of ice during the Pleistocene.

Stop 1. Finavarra Head (UTM 493629 m E, 5889993 m N; Map 51, 126784 m E, 212615 m N)

The oldest exposed limestones of the north Burren succession are developed for over 1.5 km along the northern shore of Finavarra headland. The beds are almost horizontal, dipping gently inland and are composed mostly of pale grey, medium-grained packstones. They have been assigned to the upper part of the Tubber Formation (Finavarra Member), which is of probable Holkerian age (Pracht *et al.*, 2004). Some 4 m of stratigraphic section is observed in three wave-cut terraces. The extensive bedding planes expose very abundant fasciculate (branching) rugose coral colonies (Fig. 5.2.1), many in growth position, although several smaller colonies are inverted (Fig. 5.2.2). The most abundant genus is *Siphonodendron*, with two species dominant, *S. martini* and *S. sociale*. The colonies are typically 25 cm in diameter but can be up to one metre in width. They can form radiating (stellar) clusters in plan view and fans in profile, with colonies often in close contact. Other corals present are small colonies of the tabulate *Syringopora* and the solitary rugose coral *Palaeosmilia murchisoni*. Such profusion of coral colonies within individual beds, forming biostromes, demonstrates optimum growth conditions in moderately turbulent conditions, as evidenced by the frequency of overturned colonies and colonies with pronounced laterally-directed growth. It is also interesting to note that the coral assemblage is dominated mainly by two fasciculate species and that solitary rugosans are a minor component.

Transfer to Stop 2

Follow the road back to New Quay and then head south for 1 km to join the main coast road (N67). Follow this road for *c*. 9 km to Ballyvaughan and then take the R477 coast road to Black Head. After 9 km, the Black

Fig. 5.2.1. Radiating clusters of *Siphonodendron* colonies in plan view from the Finavarra Member (Tubber Formation) at Finavarra headland (Stop 1). The hammer is 40 cm long.

Fig. 5.2.2. Fasciculate *Siphonodendron* colonies in profile showing upright *in situ* fans. Note the small inverted colony next to the hammer. Finavarra Member (Tubber Formation) at Finavarra headland (Stop 1). The hammer is 40 cm long.

Head lighthouse at the northern point of the Burren will come into view. Please note that there is very limited parking for vehicles at this locality.

Stop 2. Black Head *(UTM 482287 m E, 5889379 m N; Map 51, 115429 m E, 212159 m N)*

The oldest beds exposed here, close to Black Head lighthouse (Fig. 5.2.3), north of and below the coastal road R477, belong to the upper part of the Tubber Formation (Finavarra Member) and are similar to those at Finavarra (Stop 1), which are approximately along-strike. These thick-bedded limestones are peloidal, algal-rich, crinoidal grainstones containing the dasycladacean alga *Koninckopora*. They represent shallow water, fairly high-energy platform facies within the photic zone. Fasciculate colonies of *Siphonodendron martini* are recorded, as well as rare cerioid *Lithostrotion* colonies and *Syringopora* close to the lighthouse. The limestones here show patchy brown, weathering dolomitization. The top of the member and the formation is marked by a dolomite marker bed (50 cm thick) that forms the wall beside the road; and the road is locally built on this horizon (Fig. 5.2.3). Stratigraphically younger beds in the cliff south of the road have been assigned to the basal Black Head Member of the overlying Burren Formation (Gallagher, 1992; 1996; Pracht *et al.* 2004; Gallagher *et al.*, 2006). The lower part of the Burren Formation in the Black Head area has been subdivided into three members, the Black Head, Fanore and Dangan Gate members (Figure 5.1.3), but only the lower two will be examined herein, both here and at Fanore Beach (Stop 3). In the upper part of the Burren Formation, two members are distinguished: the Maumcaha and Aillwee members (Figs 5.1.2 and 5.1.3), both of which are exposed at Stop 4.

The Black Head Member

The Black Head member (88 m thick) comprises thick-bedded to very thick-bedded (1 to 10 m), grey, uniform and homogeneous limestone with bioturbated horizons. Cerioid *Lithostrotion araneum* and smaller *L. decipiens* coral bands appear in the lower part of the Black Head Member, a few metres above the dolomite marker bed. Other macrofauna consists of rare fasciculate *Siphonodendron* and solitary rugose corals, with gastropods and bivalves. The microfacies consists of coarse-grained, moderately-sorted to well-sorted, skeletal peloidal grainstone with common dasyclad algae (*Koninckopora*), crinoids, micrite-coated bioclasts and foraminifers of early Asbian age. Fenestrate bryozoans are mostly absent. The Black Head Member was deposited in an open-marine, shallow-water subtidal

Fig. 5.2.3. Limestone pavement in the upper part of the Tubber Formation at Black Head lighthouse (Stop 2). The dolomitized band (D) at the top by the road forms the resistant cap.

environment in the zone of normal wave action, inferred from the presence of green algae and *in situ* cerioid colonial colonies. Constant reworking of allochems at shallow photic depths is suggested by the well-sorted nature of the limestone and the presence of micritic coatings on the bioclasts.

Transfer to Stop 3

Follow the R477 coast road south for 4 km and then turn right on to the minor road leading down to the Fanore beach and large car park.

Stop 3. Fanore (UTM 480710 m E, 5885699 m N; Map 51, 113800 m E, 208500 m N)

Descend the wooden steps onto the beach and proceed for 100 m northwards to low limestone pavements amid the sand that are exposed at low tide. These almost flat bedding plane surfaces reveal dark grey to black, fine-grained, mottled limestones (bioturbated wackestones) rich in fossils, which are typical of the Fanore Member. The most conspicuous fossils are small cerioid colonies of *Lithostrotion decipiens* and *L. vorticale* up to

Fig. 5.2.4. *Lithostrotion* colony in dark grey crinoidal wackestone from the Fanore Member (lower Burren Formation) at Fanore strand (Stop 3). The coin is 2.3 cm in diameter.

25 cm in diameter (Fig. 5.2.4). Also present are the solitary corals *Siphonophyllia samsonensis* and *Axophyllum* and occasional fasciculate colonies of *Siphonodendron* and *Syringopora*. In addition, the limestones contain crinoids, gastropods and small brachiopods.

Fanore Member

The Fanore Member (46 m thick) is distinguished from the underlying member by the presence of thinner beds (0.3 to 2 m) of darker grey limestone with thin shale interbeds and an overall distinctive wavy bedding. The fauna is generally sparse with *in situ* fasciculate *Siphonodendron* and *Solenodendron furcatum* coral thickets and locally abundant *Lithostrotion* colonies. The microfacies consists of fine-grained to medium-grained skeletal peloidal wackestone, packstone and grainstone with abundant fenestrate bryozoans (sheets and fragments). Crinoids, trepostome bryozoans, sponge spicules and foraminifers are also common. Occasional bioturbated wackestone intraclasts and bored bioclasts are present. *Koninckopora* is absent or rare. The Fanore Member was deposited in subtidal, open-marine conditions, below normal wave-base but still in the photic zone. The preservation of fenestrate bryozoan sheets and the presence of trepostome bryozoans suggest quiet water, lower-energy conditions.

Transfer to Stop 4

Continue south-west along the R477 coast road for 7 km, passing a large 2 m high limestone erratic boulder (at *c*. 6.5 km) on the right (seaward) side of the road. Just past the erratic are written inscriptions (25 to 51) on the limestone cliff on the landward side of the road, which indicate the location of rock anglers' fishing points. Park vehicles by the road (there are several lay-by areas further to the south) and head down to the coast (50 m below), following the numbers on the rocks to the flat platform above the high water mark.

Stop 4. Ailladie (The Cliff of the Blind Man - UTM 475987 m E, 5880336 m N; Map 51, 109000 m E, 203200 m N)

Looking north-east from the top of the cliff towards Slieve Elva, a marked change in topography can be seen on the flanks of the hill. The lower slope forms a relatively smooth topography and corresponds to the massive limestones of the Maumcaha Member, whereas above, the upper part of the slope is distinctly terraced with well-bedded limestones of the Aillwee Member (see Fig. 5.1.2). The contact between the two members is a pronounced palaeokarst surface, which can be seen at the foot of the cliff below at Ailladie. The step-like topography of the Aillwee Member results from differential erosion of clay horizons and recent karst erosion (Gallagher, 1992; Simms, 2006; Gallagher *et al*., 2006).

The massive limestones near the top of the Maumcaha Member form the wave-cut platform at the foot of the cliff at Ailladie. They comprise pale grey, coarse-grained crinoidal packstones with *Siphonodendron* colonies and a concentration of large gigantoproductid brachiopods near the top of the bed and an upper irregular surface (low-relief palaeokarstic horizon) draped by a recessive, thin, laminated yellow shale. The overlying 60 cm thick limestone bed is a darker grey, fine-grained wackestone and has a concentration of mostly overturned *Gigantoproductus*. This bed also has a marked upper irregular palaeokarstic surface, often forming occasional hollows that can reach a metre in width and 30 cm in depth (Fig. 5.2.5). The topography of this undulose surface is locally filled by coarse dark limestone pebbles with occasional granite pebbles (Fig. 5.2.6). In places, this breccia bed also cuts up into the overlying limestone bed and is interpreted to be a cave fill, probably of Pleistocene age – the granite pebbles presumably being derived from an overlying glacial boulder clay that is no longer present on the Burren surface. The basal dark grey limestone beds of the overlying Aillwee Member (Terrace 1 of Gallagher *et al*., 2006) are much thicker (1 to 2 m thick) and contain gigantoproductid brachiopods at the base, many in growth position (concave up). This same

Fig. 5.2.5. Two palaeokarstic surfaces (white arrows) and palaeosol (yellow arrows) at the Maumcaha/Aillwee Member boundary at Ailladie cliff (Stop 4). M = Maumcaha Member, A = Aillwee Member.

Fig. 5.2.6. Close-up of palaeokarst surface (arrowed) showing upper thick palaeosol at the top of the Maumcaha Member (M) with angular dark grey-black limestone pebbles in clay matrix (see Fig. 5.2.5). This is overlain by a limestone bed of Terrace 1 (Aillwee Member, A) at Ailladie (Stop 4). The hammer is 40 cm long.

horizon is also well-exposed at Aillwee show cave (UTM 490256 m E, 5882134 m N; Map 51, 123300 m E, 204800 m N; Fig. 5.1.1) where the horizontal cave passage is developed on the impermeable clay layer below the limestone bed at the base of Terrace 1 (Simms, 2006).

Maumcaha Member

The massive, or poorly bedded, homogeneous pale grey limestones of the Maumcaha Member (80 m thick) form a distinctive smooth landscape feature and are characterised by a paucity of macrofauna, except in the upper metre, which has a concentration of brachiopods. The microfacies comprises medium-grained to coarse-grained, skeletal peloidal packstones to grainstones with abundant *Koninckopora*, kamaenid algae (*Kamaena* and *Kamaenella*), crinoids and foraminifera. Fenestrate bryozoan fragments are rare or absent. The Maumcaha Member was deposited in a shallow-marine subtidal environment, in the photic zone.

Aillwee Member

The well-bedded limestones of the Aillwee Member (152 m thick) form the terraces of the Burren landscape. These terraces result from episodes of glacio-eustatically controlled cyclic sedimentation, with each shallowing-upward cycle culminating in subaerial exposure, witnessed by the presence of palaeokarstic surfaces and occasionally bentonitic clay bands. These are best observed at coastal sites and at Aillwee Cave (Simms, 2006; Gallagher *et al.*, 2006). Nine terraces (T1 to T9) can be distinguished in the Aillwee Member (Fig. 5.1.3) and traced throughout the Burren (Fig. 5.1.2; Gallagher *et al.*, 1996; Gallagher & Somerville, 2003; Pracht *et al.*, 2004; Gallagher *et al.*, 2006). The microfacies of the Aillwee Member are variable: fenestrate bryozoans are present near the base and middle of each cycle but are often absent near the top. The alga *Koninckopora* occurs typically near the top of cycles in packstone to grainstone facies. In the lower cycles (T1 to T4), the alga *Kamaenella* is abundant but is replaced by the problematic red alga *Ungdarella* in higher cycles (T5 to T9), especially in the upper part of individual cycles. The macro-facies and micro-facies characteristics of the Aillwee Member are typical of shallowing-upward cycles in late Asbian platform rocks elsewhere in southern Ireland and on the Aran islands offshore from the Burren (Gallagher, 1996; Gallagher & Somerville, 1997, 2003; Somerville, 1999; Cózar & Somerville, 2005). The initial transgressive events of these cycles were shallow subtidal in nature. Subsequently, deeper subtidal conditions with occasional storm deposition of crinoid-rich facies prevailed. Shallowing to higher-energy subtidal conditions and emergence with pedogenesis concluded cycle deposition.

Transfer to Stop 5

From Ailladie, proceed south and south-east for 3 km along the R477 to the junction with the R479 road and then follow the R479 south-west for 2 km heading towards Doolin. Here, take a sharp right turn onto a minor road in the townland of Ballyvoe. Proceed for 1 km in a north-west direction until the road ends. Park the vehicle (note there is very limited parking here) and then head north-westwards across mostly grass-covered limestone pavements for 600 m (passing a large limestone erratic with an arrow pointing down towards the coast) to reach the coastal cliffs at Lackglass. Descend carefully onto the wave-cut platform. **N.B. This section is best avoided near high tides, especially when there is a strong westerly wind blowing, as the swell of the Atlantic Ocean can cause very high waves that are particularly dangerous.**

Stop 5. Lackglass (UTM 474191 m E, 5876720 m N; Map 51, 107153 m E, 199608 m N)

At Lackglass cliffs, three stepped terraces of the Aillwee Member (T7 to T9 in Gallagher *et al.*, 2006) are developed in gently-dipping limestones, with the contacts between terraces occasionally showing evidence of subaerial exposure. The uppermost terrace (T9) marks the boundary between the top of the Asbian Burren Formation and the base of the overlying Brigantian Slievenaglasha Formation. This contact shows a poorly-developed irregular palaeokarst surface (Fig. 5.2.7). The hollows of this surface are filled with orange bentonitic clay representing the breakdown product of volcanic ash. At the seaward end of this terrace, a more pronounced palaeokarstic surface with amplitude of up to 50 cm is visible in the cliff face (Fig. 5.2.8) (care must be exercised here). The limestone bed immediately above the clay and palaeokarst is a very distinctive pale grey, coarse-grained crinoidal limestone (grainstone) 20 cm thick. This bed, in turn, is followed by another thin orange clay band. The lower part of the Slievenaglasha Formation is composed of well-bedded, pale grey crinoidal limestone, which is much thinner-bedded than in the underlying Burren Formation (Fig. 5.2.8). Some 4 m above the base, there is a band of large unilocular foraminiferans, *Saccamminopsis*, which has been identified at the same level in other sections in the Burren (Gallagher *et al.*, 2006). Higher beds near the top of the cliff are crinoidal-rich wackestones and packstones, with chert nodules and bands, *c.* 20 m above the base of the Slievenaglasha Formation, which is *c.* 95 m in total thickness (Pracht *et al.*, 2004; Gallagher *et al.*, 2006). Occasionally, partially silicified fasciculate colonies of *Siphonodendron* and *Diphyphyllum* can be found, as well as very rare cerioid colonies of *Palastraea regia* and *Actinocyathus floriformis*,

Fig. 5.2.7. A palaeokarstic surface (arrowed) marking the boundary between the Asbian Burren Formation (top of Aillwee Member, labelled A) and Brigantian Slievenaglasha Formation (labelled S) – top of Terrace 9, Lackglass (Stop 5). The hammer is 40 cm long. Note the presence of an orange clay band at the contact (C) and the coarse crinoidal limestone bed above, representing the basal lithology of the Slievenaglasha Formation.

Fig. 5.2.8. Lackglass (Stop 5), showing the same irregular palaeokarstic surface in the cliff face (arrowed), as seen in Fig. 5.2.7, but here showing a greater erosive relief of up to 0.50 m. A - Asbian Burren Formation, S - Brigantian Slievenaglasha Formation.

which are zonal corals for the Brigantian (Gallagher, 1992; Jones & Somerville, 1996; Gallagher *et al.*, 2006; Rodríguez & Somerville, 2007).

The lower two terraces (T7 and T8) are composed of very thick posts (up to 10 m thick) of dark grey, fine-grained limestone containing rich concentrations of brachiopods. There is a particularly conspicuous band, 25 cm thick, near the top of Terrace 8 (Fig. 5.2.9), with abundant *Gigantoproductus* and rare *Davidsonina septosa* (identifiable by the presence of a median septum). In addition, this limestone bed contains occasional colonies of *Siphonodendron* and *Syringopora* showing an unusual ramose (cylindrical bush-like) growth form, up to 25 cm long (Fig. 5.2.10). Blocks of this unit occur strewn along the storm-affected coast. A similar rich concentration of *Gigantoproductus* is developed at the top of Terrace 9 below the palaeo-karst and this is accompanied by spherical colonies (up to 25 cm in diameter) of the sponge *Chaetetes* and *Syringopora*. The top of T7 forms an extensive flat bedding plane and coincides with the high water mark, but evidence of subaerial exposure is not very well developed at this horizon.

Transfer to Stop 6

From Ballyvoe, travel south along the R479 for 3 km, passing through Doolin village and then west on the R459 to Doolin Quay, where the ferries depart for the Aran Islands.

Fig. 5.2.9. Concentration of gigantoproductid brachiopods (showing thick convex pedicle valve and thin concave brachial valve), with occasional *Davidsonina septosa* (D) at the top of the limestone bed in Terrace 8. Lackglass (Stop 5). The hammer head is 7 cm long.

Fig. 5.2.10. Unusual cylindrical ramose growth form of *Syringopora* colony from Terrace 8 of the Burren Formation. Lackglass (Stop 5). Coin is 2.2 cm in diameter.

Stop 6. Doolin Quay (UTM 472885 m E, 5874157 m N; Map 51, 105811 m E, 197062 m N)

Around the harbour wall and on the foreshore south of the harbour are exposures of horizontally bedded limestones (1 to 2 m thick) of the Slievenaglasha Formation. The pale grey limestones are mostly fine-grained to medium-grained, crinoidal packstones and grainstones, with a very sparse macrofauna of small brachiopods. These limestones are inter-bedded with fine-grained limestones poor in crinoids and weakly biotur-bated, in a cyclic pattern, but there is no evidence of subaerial exposure. The top of the formation is not exposed here as it passes under the recent drift below the golf course and around the mouth of the Aille River. The cliffs further to the south-east are mostly shales of the succeeding Clare Shale Formation (see Fisherstreet Bay Stop 2, Chapter 6.2). The contact between these two formations is best exposed at Stop 7.

Transfer to Stop 7

Head back to Doolin village, then turn east onto the R479 and proceed for 2 km to the junction with the R478 road and follow this road north-east for 5 km into the town of Lisdoonvarna. At the cross-roads in the middle of the town, turn right on to a minor road and travel due east for one kilome-tre to the bridge which crosses the Gowlaun River. Cross the bridge and park in the lay-by beyond the bend on the south side of the road. From here, descend the steep track cut into blue-grey shales down to the river. Follow the river east and then north around the bend to the scarp on the left side of the river (west bank). The Gowlaun River (a tributary of the Aille River) is normally a very shallow river (in summer, the river bed is usually dry) and it is easily forded on foot. This section may be inaccessible during rainy periods and high river flows.

Stop 7. St. Brendan's Well-Gowlaun River section (UTM 481620 m E, 5875618 m N; Map 51, 114570 m E, 198402 m N)

The Gowlaun River flows over very gently dipping limestones at the top of the Slievenaglasha Formation (Fig. 5.2.11), which are overlain by a very thin unit (6 cm thick) of phosphatic conglomerate called the St. Brendan's Well Phosphate Bed by Wignall & Best (2000). Some studies consider this to be the northern-most development of the Magowna Formation, a 3 m thick unit in southern County Clare (Sleeman & Pracht, 1999; Pracht *et al.*, 2004; Fig. 5.1.1), although this is a limestone unit lacking phosphate.

The limestones in the river bed are pale grey, medium-grained crinoidal grainstones, similar to those seen at Doolin Quay, and belong to the upper part of the Slievenaglasha Formation. They contain abundant late Brigantian conodonts dominated by ornamented species of *Lochriea* (Barham *et al.*, 2015). The youngest member (Lissylisheen Member),

Fig. 5.2.11. Limestones in the bed of the Gowlaun River from the top of the Brigantian Slievenaglasha Formation (S) and black shales of the Clare Shale Formation (C) forming the cliff (Stop 7). Arrow mark shows the boundary of the formations.

which is a micrite with intraclasts and ooids (Pracht *et al.*, 2004; Gallagher *et al.*, 2006) up to 4 m thick is not developed here. However, this member is present further south at Vigo Cave near Kilfenora (UTM 493215 m E, 5867918 m N; Map 51, 126062 m E, 190537 m N), 15 km SE of Lisdoonvarna, where there is evidence of sudden shallowing with subaerial exposure (Gallagher *et al.*, 2006; Sevastopulo & Wyse Jackson, 2009). The cliff section above the river comprises black shales of the Clare Shale Formation (*c.* 15 m thick) that are described in Stop 1 of Chapter 6.

Chapter 6
The Clare Shales

PAUL B. WIGNALL, IAN D. SOMERVILLE
& KAREN BRAITHWAITE

6.1 Introduction

The Clare Shale Formation is a manifestation of a widespread phase of black shale deposition seen in the mid-Carboniferous basins of NW Europe. The shales separate Lower Carboniferous carbonates from the clastic basin fill of the Upper Carboniferous. The Clare Shale Formation shows dramatic thickness changes from a maximum of 230 m in the Shannon Estuary and passes northwards into sections only a tenth of this value north of Kilfenora. On the southern margin of the basin, in NW Co. Cork, the shales are up to 40 m thick and disappear southwards (Morton, 1965).

The first detailed study of the Clare Shale Formation was undertaken by Frank Hodson and his student Gillian Lewarne, who identified numerous goniatite-bearing marine horizons (Hodson, 1954; Hodson & Lewarne, 1961). Each band has its own unique and diagnostic fauna that have allowed the establishment of a high-resolution biostratigraphic scheme that is still in use (Table 6.1.1). This work showed that the base of the formation is diachronous, being of latest Brigantian age at its southernmost outcrops in Ballybunnion whilst at St. Brendan's Well, in northern County Clare, black shale deposition began three stages later in the Chokierian (Hodson & Lewarne 1961; Fallon & Murray, 2015; Fig. 1.7.2). The termination of black shale deposition was similarly diachronous: the topmost Clare Shale in northern County Clare is the lateral equivalent of the Ross Sandstone Formation in the south of the county.

The Clare Shale Formation is well exposed at several coastal localities and a few inland sites in both northern County Clare and along the

A Field Guide to the Carboniferous Sediments of the Shannon Basin, Western Ireland, First Edition. Edited by James L. Best and Paul B. Wignall.
© 2016 International Association of Sedimentologists.
Published 2016 by John Wiley & Sons, Ltd.
Companion website: www.wiley.com/go/best/shannonbasin

Table 6.1.1. Goniatite zones of the early Namurian stages of western Europe. Those present in the County Clare sections are denoted by lines adjacent to the stratigraphic columns.

Stage	Goniatite zones	
Kinderscoutian	*Reticuloceras dubium*	$R_{1a}5$
	Reticuloceras todmorense	$R_{1a}4$
	Reticuloceras subreticulatum	$R_{1a}3$
	Reticuloceras circumplicatile	$R_{1a}2$
	Hodsonites magistrorus	$R_{1a}1$
Alportian	*Homoceratoides prereticulatum*	$H_{2c}2$
	Vallites eostriolatus	$H_{2c}1$
	Homoceras unulatum	$H_{2b}1$
	Hudsonoceras proteum	$H_{2a}1$
Chokierian	*Isohomoceras* sp./*Dimorphoceras*	$H_{1b}2$
	Homoceras beyrichianum	$H_{1b}1$
	Isohomoceras subglobosum	$H_{1a}3$
	Isohomoceras subglobosum	$H_{1a}2$
	Isohomoceras subglobosum	$H_{1a}1$
Arnsbergian	*Nuculoceras nuculum*	$E_{2c}4$
	Nuculoceras nuculum	$E_{2c}3$
	Nuculoceras nuculum	$E_{2c}2$
	Nuculoceras stellarum	$E_{2c}1$
	Cravenoceratoides nititiodes	$E_{2b}3$
	Cravenoceratoides spp.	$E_{2b}2$
	Cravenoceratoides edalensis	$E_{2b}1$
	Eumorphoceras bisulcatum	$E_{2a}1$
Pendleian	*Cravenoceras malhamense*	E_{1c}
	Eumorphoceras pseudobilingue	E_{1b}
	Cravenoceras leion	E_{1a}

Goniatite records around Shannon Estuary

Ross Sst.

Clare Shale

Goniatite records in northern Co. Clare

Clare Shale

Shannon Estuary. However, there are few exposures in the intervening ground, except for the occurrence of a few small stream sections in steep-sided gullies that are the preserve of die-hard black shale enthusiasts. This chapter describes localities that are widely separated and you may

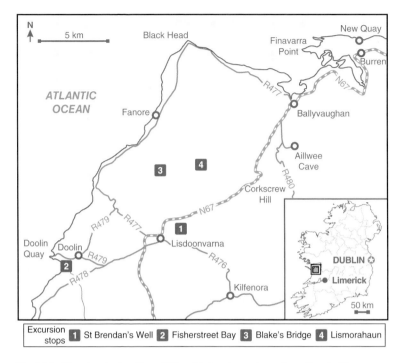

Fig. 6.1.1. Map showing location of Clare Shale outcrops in northern County Clare.

wish to combine visits to the Clare Shale sites with nearer-at-hand localities described in other chapters in this field guide (see Fig. 1.8.1). We start with the best-known and best-studied outcrop near Lisdoonvarna in northern County Clare (Fig. 6.1.1).

6.2 Clare Shale in Northern County Clare

Directions to Stop 1

The St. Brendan's Well section can be reached by taking the road that runs eastwards from the central crossroads in Lisdoonvarna. Cross the small bridge over the equally small Gowlaun River and, after ~100 m, park on the right of the road. The section can be reached through a gate and descending a farmer's track (frequently muddy and overgrown) down to the river. At most times the riverbed is dry and it is possible to walk up a gently dipping limestone pavement to examine the 15 m high cliff of the Clare Shale Formation. Only after extremely wet weather (even by the standards of western Ireland it has to be very wet) is there more than a trickle of water at the base of the cliff. Note that the cliff section is unstable, as testified by the fallen material at its base, and it is necessary to wear a hard hat at this location.

Stop 1. St. Brendan's Well (UTM 481611 m E, 5875654N m N; Map 51, 114561 m E, 198438 m N)

The limestones in the riverbed are pale grey, crinoidal grainstones similar to those seen at Doolin Quay (see Chapter 5, Stop 6) and belong to the upper part of the Slievenaglasha Formation. They are overlain by a half metre-thick bed of crinoidal limestone belonging to the Magowna Formation (see Fig. 5.1.3). A younger limestone unit (the Lissylisheen Member) is developed in the region but it is not developed at this location (Gallagher *et al.*, 2006; Sevastopulo & Wyse Jackson, 2009).

The St. Brendan's Well section has long been regarded to display a disconformable contact between the Viséan limestones and overlying Namurian Clare Shale (Douglas, 1909; Hodson, 1954). Evidence for emergence includes the gently undulatory top surface of the Magowna Limestone and the presence of vertical, dolomitised pipes that may record rootlets (Fig. 6.2.1).

The limestone is sharply overlain by a 6 cm thick bed of phosphate known as the St. Brendan's Well Phosphate Bed that, to the south of St. Brendan's Well, divides into several thin beds interleaved in the basal metres of the Clare Shale (Wignall & Best, 2000). Pracht *et al.* (2004) suggest that this bed may be the northern-most, feather-edge development of the Magowna Formation but it rests on a sharp, irregular contact and is best considered a part of the Clare Shale.

Please do not hammer the phosphate bed indiscriminately because this is one of the few places where this intriguing horizon can be readily located. There are usually loose blocks of material lying around for examination. The main feature of the phosphate bed is that it is a matrix-supported conglomerate with centimetre-sized clasts of rounded phosphate, primarily collophane (carbonate fluoroapatite) partly replaced by apatite and abundant crystals of pyrite (O'Brien, 1953; Fig. 6.2.2). The matrix consists of small crystals of collophane, calcite and dolomite and fine-grained pyrite. Some of the collophane clasts show a micro-stromatolitic structure typical of oncoid grains. Other clasts are reworked grains of the underlying limestone replaced by collophane (although columnals of crinoids often remain unaltered and thus stand out as white discs in a dark matrix). Rarer clasts include shark teeth and very rare well-polished grains of quartz around 5 to 6 mm in diameter. The top surface of the bed shows gentle undulations and is burrowed by *Rhizocorallium*.

Fig. 6.2.1. The St. Brendan's Well outcrop. The foreground shows the uneven (eroded?) top surface of the Magowna Formation and the Clare Shales form the cliff behind. The person is standing on the thin bed of phosphatic pebbles that separates these two formations.

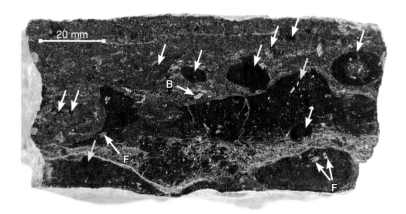

Fig. 6.2.2. Polished block of the St Brendan's Well Phosphate Bed from its type locality showing rounded pebbles (white arrows) and tabular slabs of phosphate (yellow arrows) in a matrix of phosphate and pyrite. Note the pyritized burrow (labelled 'B') and small fossil fragments, probably crinoid columnals (labelled 'F'). Photo courtesy of Jim Best.

The diverse range of features displayed by the St. Brendan's Well Phosphate Bed is testimony to a prolonged and complex depositional history and it is possible that the Early Namurian interval missing at the hiatus is represented by this bed (Wignall & Best, 2000). The phosphatic clasts probably formed as rounded, diagenetic concretions, in which case their smooth surfaces need not necessarily indicate prolonged abrasion. However, the presence of phosphatic oncoids does suggest current strengths sufficiently strong to regularly roll these cm-sized pebbles; they also imply formation within the photic zone. Phosphate precipitation occurs at redox boundaries and this, together with the abundant pyrite content, indicates predominantly oxygen-restricted deposition, although the burrows suggest intervals of more elevated sea floor oxygenation.

The primary source of the phosphate is unclear. Phosphate-rich horizons are often attributed to upwelling of nutrient-rich, deep ocean water and this model was proposed in the first study of this horizon (O'Brien, 1953), but it is unlikely this scenario applies to deposits like these, which formed within an epicontinental basin. There is clearly an element of sediment condensation involved in the formation of this phosphate bed, as indicated by its complex depositional history and its development at a hiatal surface (see discussion of the biostratigraphy below). Even more intriguing is the origin of the polished quartz pebbles; perhaps they were dropped from the roots of floating driftwood?

The sharp transition from shallow-water limestone to deep-water black shale at St. Brendan's Well is one of the most spectacular lithological changes to be seen in the geology of the Shannon Basin, and it has long been recognised that the contact between these two facies represents a considerable hiatus that may in part be recorded in the phosphate bed (Hodson, 1954; Lewarne & Hodson, 1961). The precise age of the topmost Slievenaglasha Formation is unclear but it is unlikely to range higher than the mid-Brigantian (Gallagher *et al.*, 2006). Conodonts confirm a Brigantian age for the topmost limestones and a Chokierian age for the overlying shales because *Declinognathodus noduliferus* and *Idiognathoides* sp. are recorded from only 2 m above the base (Barham *et al.*, 2014), marking the approximate position of the base of the Upper Carboniferous. This indicates that there is a substantial hiatus of around 10 million years, which spans the Pendleian and Arnsbergian stages.

The age of the Clare Shale is much easier to ascertain than the age of the St Brendan's Well Phosphate Bed because of the abundant biostratigraphically-useful goniatites present at several levels within the cliff at St. Brendan's Well. These marine horizons are associated with carbonate nodules (bullions) that contain beautiful, uncrushed examples of goniatites. The lowest horizon, around 2.0 m above the base of the formation, contains *Homoceras beyrichianum*, indicating the $H_{1b}1$ Subzone and the *Homoceratoides prereticulatum* Subzone ($H_{2c}2$) is 2.0 m higher (Fig. 6.2.3A), although somewhat

(A)

(B)

Fig. 6.2.3. A. Uncrushed specimen of *Homoceras* from a concretion low in the St. Brendan's Well section. B. *Dunbarella* from the Clare Shale at St. Brendan's Well.

difficult to access in the steep, precarious cliff face. At the nearby Aille River section at Roadford, Hodson (1954) recorded the presence of *Isohomoceras subglobosum* (H_{1a} Zone), which defines the base of the Late Carboniferous, but this goniatite has not been found at St. Brendan's Well. Therefore, the hiatus between limestone and shale at St. Brendan's Well spans the late Brigantian to Arnsbergian stages and possibly the early part of the Chokierian stage too.

Other than goniatites, bivalves are common fossils in the Clare Shale at St. Brendan's Well, with *Dunbarella* (with its numerous fine radial ribs) being the most distinctive (Fig. 6.2.3B). Nautiloids and the bivalve *Caneyella* are the only other common fossils, whilst specimens of trilobites and small, articulated crinoids occur occasionally.

Transfer to Stop 2

From St. Brendan's Well, drive back to Lisdoonvarna and then head towards Doolin and the coast. There is plenty of roadside parking in this strung-out village, although it can get very busy at peak tourist season. From the bridge over the Aille River, take the road that heads steeply uphill, signposted 'The Burren Way'. After ~250 m, a step stile appears in the wall on the right, opposite a 'B & B' sign. Take this and head across a field to a steep track that takes you down to the beach.

Stop 2. Fisherstreet Bay (UTM 473709 m E, 5873457 m N; Map 51, 106625 m E, 196350 m N)

The sea cliff on the south side of Fisherstreet Bay displays an excellent profile in the upper part of the Clare Shale Formation. The *R. subreticulatum* Marine Band ($R_{1a}3$), of the mid-Kinderscoutian Stage, is seen at the base of the cliff in a wave-cut hollow (with dangerous overhang) and the underlying *R. circumplicatile* Marine Band ($R_{1a}2$) is beautifully displayed on wave polished outcrops at low tide, although in recent years this surface has been covered in coarse cobbles. The marine bands have a fauna dominated by goniatites and the bivalves *Dunbarella* and *Caneyella* once again. The majority of the goniatites are a centimetre or so in size, typical values for black shale fauna of this age, but occasionally giant specimens approaching 20 to 25 cm are encountered. Nautiloids, both planispiral and orthocone varieties, are also relatively common and they too can reach decimetre sizes. The low diversity fauna of the black shales is typical of impoverished dysaerobic assemblages in which very low sea floor oxygenation values restricted the fauna to a few hardy, opportunistic forms.

The upper part of the cliff shows a gradual coarsening upwards as silt-laminated shales and siltstones are developed. These are inaccessible but the abundant fallen blocks on the beach reveal that these lithologies occasionally contain exquisitely preserved, delicate plant fragments. They have also yielded a single specimen of a damsel fly-like insect, one of the oldest flying insects ever found.

Note that after visiting this section you may want to see the spectacular slump outcrop at the nearby exposure described in the Chapter 8 (Stop 2). To reach this Fisherstreet Slump, you must return to the road and then turn right to continue along the dirt road – the Burren Way.

Transfer to Stop 3

From Doolin, head back to Lisdoonvarna and take the main road (N67) north out of town (Figure 6.1.1). After ~1 km, turn left onto a small road that takes you over Blake's mountain, a forested hilltop capped by the Clare Shale and towards Slieve Elva, the highest point in northern County Clare.

Stop 3. Blake's Bridge (UTM 480350 m E, 5879973 m N; Map 51, 113360 m E, 202776 m N)

The higher parts of the Clare Shale Formation are best seen at Blake's Bridge near the top of the southern slopes of Slieve Elva. Recently excavated forest tracks to the north of the bridge, together with older natural exposures, reveal several black shale sections with septarian concretions. Most levels are barren of fossils but occasional goniatite-rich levels yield *Reticuloceras* spp., indicating stratigraphic levels belonging to the Kinderscoutian Stage, along with the bivalve *Dunbarella*.

Transfer to Stop 4

The most northerly Clare Shale outcrops occur on the northern slopes of Slieve Elva and can be reached by heading northwards from Blake's Bridge to reach the coast road at Fanore, where you turn right. After 1 km, turn right again onto the small road that takes you up through the beautiful limestone scenery of the Caher Valley. After 2 km, turn right onto another tiny road. Note, that for a large part of its length this road runs approximately along the contact between the Viséan limestones (on the left) and the Clare Shale (on the right), with the result that the terrain on either side of the road is fundamentally different. After 3 km you reach the hamlet of Lismorahaun.

Stop 4. Lismorahaun (UTM 483089 m E, 5880357 m N; Map 51, 116106 m E, 203122 m N)

Hodson (1954) considered that the base of the Clare Shale becomes progressively younger northwards, between Lisdoonvarna and the northern slopes of Slieve Elva. However, this is a contentious claim because the northernmost outcrops provide only poor exposures of the basal metres of the Clare Shale and their contained goniatites are difficult to identify. Hodson (1954) reported that the lowest goniatite found on Slieve Elva could possibly be a poorly preserved *H. beyrichianum*, indicating the late Chokierian H_{1b} zone. However, it is also possible that the Slieve Elva sections are as biostratigraphically complete as the more southerly sections around Doolin, Lisdoonvarna and Kilfenora, and that older Chokierian strata are present in the north. At Slieve Elva, Hodson & Lewarne (1961) reported the presence of nearly 2.5 m of black shale between the top of the limestone and shales with Alportian goniatites. This black shale could be of Chokierian age, albeit a thinner development than the 4.5 m thick Chokierian shales seen at St. Brendan's Well. In this alternative view, the Clare Shale of northern County Clare records the sudden, widespread onset of black shale deposition in the early Chokierian following a prolonged hiatus that was initiated in the Brigantian Stage.

The northernmost Clare Shale outcrops can be seen in a series of small gullies incised into the eastern flank of Slieve Elva, the best being in a stream section at Lismorahaun. The Lismorahaun stream section reveals a small section of black shale with only the 'ghosts' of goniatite impressions. Hodson's difficulties in determining the age of these shales can be appreciated here!

6.3 Clare Shale around the Shannon Estuary

This location is a long drive from any others in the field excursion, but it is worth visiting because it displays one of the best (most expanded) mid-Carboniferous boundary sections to be seen anywhere in the world (Hodson & Lewarne, 1961; Braithwaite, 1993; Fallon & Murray, 2015). As an added bonus, the lower part of the Ross Sandstone Formation is also well displayed, showing a somewhat different character from the sections seen around Loop Head and north of Ballybunnion (described in Chapter 7).

Directions to Stop 1

Inishcorker is reached by taking the R473, which snakes its way along the north bank of the Shannon Estuary (Fig. 6.3.1). Park at the top of a farm track (UTM 492640 m E, 5835210 m N; Map 64, 125032 m E,

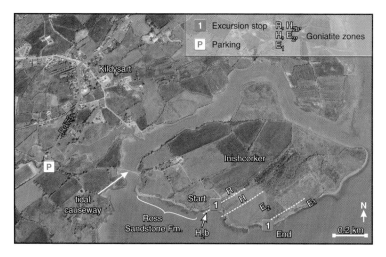

Fig. 6.3.1. Google Earth image of Inishcorker Island, showing access and principal outcrops. The goniatite zones are taken from figure 2 of Hodson & Lewarne (1961). Image courtesy GoogleEarth; © DigitalGlobe.

157824 m N) and head towards the estuary where you meet a tidal causeway (Fig. 6.3.1). Inishcorker is an island in the Shannon Estuary that is connected to the mainland at low tide. Inevitably, this means that **it is possible to become cut off** and it is always worth being somewhat paranoid about the state of the tide when on Inishcorker! The returning tide moves fast across the estuarine mudflats – you have been warned. Once you have crossed the causeway, turn right and follow the beach to the south.

**Stop 1. Inishcorker (UTM 493774 m E, 5835038 m N
(Map 64, 126164 m E, 157636 m N) to UTM 494137 m E,
5834950 m N (Map 64, 126526 m E, 157543 m N))**

The Ross Sandstone Formation is displayed for 500 m as a series of ridges that run across the foreshore (Fig. 6.3.1). Sedimentary style alternates between thin-bedded intervals and 4 to 5 m thick sandstones that are the result of bed amalgamation. The base of the Ross Sandstone Formation is especially well exposed in cliffs on the north side of a small bay (Fig. 6.3.2), which displays several thick sandstone beds, together with thinner beds. The development of the Ross Sandstone Formation was clearly not a story of gradual progradation of a turbidite fan but rather deposition began with the abrupt arrival of several, major turbidity currents.

Fig. 6.3.2. Basal beds of the Ross Sandstone Formation on Inishcorker showing presence of several thick sandstone beds, even at the onset of turbidite deposition.

South of the Ross Sandstone Formation cliff is a short stretch of poorly exposed shales, approximately 15 m thick, followed by a continuous, low cliff line that stretches for a further 500 m around the south point of the island (Fig. 6.3.3). This >200 m thick succession provides the most complete and continuous Clare Shale section of County Clare (Fig. 6.3.4). Goniatites occur at many levels and they record a succession of goniatite zones ranging from the Pendleian to the Chokierian Stage. The marine bands are thicker and more closely spaced in the Chokierian Stage where they frequently contain large carbonate concretions. This is the interval when Clare Shale deposition was established in northern County Clare (e.g. St. Brendan's Well) and the increased proportion of marine strata (and concretions) in the Inishcorker succession may reflect condensation and/or overall improved marine connections at this time. Older (Arnsbergian and Pendleian stages) shales occur along the southeast shore of the island (Figure 6.3.1). The marine fauna in these levels is much more difficult to find, often being confined to just a few centimetres of strata separated by barren shales, and large concretions are no longer present. A curious oddity is the occasional presence of vertical, paired pyrite tubes resembling pyritized *Arenicolites*.

Fig. 6.3.3. Topmost beds of the Clare Shale exposed in low cliffs at Inishcorker. The cliffs here are *c.* 5 m in height.

Like the Clare Shale in northern County Clare, the Shannon Estuary outcrops record persistent oxygen-poor deposition; however, they are for the most part older. Thus, whilst black shale deposition was occurring at Inishcorker, a highly condensed-style of phosphate deposition was occurring in the north of the basin. With the arrival of turbidite deposition in the Shannon Estuary area, the main locus of black shale deposition shifted to the north.

Transfer to Stop 2

If visiting the Ballybunnion section immediately after Inishcorker, then the best route is to head westwards along the R473 and, taking the ferry at Killimer, cross the Shannon Estuary to Tarbert. However, the Clare Shale outcrops at Ballybunnion are immediately to the north of some marvellous Viséan strata and it is anticipated that they will be visited after a short stroll across the beach from the final stop described in Chapter 5. The overlying Ross Sandstone Formation is also well displayed in the cliff line to the north of Ballybunnion (see descriptions in Chapter 7, Stop 6) and these too can be visited, although it requires more than a day in the field to do justice to all the sections in this stretch of the County Kerry coastline.

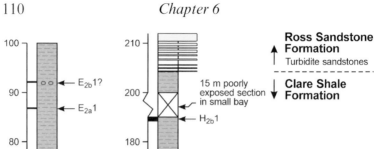

Fig. 6.3.4. Log of Clare Shale at Inishcorker (redrawn from Braithwaite, 1993). Published with the permission of the Société géologique de Belgique.

Stop 2. Ballybunnion (UTM 454073 m E, 5818461 m N; Map 63, 086216 m E, 141602 m N)

On the northern side of the main beach in Ballybunnion is an impressive cliff line of black shales from near the base of the Clare Shale, dipping gently to the south (towards the beach). Fossils are exceptionally rare but

Fig. 6.3.5. Cliff of Clare Shales on the northern side of Ballybunnion beach. The large hollows (e.g. example in lower left of photograph with bleached, white seaweed) may record the former presence of giant pyrite concretions, now oxidised.

late Brigantian (P_2) ammonoids have been recorded from near the base (Kelk, 1960). The shales show alternating hard and softer lithologies, the former being more siliceous and thus resistant to weathering. The presence of abundant pyrite in the shales is indicated by their rusty and sulphurous weathering appearance and the presence of hollows in the cliff is probably testimony to the former presence of large pyrite concretions now lost due to oxidation (Fig. 6.3.5).

Chapter 7
Architecture of a Distributive Submarine Fan: The Ross Sandstone Formation

DAVID R. PYLES & LORNA J. STRACHAN

7.1 Introduction to the Ross Sandstone Formation

The Namurian Ross Sandstone Formation crops out on sea cliffs and wave-cut platforms of the Loop Head Peninsula, County Clare and near Ballybunnion, County Kerry (Figs 7.1.1 and 7.1.2). The formation contains some of the most laterally continuous exposures of turbidites to be found anywhere, and is an excellent outcrop analogue for distributive submarine fans, which form important oil and gas reservoirs around the world. The Ross Sandstone Formation has attracted considerable attention due to its use as an analogue for turbidite hydrocarbon reservoirs, and the chapter begins with an overview of the geological setting of the Formation, together with its lithofacies and main characteristics. This summary provides the background from which the Ross Sandstone Formation outcrops are then detailed. The excursions described in this chapter provide an opportunity to visit the principal outcrops of the Ross Sandstone Formation and examine the wide array of sedimentary structures and facies that comprise this beautifully exposed ancient distributive submarine fan.

Geologic setting

Biostratigraphy

The biostratigraphic zonations of Namurian strata of western Ireland were defined by Hodson (1954a,b), who identified thin (<1.0m thick), black, goniatite-bearing, organic-rich shale beds in north County Clare (Fig. 7.1.3) and termed them "marine bands". Hodson (1954a, b) recognised that successive marine bands contain a unique goniatite assemblage

A Field Guide to the Carboniferous Sediments of the Shannon Basin,
Western Ireland, First Edition. Edited by James L. Best and Paul B. Wignall.
© 2016 International Association of Sedimentologists.
Published 2016 by John Wiley & Sons, Ltd.
Companion website: www.wiley.com/go/best/shannonbasin

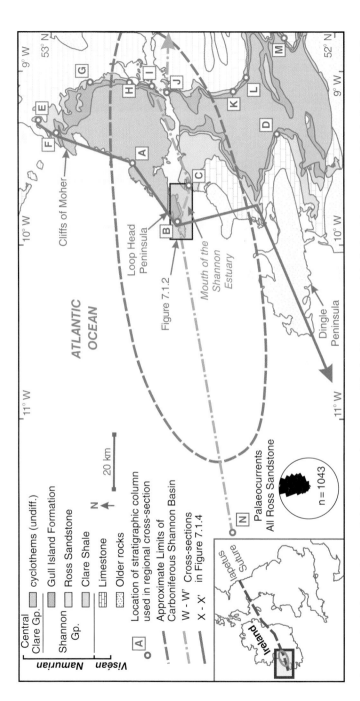

Fig. 7.1.1. Geologic map of western Ireland documenting outcrops of Carboniferous strata and the limits of deepwater deposition in the Shannon Basin (modified after Hodson & Lewarne, 1961; Holland, 1981; Martinsen et al., 2000; Wignall & Best, 2000; and Pyles, 2008). Figure modified from Pyles (2008, AAPG © used by permission of the AAPG whose permission is required for further use.

Chapter 7

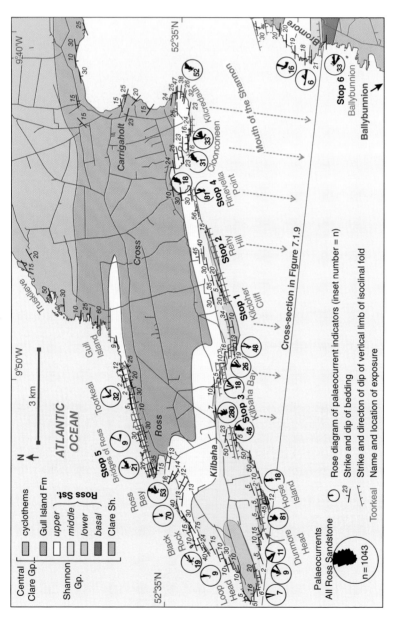

Fig. 7.1.2. Geologic map of Loop Head Peninsula documenting the location of the contacts for the Namurian lithostratigraphic units, all fold limbs in the area, strike and dip of bedding, names of exposures around the peninsula, locations of field stops described in this chapter and palaeocurrent measurements (modified after Gill, 1979; Ordnance Survey of Ireland, 2001, and Pyles, 2008 (AAPG ©, used by permission of the AAPG whose permission is required for further use).

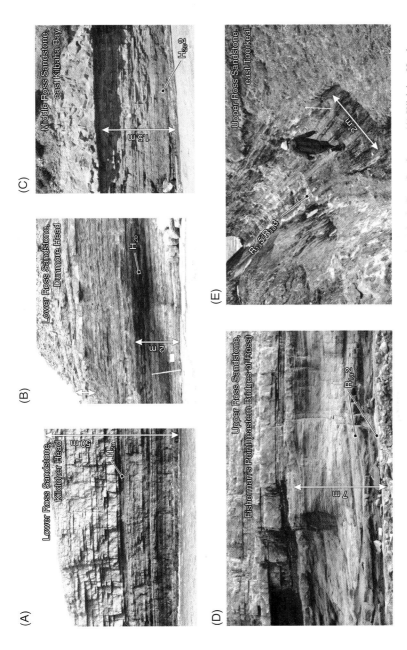

Fig. 7.1.3. Photographs of organic-rich, goniatite bearing marine bands (condensed sections) in the Ross Sandstone: (A) Kilcloher Head, (B) Dunmore Head, (C) Kilbaha Bay, (D) Fisherman's Point and (E) Toorkeal. The locations of the outcrops are shown in Fig. 7.1.2.

and that they are laterally persistent, allowing them to be used as correlation markers. More recently, Hampson *et al.* (1997) and Pyles (2008) interpreted these regionally extensive beds as condensed sections recording intervals of time when clastic sedimentation rates in the basin were low. A summary table of the lithostratigraphy and biostratigraphy of the Shannon Basin is given in Best & Wignall (Chapter 1, Fig. 1.7.2).

Lithostratigraphy

The Ross Sandstone Formation, shortened herein to Ross Sandstone, contains interbedded sandstones and siltstones interpreted by Rider (1974) as turbidites deposited in a submarine fan. Regional studies of the Ross Sandstone and Clare Shale by Hodson & Lewarne (1961), Collinson *et al.* (1991), Chapin *et al.* (1994), Wignall & Best (2000), Martinsen *et al.* (2003) and Pyles (2008) documented the Ross Sandstone to be up to 600 metres in thickness and to overlie, and laterally interfinger with, the Clare Shale on a regional scale. The base of the Ross Sandstone is therefore time-transgressive (Fig. 7.1.4). The contact between the Ross Sandstone and overlying Gull Island Formation was defined as a sharp contact following the R1a5 (*R. dubium*) marine band (condensed section) (Fig. 7.1.4; Gill, 1979; Collinson *et al.*, 1991), although this boundary was not clearly mapped. Pyles (2008) and Strachan & Pyles (Chapter 8) describe the lithological character of the upper Ross Sandstone and lower Gull Island Formation as being essentially the same – a mixed association of slumps, turbidite channels and lobes. The boundary between the formations is best described as gradational.

There is general consensus that the Ross Sandstone was deposited in a submarine-fan. The one exception was proposed by Higgs (2004, 2009) who interpreted the strata as being deposited in a ~5 m deep freshwater lake – an interpretation based on the similarity of lithological character between the Ross Sandstone and the Bude Formation (Cornwall, England).

Palaeocurrent patterns

Palaeocurrent directions reported for the Ross Sandstone vary spatially, with vector means ranging from 017° to 048° (Rider, 1974; Collinson *et al.*, 1991; Chapin *et al.*, 1994; Elliott 2000a; Martinsen *et al.*, 2000; Wignall & Best, 2000; Lien *et al.*, 2003; Pyles, 2007, 2008; Pyles & Jennette, 2009). Pyles (2008) used ripples, flutes, megaflutes and orientations of channel margins to calculate a modal palaeocurrent direction of 005° and a vector mean of 017° (Figs 7.1.1 and 7.1.2). The measurements of Pyles (2008) are locally consistent with those reported in earlier studies.

Pyles (2008) recorded a bimodal palaeocurrent pattern (Fig. 7.1.2) in the upper part of the Ross Sandstone, which he interpreted to indicate a

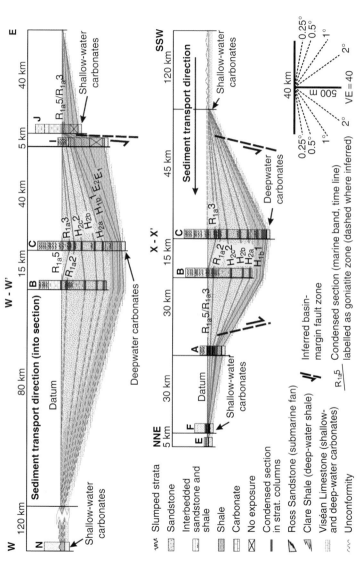

Fig. 7.1.4. Stratigraphic cross-sections documenting regional stacking patterns of the Ross Sandstone and Clare Shale in western Ireland (modified from Pyles, 2008). The locations of the cross-sections are annotated in Fig. 7.1.1. Datum for the cross-sections is the R1a5 condensed section. Data included in cross-sections are from Hodson (1954a, b), Lewarne (1959), Hodson & Lewarne (1961), Philcox (1961), Brennand (1965), Morton (1965), Ramsbottom *et al.* (1978), Tate & Dobson (1989), Collinson *et al.* (1991), Braithwaite (1993), Croker (1995), Lien *et al.* (2003) and Pyles (2008). Figure modified from Pyles, 2008 (AAPG ©, used by permission of the AAPG whose permission is required for further use).

dominant source from the south-south-west and a minor western mud-rich source that became active at that level. However, most of the slumps in the upper Ross Sandstone and lower Gull Island Formation have an east-north-eastward movement direction (Wignall & Best, 2000; Strachan, 2002a,b; Strachan & Alsop, 2006; Strachan & Pyles, Chapter 8).

Regional stacking patterns

Regional cross-sections by Hodson & Lewarne (1961), Collinson *et al.* (1991), Chapin *et al.* (1994), Wignall & Best (2000), Martinsen *et al.* (2003), Pyles (2008) and Pyles & Jennette (2009) documented a trough in sedimentation, or depocentre, located over the present-day Shannon Estuary. The trough, or basin, was referred to as the Shannon Trough (Sevastopulo, 1981a,b; Strogen, 1988), Foynes Basin (Haszeldine, 1984), Western Irish Namurian Basin (Collinson *et al.*, 1991; Wignall & Best 2000, 2002) and the Carboniferous Shannon Basin (Martinsen *et al.*, 2000; Martinsen & Collinson, 2002; Martinsen *et al.*, 2003; Pyles, 2007, 2008; Pyles & Jennette, 2009). However, there remain disagreements about the geometry and subsidence history of the Carboniferous Shannon basin during deposition of the Ross Sandstone (see Chapter 3).

Several authors interpreted the axis of the basin, indicated by the site of maximum stratal thickness, to be located along the present day Shannon Estuary (Collinson *et al.*, 1991; Martinsen *et al.*, 2000, Martinsen & Collinson, 2002; Pyles, 2008) and to overlie the interpreted trend of the Iapetus Suture (Phillips *et al.*, 1976). Others interpreted the Ross Sandstone to be part of a north-eastward prograding depositional system with a basin axis oriented south-west to north-east (Wignall & Best 2000, 2002, Chapter 3).

There are multiple interpretations of how the Ross Sandstone fits into a basin-fill succession. Rider (1974) interpreted the upward transect through Namurian strata in the basin to record eastward progradation of a linked shelf-slope-basin system. The upward succession was interpreted to be Waltherian, meaning upward changes in strata record laterally adjacent environments during deposition. Collinson *et al.* (1991) used regional cross-sections to document the basin as confined, and interpreted the Ross Sandstone to record turbidites deposited on the basin floor that were largely sourced from the north-north-west. Collinson *et al.* (1991) interpreted the upward succession of formations to record east-south-east progradation of an unstable slope (Gull Island Formation) and deltaic system (Central Clare Group) over the basin. Wignall & Best (2000, 2002) reinterpreted the upward succession to be associated with the north-eastward progradation of a related shelf-slope-basin system across the basin with the deepest part of the basin located to the north-east.

Pyles (2008) used marine bands (condensed sections) to correlate out-crops around the entire basin and cores from the Irish Shelf west of the Loop Head Peninsula to constrain three regional cross-sections through the Ross Sandstone and Clare Shale, that collectively document the three-dimensional geometry of the basin (Fig. 7.1.4). The cross-sections document the depocentre of each marine-band (condensed-section) bounded unit to be located at the Loop Head Peninsula and the regional stacking pattern to therefore be aggradational, although the depositional area increased through time. Pyles (2008) interpreted the fixed depocentre to reflect continued and focused subsidence during deposition of the Ross Sandstone, and the progressive outward onlap of strata onto the margins of the basin was interpreted to reflect that sediment supply exceeded the rate of accommodation created by subsidence. Regional palaeocurrent data from the Ross Sandstone reveals a fan-like, radially dispersive, pattern that Pyles (2008) interpreted to record a distributive submarine fan that was sourced from the south-south-west. Pyles (2008) also interpreted the upward succession of formations to record an evolving landscape that was not Waltherian (i.e. the vertical facies successions do not record laterally equivalent facies). The Ross Sandstone was interpreted to record a pon-ded, distributive submarine fan sourced from the south-south-west. The upper Ross Sandstone and lower Gull Island Formations were interpreted to represent the transition from a sand-rich southerly source to a mud-rich westerly source. The upper Gull Island Formation and Central Clare Group were then interpreted to record deltas that prograded over the basin. Pyles (2008) further interpreted the organic-rich Clare Shale located on the inferred perimeter of the basin in north-east County Clare and south-east County Kerry to be deposited in a shallow (but sub wave-base), anoxic environment because: (1) the shale abruptly overlies shallow-water Viséan carbonates in northern County Clare, whereas the coevally-deposited Ross Sandstone overlies deep-water carbonates that contain sediment gravity-flow deposits; and (2) the shale on the margin is deposited coevally to the Ross Sandstone, which thins and laps out onto the margins of the basin (Fig. 7.1.4). However, see an alternative interpretation for the basin margin presented in Chapter 6.

Local stratigraphic features

Lithofacies

Pyles (2004, 2007) described twelve lithofacies in the Ross Sandstone, that are characterised by bed thickness, grain size, internal stratification and nature of lower bounding surfaces of beds. Photographs and descriptions of each are included in Fig. 7.1.5.

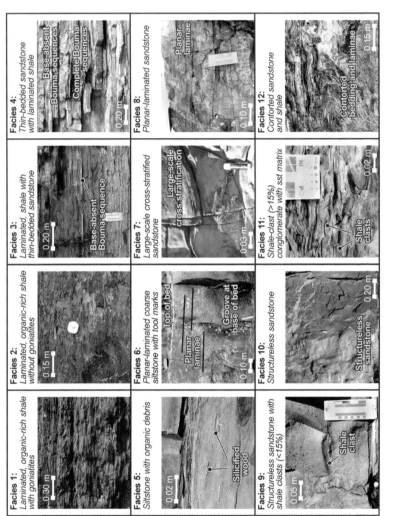

Fig. 7.1.5. Photographs and a description of the twelve lithofacies documented in the Ross Sandstone (modified after Pyles, 2007, AAPG ©, used by permission of the AAPG whose permission is required for further use).

Megaflute erosional surfaces

Megaflutes are common bedding-plane features in the Ross Sandstone (Fig. 7.1.6) and have been documented by Chapin *et al.* (1994), Elliott (2000b), Lien *et al.* (2003), Pyles (2004) and Macdonald *et al.* (2011). The megaflutes are parabolic-shaped erosional scours located on the upper surfaces of thick, amalgamated turbidite sandstone beds, and range from 1 to 45 m in width, 0.5 to 3 m in depth and 5 to > 25 m in length (Chapin *et al.*, 1994; Elliott, 2000b).

Using correlation panels from Kilbaha Bay, Elliott (2000b) proposed that megaflutes are located adjacent to channels and developed a process model for the formation of turbidite channels. Elliott (2000b) suggested turbidite channels formed by single, high-magnitude, low-frequency, fully turbulent gravity flows. Later work provides additional observations that collectively support an alternate interpretation. Firstly, Lien *et al.* (2003) and Pyles (2004) documented multiple megaflute surfaces in relatively thin (~1 m thick) intervals of strata. Secondly, Pyles (2004), Pyles & Jennette (2009) and Straub & Pyles (2012) documented megaflutes in multiple stratigraphic levels within channels. Thirdly, Pyles (2004), Pyles & Jennette (2009), Macdonald *et al.* (2011) and Pyles *et al.* (2014) demonstrated megaflutes to be common in lobes. Collectively, these observations show that megaflutes are abundant and are not restricted to regions that are laterally adjacent to channels. Consequently, they do not record relatively infrequent or high-magnitude events.

Pyles (2004), Pyles & Jennette (2009) and Pyles *et al.* (2014) documented megaflutes to be clustered in areas underlying the axial, or thickest parts, of thickening-upward beds interpreted as lobes. The upward association was interpreted to reflect a genetic relationship between megaflutes and overlying strata. Macdonald *et al.* (2011) interpreted the megaflutes to form in the late stage of lobe element progradation as flow bypass becomes increasingly prevalent.

Although some observations and interpretations are debated, all previous workers agree megaflutes are well-exposed in the Ross Sandstone and were created by fully turbulent gravity flows.

Architectural elements

Pyles (2007) interpreted the Ross Sandstone to be constructed of four architectural elements: channels, lobes, slumps and mudstone sheets (Fig. 7.1.7). The cross-sectional shape, internal characteristics and size of each element are distinctive (Fig. 7.1.8).

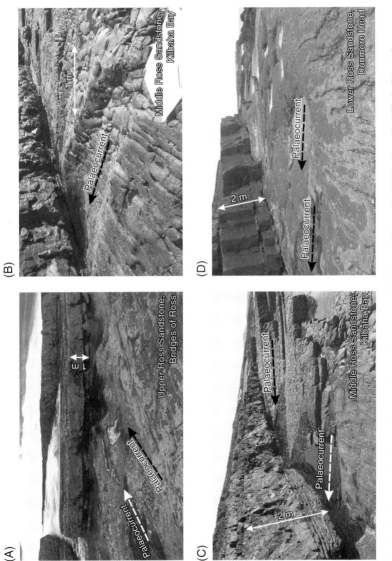

Fig. 7.1.6. Photographs of megaflutes in the Ross Sandstone: (A) Bridges of Ross, (B) Kilbaha Bay, (C) Kilbaha Bay and (D) Dunmore Head. The locations of the outcrops are annotated in Fig. 7.1.2.

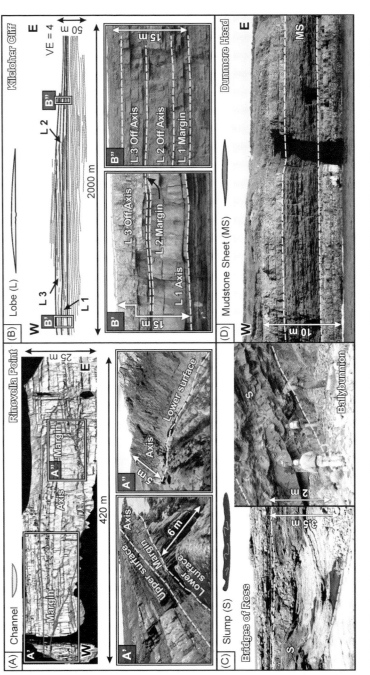

Fig. 7.1.7. The Ross Sandstone contains four different architectural elements (modified from Pyles, 2007, AAPG ©, used by permission of the AAPG whose permission is required for further use): (A) channels, (B) lobes, (C) slumps and (D) mudstone sheets. Each is distinct in terms of cross-sectional shape in strike view and internal characteristics. Locations of outcrops are shown in Fig. 7.1.2.

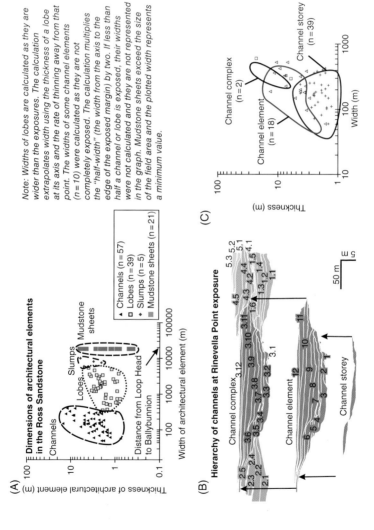

Fig. 7.1.8. Plots and diagrams of architectural elements in the Ross Sandstone (modified from Pyles, 2007. AAPG ©, whose permission is required for further use). A) Plot of thickness and width for architectural elements in the Ross Sandstone. Each architectural element has a unique width and aspect ratio. B) Diagram documenting hierarchical stacking of channels from Rinevella Point. Hierarchical levels are: channel storey, channel element, channel complex. C) Plot of thickness and width for all hierarchical levels of channels in the Ross Sandstone.

Channels

The lower bounding surfaces of channels are erosional into older strata, have a concave-upward shape and are mantled by megaflutes (Fig. 7.1.7A). The upper bounding surfaces of channels are conformable and typically planar, except where in erosional contact with younger channels. Channel fills have an average thickness and width of 4 m and 170 m, respectively, yielding an average aspect ratio of 58 (Fig. 7.1.8A). Channels contain distinctive axis-to-margin changes in lithofacies. The axes of channels often, but not always, contain a basal shale-clast conglomerate (Facies 11; Fig. 7.1.5) overlain by thick-bedded, amalgamated, structureless sandstone (Facies 10; Fig. 7.1.5) and to a lesser degree structureless sandstone with shale clasts (Facies 9; Fig. 7.1.5) and large-scale cross-stratified sandstone (Facies 7; Fig. 7.1.5). These beds thin laterally, de-amalgamate and transition to thin-bedded sandstones and laminated shales (Facies 4; Fig. 7.1.5) and laminated shale with thin-bedded sandstone (Facies 3; Fig. 7.1.5) toward the margin. The surfaces within channels are commonly locally erosional and many of the channels are asymmetric both with regard to cross-sectional shape and internal facies patterns. Furthermore, many channels in the Ross Sandstone contain inclined structureless sandstone (Facies 10; Fig. 7.1.5) units encased in shale-clast conglomerate (Facies 11; Fig. 7.1.5), which are interpreted as lateral accretion packages (LAPs) that are asymmetrically positioned on one side of the channel (Elliott, 2000a; Abreu *et al.*, 2003; Pyles, 2004).

Lobes

The lower bounding surfaces of lobes are planar and conformable with older strata, although in some examples the surfaces are locally erosional into underlying strata by up to 1 m. Erosion is related to megaflutes and the location of the maximum amount of erosion underlies the axis, or thickest part, of the lobe (Fig. 7.1.7B). In contrast to channels, the amount of erosion at the base of lobes does not equal the thickness of the lobe. The upper bounding surfaces of lobes are typically convex upward on a broad scale (> 500 m) except where in erosional contact with younger lobes. Lobes are thickest at their axis and thin toward their lateral and distal margins. The average aspect ratio of lobes is 1100 (Fig. 7.1.8A).

Lobes contain distinctive upward and axis-to-margin changes in lithofacies. The axes of lobes most commonly contain thickening-upward and coarsening-upward beds that commonly locally overlie megaflutes, with the lower beds containing thin-bedded, laminated shale with thin-bedded sandstone (Facies 3; Fig. 7.1.5), thin-bedded sandstone with laminated shale (Facies 4; Fig. 7.1.5) and the upper beds contain thick-bedded, amalgamated, structureless sandstone (Facies 10; Fig. 7.1.5) and to a lesser degree structureless sandstone with shale clasts (Facies 9) and planar-laminated sandstone (Facies 8; Fig. 7.1.5). The thick, amalgamated beds in the upper, axial parts of lobes become thinner laterally, de-amalgamate and transition into thin-bedded sandstone with laminated shale (Facies 4; Fig. 7.1.5) and eventually laminated shale with thin-bedded sandstone (Facies 3; Fig. 7.1.5) toward the margins of the lobe. The facies overlying megaflutes, when present, is most commonly siltstone with organic debris (Facies 5; Fig. 7.1.5). Beds of this facies type are only located in the lower, thin-bedded intervals of lobes and in strata deposited on the lateral and distal margins of the basin.

Slumps

The lower bounding surfaces of slumps are planar décollement and/or detachment faults (Fig. 7.1.7C), whereas the upper surfaces are conformable with overlying strata except where in erosional contact with an overlying architectural element. Slumps have an average thickness and width of 4 and 7500 m, respectively (Fig. 7.1.8A), although this measurement is conservative as both margins of slumps rarely crop out in the field area, and the average aspect ratio of slumps is 2100. Some slumps thin toward their margins, whereas others maintain a relatively constant thickness and have an abrupt lateral termination against the detachment fault.

Stratification within slumps is variable, with contorted sandstone and shale, fluidized sandstone, sandstone dykes and sandstone volcanoes (Strachan, 2002a, b). Contorted bedding in slumps is most abundant in interbedded sandstone and mudstone beds (Facies 3 and 4; Fig. 7.1.5) and thin-bedded sandstone with laminated shale beds (Facies 4; Fig. 7.1.5).

Mudstone sheets

The lower bounding surfaces of mudstone sheets are conformable with older strata and drape underlying topography (Fig. 7.1.7D), whereas the upper bounding surfaces are planar. Mudstone sheets have an average thickness of 2 m and their width exceeds the outcrop extent so cannot be measured. Mudstone sheets are the largest type of architectural element in the Ross Sandstone (Fig. 7.1.8A) and they are composed predominantly of laminated, organic-rich shale (marine bands/condensed sections; Facies 1 and 2; Fig. 7.1.5) and laminated shale with thin-bedded sandstones (Facies 3; Fig. 7.1.5).

Hierarchy of channels

Pyles (2007) and Straub & Pyles (2012) described a three-level hierarchy for channels that is (from smallest to largest): channel storey, channel element and channel complex (Fig. 7.1.8B and C). The channel element (or channel), the fundamental unit, records a series of flows that deepen and widen the channel followed by a series of flows that aggrade, or fill, the channel (Pyles *et al.*, 2010, 2012). The boundary between stratigraphically adjacent channels correlates to an abrupt shift in the location of erosion and sedimentation and was interpreted to record abandonment and avulsion. Individual channels contain systematic upward and lateral associations of strata and contain sub-units composed of beds and bedsets that are bounded by fine-grained intervals and minor erosional surfaces mantled by megaflutes. These intervals record a phase of abandonment and re-initiation within the fill of the channel. These sub-element packages are referred to as channel storeys (*sensu* Friend *et al.*, 1979). Channel complexes contain two or more genetically related channels of similar grain-size, lithofacies and architectural style that stack in a narrow geographic area.

Stratigraphic position of Ross Sandstone outcrops

Pyles (2008) constructed a composite stratigraphic cross-section of the formation whereby all coastal exposures of the Ross Sandstone near the mouth of the Shannon Estuary are projected onto a plane oriented normal to the modal palaeocurrent direction (Figs 7.1.2, 7.1.9). The exposures were correlated by mapping marine bands across the field area (Fig. 7.1.2) and on photo-panels of sea cliffs. The spatial positions of architectural elements and palaeocurrent measurements are presented in their correct position and scale, although all lobes are presented as 2 km wide, which is the average calculated width of this architectural element (Fig. 7.1.8). The cross-section documents the Ross Sandstone as 500 m thick in the Loop Head area, +/− 5%, a greater estimate than those by Collinson *et al.* (1991) and Lien *et al.* (2003) who reported thicknesses of 380 m and 460 m, respectively.

Upward changes in stratigraphic architecture

Upward changes in stratigraphic architecture in the Ross Sandstone are evident (Fig. 7.1.10). Gill (1979), Martinsen & Bakken (1990) and Strachan (2002a, b) qualitatively described an upward increase in the proportion of slumps. Elliott (2000a) and Sullivan *et al.* (2000) also noted an upward increase in the proportion of channels, which they used to informally

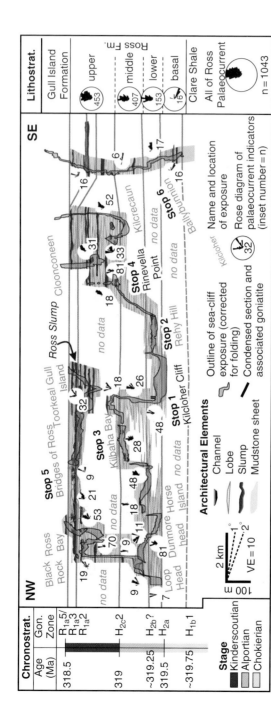

Fig. 7.1.9. Stratigraphic cross-section through the Ross Sandstone at Loop Head Peninsula showing the stratigraphic position of all sea-cliff exposures and rose diagrams of palaeocurrents (see Fig. 7.1.2 for location). The size and position of architectural elements are accurately represented, although most lobes are wider than the outcrops and the average measured width of 2 km was used to convey their dimensions in the cross-section. The datum for the cross-section is the R 1a5 condensed section. Modified after Pyles, 2008 (AAPG ©, used by permission of the AAPG whose permission is required for further use).

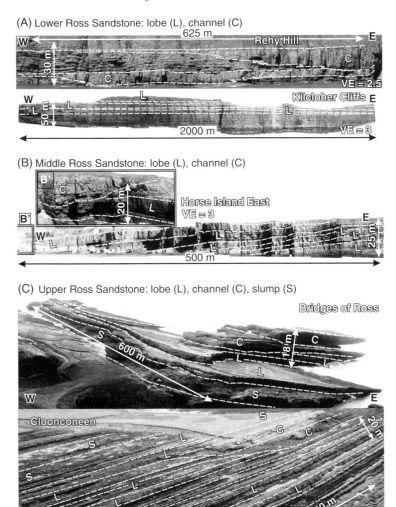

Fig. 7.1.10. Photographs of selected (A) lower, (B) middle and (C) upper Ross Sandstone exposures showing upward trends in stratigraphic architecture (modified from Pyles, 2008). The locations of the outcrops are documented on Figs 7.1.2 and 7.1.9. (L=lobes, C=channels, S=slumps). Figure modified from Pyles, 2008 (AAPG ©, used by permission of the AAPG whose permission is required for further use).

divide the Ross Sandstone into lower, middle and upper units, with the lower Ross Sandstone being predominantly lobes and the upper Ross Sandstone being predominantly channels. Subsequent work by Pyles (2008) identified seven upward patterns: (1) lobes are the most abundant architectural element by area at all stratigraphic positions in the Ross Sandstone (Figs 7.1.9, 7.1.10, 7.1.11A); (2) channels account for ~5% of the total cross-sectional area of the Ross Sandstone Formation with a subtle increase from 3% (by area) in the lower Ross Sandstone to 7% (by area) in the upper Ross Sandstone, at the expense of lobes (Figs 7.1.10, 7.1.11A); (3) channel complexes are only present in the upper Ross Sandstone (Fig. 7.1.10), although only two complexes are documented in this interval (Rinevella Point and Bridges of Ross); (4) slumps increase in proportion upward from 0% in the lower Ross Sandstone to 23% in the upper Ross Sandstone (Figs 7.1.9, 7.1.10, 7.1.11A); (5) the percentage of sandstone decreases from 70% in the lower Ross Sandstone to 51% in the upper Ross Sandstone, largely due to an increase in slumps (Fig. 7.1.11A); (6) although the average size of channels does not change upward, there is an upward increase in the range of channel sizes (Fig. 7.1.11B); and (7) the average size of lobes does not change upward (Fig. 7.1.11C).

7.2 Ross Sandstone Formation Outcrops

Axis of the basin

Five outcrops documenting an upward transect through the Ross Sandstone in the axis of the basin are included below. Stops 1 and 2 can only be viewed from a boat. A charter boat (Dolphinwatch, see companion website for information) can be hired from Carrigaholt Harbour (Fig. 7.1.2).

Stop 1: Kilcloher Cliff (UTM 445650 m E, 5824050 m N; Map 63, 077866 m E, 147311 m N)

Stop 1 is located on the southern coast of Loop Head Peninsula and contains strata from the lower Ross Sandstone (Figs 7.1.2, 7.1.9) and requires about 1 hour to view from the boat.

This outcrop has the best-exposed, most laterally persistent exposures of lobes in the Ross Sandstone, although internal facies and bed patterns are not resolvable from the boat or photographs. The outcrop is 2000 m wide and up to 50 m high (Fig. 7.2.1), with the strata being nearly horizontal. The modal palaeocurrent direction measured from megaflutes and ripples at the western part of the exposure is 010° and the outcrop is oriented roughly perpendicular to this direction (Fig. 7.1.2).

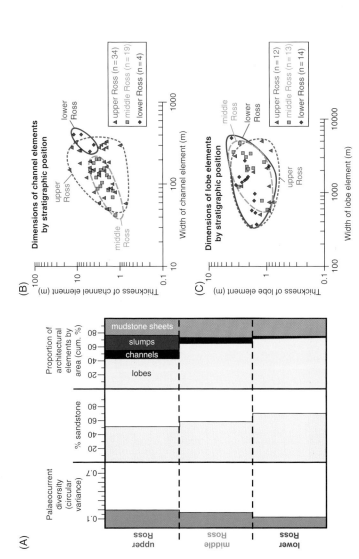

Fig. 7.1.11. A) Graphs documenting upward changes in palaeocurrent patterns, percentage sandstone and proportions of architectural elements (modified after Pyles, 2008, AAPG ©, used by permission of the AAPG whose permission is required for further use). B) Graph documenting the size of channels by stratigraphic position: t-tests and f-tests conducted at the 95% confidence interval document the average width and variance of channels in each stratigraphic level to be statistically different (modified after Pyles, 2004). The average size and variance of channels increases upward. C) Graph documenting the size of lobes by stratigraphic position: t-tests and f-tests conducted at the 95% confidence interval document the average width and variance of lobes in each stratigraphic level to be statistically similar (modified from Pyles, 2004).

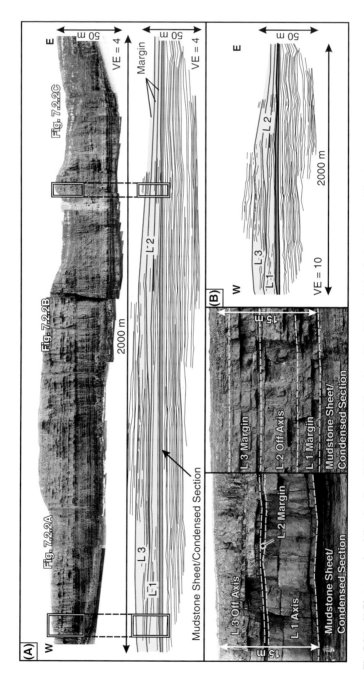

Fig. 7.2.1. A) Photo-panel and line drawing of Kilcloher Cliff (Stop 1; see Figs 7.1.2 and 7.1.9 for location; modified from Pyles, 2004). B) Line drawing of Kilcloher cliffs presented at 10x vertical exaggeration.

Fig. 7.2.1 contains a vertically exaggerated, flattened photo-panel of the outcrop and a line drawing documenting boundaries between bedsets and elements, whereas Fig. 7.2.2 contains interpreted photo-panels of the cliff. As the cliff is wide, it is imperative to request the boat captain to repeatedly troll the boat northward and southward along the outcrop to fully examine the bedding features and stratal architecture.

The lower part of the cliff contains laterally continuous, thick-bedded and thin-bedded strata that can be correlated across the width of the cliff. Chapin *et al.* (1994) referred to these deposits as "layered sheets." These beds are silty sandstones (Facies 6; Fig. 7.1.5) and have characteristics similar to those reported in co-genetic debrite-turbidite beds (Pyles & Jennette, 2009; *sensu* Talling *et al.*, 2004) and hybrid beds (Haughton *et al.*, 2009). The H_{2a} marine band (condensed section) is interpreted to be located in the middle of the cliff (Pyles, 2008) and is composed of black, organic-rich shale (Facies 1, 2; Fig. 7.1.5) that is locally recessive compared to stratigraphically adjacent lithologies.

Strata above the condensed section are sand-rich, thick-bedded sandstones that are interpreted by Collinson *et al.* (1991), Chapin *et al.* (1994) and Sullivan *et al.* (2000) as amalgamated sheets (or lobes) that do not change in character across the width of the outcrop. An alternate interpretation is described below. The western part (UTM 444710mE, 5824206mN; Map 63, 076928mE, 147480mN) of the cliff contains a thick (~10m) unit of amalgamated sandstone beds overlying the condensed section (Fig. 7.2.1B). This unit is interpreted as the axis of Lobe 1, which thins to the east where beds laterally de-amalgamate and become thinner and finer-grained. In the east-central part of the cliff, near the 'kink' in the cliff face (Fig. 7.2.1), a thick unit of amalgamated sandstone overlies the margin of Lobe 1 and is interpreted as the axis of Lobe 2, which thins to the east and west. As the lobe thins, amalgamated sandstone beds laterally de-amalgamate and become thinner and finer grained. The lobe thins to a feather edge onto the axis of Lobe 1. Above this site, there is another thick-bedded sandstone unit interpreted as the axis of Lobe 3, which thins and becomes finer grained toward the east and west. The three lobes stack compensationally, meaning their axes are offset due to depositional topography created from deposition of the underlying lobe(s). Therefore, the stratigraphy of Kilcloher Cliffs is more variable than previously recognised.

Stop 2: Rehy Hill (UTM 447250mE, 5824750mN; Map 63, 079477mE, 147989mN)

Stop 2 is located on the southern coast of Loop Head Peninsula, approximately 1 km east of Kilcloher Cliff (Stop 1) and contains strata from the lower Ross Sandstone (Figs 7.1.2 and 7.1.9). This exposure can only be viewed from a boat and requires about 1 hour.

Fig. 7.2.2. Interpreted photo-panels of Kilcloher Cliff (Stop 1). The locations of the photographs are documented in Fig. 7.2.1.

This outcrop has the best-exposed and most laterally persistent outcrops of channels in the Ross Sandstone, although internal facies and bed patterns are not resolvable from the boat or photographs. The outcrop is 1500 m wide and ~30 m thick (Figs 7.2.3 and 7.2.4) and strata are nearly horizontal. The modal palaeocurrent direction measured from flutes and grooves on overhanging beds is 025° and the outcrop is oriented roughly perpendicular to this direction (Fig. 7.1.2). This stop is divided into two sections: Stop 2a (Fig. 7.2.3) and Stop 2b (Fig. 7.2.4), which includes strata of the eastern and western parts of the cliff, respectively.

Stop 2a (UTM 447425 m E, 5824800 m N; Map 63, 079652 m E, 148036 m N), the eastern half of the exposure, contains two lenticular, amalgamated, sandstone units that are ~8 and ~12 m thick, respectively (Fig. 7.2.3). Chapin *et al.* (1994), Elliott (2000a), Sullivan *et al.* (2000) and Pyles (2004) interpreted the units as Channels 1 and 2. The channels are thickest at their axes and become thinner toward their margins, although the right (eastern) margin of Channel 2 is folded (Fig. 7.2.3B). Elliott (2000a), Chapin *et al.* (2004) and Chapin (2007) document the lower surfaces of the channels as a single erosional surface that is concave upward. Sullivan *et al.* (2000) and Pyles (2004) interpreted the margins to contain several weakly erosional surfaces for each channel (Fig. 7.2.3B').

Stop 2b (UTM 447035 m E, 5824720 m N; Map 63, 079261 m E, 1457962 m N), the western part of the exposure (Fig. 7.2.4B), contains a laterally persistent unit that is ~4 m thick in the middle part of the cliff. Elliott (2000a) interpreted the unit as a channel, whereas Chapin *et al.* (1994), Pyles (2004) and Chapin (2007) interpreted this unit as a lobe. The unit is thickest in the western part of the cliff and it thins laterally to the east where bedding de-amalgamates and the beds become thinner and finer grained and the lower bounding surface is flat. The lobe interpretation is preferred here as: (1) there is no erosion at the base of the element, (2) the width of the element far exceeds the width of channels (Fig. 7.1.8) and (3) the element contains similar internal and external characteristics to other lobes in the Ross Sandstone. Lobe 1 is overlain by Channel 3 (Fig. 7.2.4B), which is distinctively different than Channels 1 and 2 (Fig. 7.2.3). The lower bounding surface of Channel 3 is well exposed (Fig. 7.2.4B) and strata in the eastern part of the channel are inclined sandstone units that are bounded by shale-clast conglomerate (Fig. 7.2.4B). Abreu *et al.* (2003) named them Lateral Accretion Packages (LAPs). The western side of Channel 3 contains sandstone beds that onlap the western margin of the channel (Fig. 7.2.4B). The axis of the channel is locally covered by grass and is interpreted as containing mudstone. The channel is asymmetric in terms of facies and cross-sectional shape.

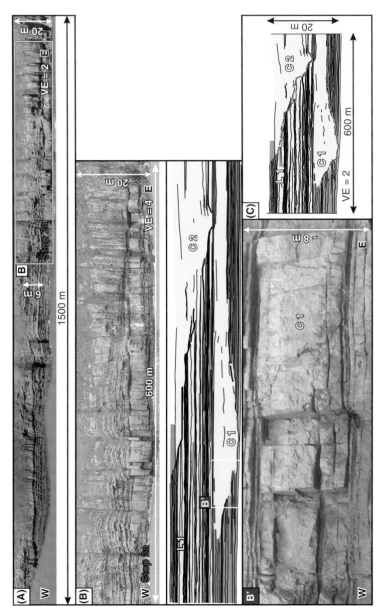

Fig. 7.2.3. A) Photo-panel of Rehy Cliff (Stop 2; see Figs 7.1.2 and 7.1.9 for location; modified from Pyles, 2004). B) Photo-panel and line drawing of the eastern part of the exposure (Stop 2a). B′) Photograph of the margin of Channel 1 (see inset in B for location). C) Line drawing of the eastern part of Rehy Hill presented at 10x vertical exaggeration. (Modified after Pyles, 2007. AAPG ©, used by permission of the AAPG whose permission is required for further use).

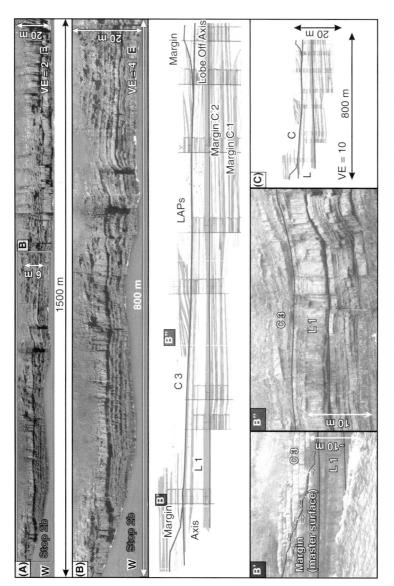

Fig. 7.2.4. A) Photo-panel of Rehy Cliff (Stop 2; see Figs 7.1.2 and 7.1.9 for location; modified after Pyles, 2004). B) Photo-panel, line drawing and photographs of the western part of the cliff (Stop 2b, modified from Chapin, 2007). C) Line drawing of the western part of Rehy Hill presented at 10x vertical exaggeration (modified after Chapin, 2007).

Transfer to Stop 3

Stop 3 is located on the southern coast of Loop Head Peninsula, near the hamlet of Kilbaha, and can be reached by driving west from Carrigaholt via the village of Cross (Fig. 7.1.2),. Parking is available near the intersection of the coastal road and the road that leads to Kilkee (UTM 442170 m E, 5824925 m N; Map 63, 074397 m E, 148234 m N). From this location, proceed to the coastal exposure where it can be accessed with ease. **This stop is only accessible during low spring tide and should be approached with caution as the rocks are very slippery when wet. This stop will require a full day to complete.**

Stop 3: Kilbaha Bay (UTM 442170 m E, 5824925 m N; Map 63, 074397 m E, 148234 m N)

This outcrop has the most accessible exposures of lobes, channels and bedding planes in the Ross Sandstone. The outcrop is 1270 m wide and a total of 30 m (vertical) of strata are exposed (Fig. 7.2.5). Strata at this exposure dip ~5 degrees to the north and several small strike-slip faults locally offset the stratigraphy, although bedding can be correlated across the faults with ease. The modal palaeocurrent direction measured from flutes, grooves, ripples and orientations of channel margins is 355° and the outcrop is oriented roughly normal to this direction (Figs 7.1.2, 7.2.5).

Several authors document the stratigraphy at Kilbaha Bay (Chapin *et al.*, 1994; Elliott 2000a,b; Sullivan *et al.*, 2000; Lien *et al.*, 2003; Pyles, 2004; Macdonald *et al.*, 2011; Straub & Pyles, 2012). Chapin *et al.* (1994) and Sullivan *et al.* (2000) document the stratal architecture of the Kilbaha Bay exposure to comprise channels and sheets (lobes) that do not change in character across the exposure. Elliott (2000a) interpreted the architecture of Kilbaha Bay to be predominantly channels and that the laterally continuous strata adjacent to the channels are "channel wings." However, Lien *et al.* (2003) interpreted the laterally continuous non-channelized strata as crevasse splays. Pyles (2004, 2008) and Straub & Pyles (2012) interpreted the outcrop to contain a mixed association of channels and lobes where channels and lobes occupy ~7% and ~93% (by area) of the outcrop, respectively. The lobe interpretation is favoured here as the units: (1) are similar in thickness to nearby channels, (2) have the same grain-size distributions as nearby channels, (3) are an order of magnitude greater in proportion than channels; and (4) display all of the characteristics described for lobes elsewhere in the Ross Sandstone where channels are not present. Straub & Pyles (2012) quantitatively document the architecture of the exposure as highly compensational.

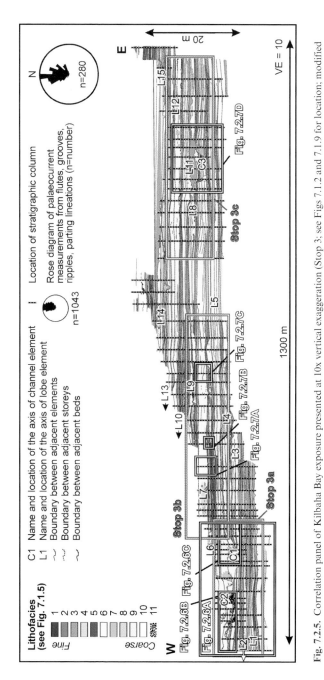

Fig. 7.2.5. Correlation panel of Kilbaha Bay exposure presented at 10x vertical exaggeration (Stop 3; see Figs 7.1.2 and 7.1.9 for location; modified after Pyles, 2004 and Straub & Pyles, 2012 (courtesy of SEPM (Society for Sedimentary Geology)).

This stop is divided into three sections (Fig. 7.2.5): (1) Stop 3a, the western part of the exposure; (2) Stop 3b, the central part of the exposure; and (3) Stop 3c, the eastern part of the exposure.

At Stop 3a (Figs 7.2.5, 7.2.6A; UTM 442075 m E, 5824868 m N; Map 63, 074301 m E, 148179 m N), the lower part of the exposure contains several thickening- and coarsening- upward units that range in thickness from 1 to 3 m and possess relatively flat lower bounding surfaces and distinctive lateral changes in facies. Pyles (2004) interpreted these units as lobes. Proceed to a unit labelled 'L1' in Fig. 7.2.5 and 'Lobe 1' in Fig. 7.2.6A, which has a thick axis of deposition located in the western part of the correlation panel (Figs 7.2.5 and 7.2.6A). The axis terminates abruptly against a fault to the west. At the axis, the lower bounding surface of Lobe 1 is exposed on an extensive bedding plane and is mantled by megaflutes (Fig. 7.2.6A). Strata overlying the megaflutes form a thickening-upward and coarsening-upward succession of beds that is ~2.5 m thick. The lower part of the succession contains laminated shale and thin-bedded sandstone (Facies 3 and 4; Fig. 7.1.5), whereas the upper part of the succession contains amalgamated, structureless sandstone (Facies 10; Fig. 7.1.5). The upper bounding surface of the lobe forms the next extensive bedding plane exposure. Bedding within the lobe de-amalgamates and becomes thinner and finer-grained laterally to the east, such that amalgamated sandstone (Facies 8; Fig. 7.1.5) passes into thin-bedded, interbedded sandstones and mudstones (Facies 3, 4; Fig. 7.1.5) over a distance of ~300 m. Critically, the thickest part of the lobe overlies megaflutes. This association was interpreted by Pyles (2004) to document a genetic linkage between the megaflutes and the overlying deposits. The lobe contains a similar axis-to-margin association as described for lobes at Kilcloher Cliff (Stop 1; Figs 7.2.1 and 7.2.2) and Rehy Hill (Stop 2; Fig. 7.2.4). The vast majority of lobes in the Ross Sandstone contain thickening-upward and coarsening-upward successions of beds and many of the walkable surfaces at Kilbaha Bay are bounding surfaces between stratigraphically adjacent lobes.

The upper part of Stop 3a (Figs 7.2.5, 7.2.6) contains two lenticular sandstone units that are approximately 5 m thick and are interpreted by Chapin *et al.* (1994), Elliott (2000a, b), Sullivan *et al.* (2000), Lien *et al.* (2003), Pyles (2004) and Straub & Pyles (2012) as channels: Channel 1 to the east (Figs 7.2.5 and 7.2.6B) and Channel 2 to the west (Figs 7.2.5 and 7.2.6B). They are distinguishable from adjacent lobes as (Figs 7.2.5 and 7.2.6A): (1) they have erosional lower bounding surfaces in which the amount of erosion is equal to the thickness of the unit; and (2) they are slightly thicker than, and not as wide as, lobes and have a significantly lower aspect ratio.

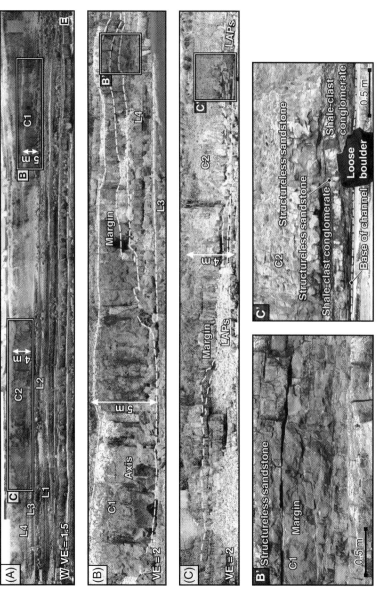

Fig. 7.2.6. A) Photo-panel of the western part of the Kilbaha Bay exposure (Stop 3a; see Fig. 7.2.5 for location). B) Photographs of the eastern part of Channel 1. C) Photographs of the western part of Channel 2.

The axis of Channel 1 (Figs 7.2.5 and 7.2.6A and B) is ~5m thick and contains amalgamated structureless sandstone (Facies 10; Fig. 7.1.5) and to a lesser degree large-scale cross-stratified sandstone (Facies 7; Fig. 7.1.5). The western part of the channel is locally covered by beach gravel and is truncated by Channel 2. The eastern part of the channel is well exposed and its lower bounding surface is locally flat at the axis and steps upward to the east. As the channel thins due to decreased erosion toward the east, bedding planes become more defined and ripple-laminated strata serve to locally de-amalgamate stratigraphically adjacent beds.

The axis of Channel 2 is well exposed (Figs 7.2.5 and 7.2.6 A and C) and is ~4m thick. The lower bounding surface is locally flat in the axis and steps upward to the eastern and western margins of the channel, which are well exposed, although they are sometimes covered by beach gravel. Strata overlying the western side of Channel 2 are laterally-dipping, structureless, sandstone (Facies 10; Fig. 7.1.5) units encased in shale-clast conglomerate (Facies 11; Fig. 7.1.5 and Fig. 7.2.6A and C). These features are similar, but smaller, to those described in Channel 3 at Rehy Hill (Stop 2; Fig. 7.2.4) and are interpreted as LAPs. Strata overlying the lower bounding surface in the eastern part of the channel are amalgamated structureless sandstone beds (Facies 10; Fig. 7.1.5) with lesser amounts of large-scale, cross-stratified sandstone (Facies 7; Figs 7.1.5, 7.2.5 and 7.2.6A). Channel 2 is asymmetric both with regard to facies and architecture.

Stop 3b focuses on the central part of the outcrop (Fig. 7.2.5). This stop emphasises Lobe 7, the axis of which is 4m thick and crops out ~200m west of Channel 1, near the upper part of the exposure (Figs 7.2.5 and 7.2.7A; UTM 442230m E, 5824915m N; Map 63, 074457m E, 148223m N). Lobe 7 is the most laterally extensive exposure of a lobe at this outcrop (Fig. 7.2.5). The lower bounding surface of Lobe 7 is locally erosional into underlying strata at the axis, where it erodes 1m into underlying strata, forming a flat surface that is ~100m wide. The eastern and western extents of this erosional feature are well exposed (Figs 7.2.5 and 7.2.7B). Although this unit has an erosional base, the amount of erosion is ~20% of the thickness of the unit and the lateral extent of the unit is 100s of metres, giving it an aspect ratio > 1000. The unit is therefore interpreted as a lobe. Strata at the axis of Lobe 7 form a thickening-upward and coarsening-upward succession. The lower part of the succession contains laminated, thin-bedded sandstone and mudstone (Facies 3 and 4; Fig. 7.1.5) and siltstone with organic debris (Facies 5; Fig. 7.1.5), whereas the upper part of the succession contains amalgamated structureless sandstone (Facies 10; Fig. 7.1.5). Approximately 350m east of the axis and across some small-scale strike-slip faults, Lobe 7 thins to 2.5m and the lower bounding surface is mantled by some of the best exposed megaflutes at Kilbaha Bay (Fig. 7.2.7C; UTM 442535m E, 5824930 N; Map 63, 074762m E, 148234m N). At this location, strata overlying the megaflutes form a thickening- and coarsening-upward succession, and bed boundaries in the upper part of the succession are more evident than at the axis. Lobe 7 continues to thin eastward to Stop 3c.

Stop 3c focuses on the eastern part of the outcrop and can be accessed from the west by correlating stratigraphy across a local embayment (UTM 442710 m E, 5824960 m N; Map 63, 074937 m E, 148262 m N) or by exiting the outcrop and re-entering it at the eastern-most part of the outcrop (UTM 443160 m E, 5824965 m N; Map 63, 075388 m E, 148261 m N). Proceed to a boulder field located in the middle of the photo-panel Fig. 7.2.7D, where Channel 3 locally incises into Lobes 7 and 8.

Lobe 7 is distinctively different here than at Stop 3b. At Stop 3c, Lobe 7 is 1.5 m thick, its lower bounding surface is flat and internal strata thicken and coarsen upward. Lobe 7 continues to thin to the eastern edge of the outcrop where it de-amalgamates and becomes finer-grained (Fig. 7.2.5; UTM 443160 m E, 5824910 m N; Map 63, 075387 m E, 148205 m N). Macdonald *et al.* (2011) studied this part of Lobe 7 and described the megaflutes at the base of Channel 3 to locally erode into the axis of Lobe 7. They used this to interpret a genetic linkage between the thickening upward successions and overlying erosional surface. This interpretation is argued herein to be implausible as: (1) the axis of Lobe 7 is located several hundred metres to the west and (2) the erosional surface cuts through Lobes 8 and 9, indicating that it is much younger than and not related to Lobe 7.

The axis of Channel 3 (UTM 442920 m E, 5824940 m N; Map 63, 075147 m E, 148239 m N) is locally covered by beach gravel, although its margins are well exposed (Fig. 7.2.7D). The lower bounding surface is locally flat at the axis and steps upward to the eastern and western margins, although the western margin is steeper than the eastern margin (Figs 7.2.5 and 7.2.7D). Strata underling the western margin (Lobes 7 and 8) are locally slumped (Fig. 7.2.5). Strata in the western part of Channel 3 are interbedded sandstone and mudstone, whereas strata in the eastern part of the channel are westward-dipping structureless sandstone (Facies 10; Fig. 7.1.5) units encased in shale-clast conglomerate (Facies 11; Fig. 7.1.5), interpreted as LAPs.

Channel 3 is overlain by Lobe 11, whose axis is in the same position as the axis of Channel 3. Lobe 11 thins substantially to its western margin (700 m) and to the west. This lobe follows a similar pattern to that described for Lobes 1 and 7 above.

Transfer to Stop 4

Stop 4 is located on the southern coast of Loop Head Peninsula, near the hamlet of Rinevella (Fig. 7.1.2; UTM 449550 m E, 5825800 m N; Map 63, 081792 m E, 1449007 m N). Public parking is available on the coastal road

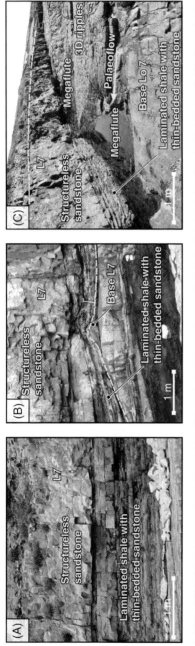

Fig. 7.2.7. A) Photograph of the axis of Lobe 7. B) Photograph of the eastern part of the erosional surface at the base of Lobe 7. C) Photograph of megaflutes on the lower bounding surface of Lobe 7 in the off-axis position. D) Photo-panel of the stratigraphy at Stop 3c. The locations of photographs are documented in Fig. 7.2.5.

at Rinevella Bay (UTM 449820 m E, 5826610 m N; Map 63, 082074 m E, 149814 m N) or along the narrow road in the hamlet of Rinevella (UTM 449375 m E, 5825965 m N; Map 63, 081620 m E, 149175 m N). After parking, walk along the rocky beach around the western limit of the bay to the tip of Rinevella Point (Fig. 7.1.2; UTM 449650 m E, 5825840 m N; Map 63, 081893 m E, 149046 m N), then proceed west past a narrow inlet in the coast line (~400 m west of the 'point'; UTM 449370 m E, 5825780 m N; Map 63, 081612 m E, 148990 m N). After crossing the inlet, enter the cliff where you will be located at the western margin of the exposure (Fig. 7.2.8). This stop can only be accessed during a low spring tide and the rocks are slippery when wet. This stop will require a half day to complete.

Stop 4: Rinevella Point (UTM 449370 m E, 5825780 m N; Map 63, 081612 m E, 148990 m N)

Rinevella Point has the most accessible exposure of a channel complex in the Ross Sandstone. The outcrop is 450 m wide and >35 m of strata are exposed, which dip ~30 degrees north. The modal palaeocurrent direction recorded from flutes, grooves, ripples and channel margin orientations is 350°, roughly perpendicular to the strike of the outcrop.

Several authors document the stratigraphy of Rinevella Point (Sullivan *et al.*, 2000; Lien *et al.*, 2003; Pyles, 2004; Straub & Pyles, 2012) and have interpreted the stratigraphy to represent channels. Pyles (2004) and Straub & Pyles (2012) document axis-to-margin changes in the channels and interpreted the stacking of channels to be hierarchical. They document five genetically related channels that stack in a narrow geographic area forming a channel complex (Fig. 7.2.8). The lower bounding surface of each channel is mantled by megaflutes (Fig. 7.2.9) and internally all of the channels contain minor erosional surfaces that have been interpreted as boundaries between adjacent storeys (Fig. 7.2.8).

This field stop focuses on the stratigraphy of Channels 3 and 4 (Fig. 7.2.8). At the western part of the outcrop (UTM 449370 m E, 5825780 m N; Map 63, 081612 m E, 148990 m N), the lower bounding surface of Channel 3 forms a small ledge that is overlain by thin-bedded sandstone and shale (Facies 3 and 4; Fig. 7.1.5). From this location there is an exceptional view of the western side of Channel 3, similar to that in the photograph in Fig. 7.2.9A. At a spring low tide, the lower bounding surface and fill of Channel 3 are entirely exposed, although the lower bounding surface is challenging to distinguish where it is underlain and overlain by sandstone beds of similar facies (e.g. below the 'C3' label in Fig. 7.2.8B). Immediately east of the narrow inlet (UTM 449375 m E, 5825760 m N; Map 63, 081617 m E, 148970 m N), the lower bounding surface is mantled by megaflutes. At this location, strata overlying the lower bounding surface

(A)

(B)

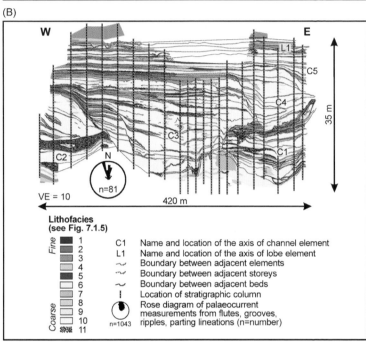

Lithofacies
(see Fig. 7.1.5)

Fine
1
2
3
4
5
6
7
8
9
Coarse
10
11

C1   Name and location of the axis of channel element
L1   Name and location of the axis of lobe element
~   Boundary between adjacent elements
~   Boundary between adjacent storeys
~   Boundary between adjacent beds
| Location of stratigraphic column
🌀   Rose diagram of palaeocurrent measurements from flutes, grooves,
n=1043   ripples, parting lineations (n=number)

(C)

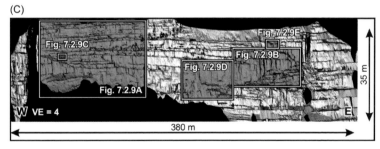

Fig. 7.2.8. A) Photo-panel of Rinevella Point (Stop 4; see Figs 7.1.2 and 7.1.9 for location). B) Correlation panel of Rinevella Point presented at 10x vertical exaggeration (modified after Pyles, 2004 and Straub & Pyles, 2012 (courtesy of SEPM (Society for Sedimentary Geology)). C) Lidar (light detection and ranging) image of Rinevella Point (modified after Pyles, 2004).

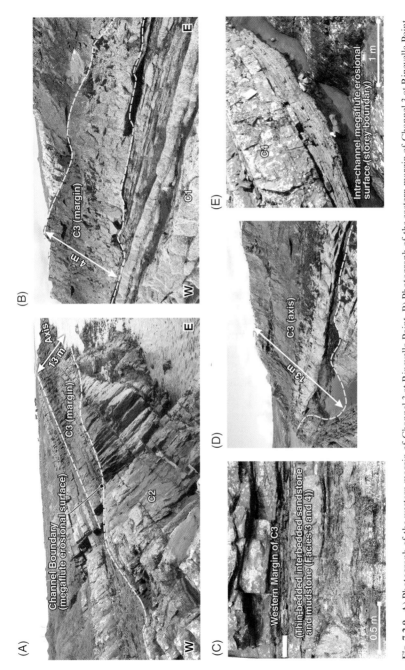

Fig. 7.2.9. A) Photograph of the western margin of Channel 3 at Rinevella Point. B) Photograph of the eastern margin of Channel 3 at Rinevella Point. C) Photograph of thin-bedded sandstones and mudstones overlying the western margin of Channel 3. D) Photograph of megaflutes located at the axis of Channel 3. E) Megaflutes at the base of a channel storey in Channel 4. The locations of all photographs are shown in Fig. 7.2.8C.

are laminated shale and thin-bedded sandstones (Facies 3 and 4; Figs 7.1.5 and 7.2.9C) overlain by a thick (>1 m) unit of structureless sandstone (Facies 10; Fig. 7.1.5). The top of this unit is the upper bounding surface of Channel 3.

The lower bounding surface erodes successively deeper into underlying strata toward the axis of Channel 3 and consequently the channel thickens and becomes coarser grained. In the off-axis position (UTM 449450 m E, 5825770 m N; Map 63, 081692 m E, 148979 m N), strata in the channel contain multiple thickening-upward and coarsening-upward bedsets, each of which is bounded by a weakly erosional surface (blue dashed lines on the correlation panel, Fig. 7.2.8B). These surfaces are interpreted as boundaries between channel storeys that record minor erosion and are then overlain by a thickening-upward succession of beds. The storeys thicken and amalgamate toward the axis of Channel 3 (UTM 449570 m E, 5825790 m N; Map 63, 081812 m E, 148997 m N), where the lower bounding surface of the channel is highly irregular (Fig. 7.2.9D). In the axis, the channel is ~10 m thick and the lower bounding surface is overlain by shale-clast conglomerate (Facies 11; Fig. 7.1.5), which is in turn overlain by amalgamated, structureless sandstone (Facies 10; Fig. 7.1.5) and lesser amounts of large-scale, cross-stratified sandstone (Facies 7; Fig. 7.1.5) and planar-laminated sandstone (Facies 8; Fig. 7.1.5, Figs 7.2.9A, B and D). Here, the strata are highly amalgamated and, as a consequence, bed and storey boundaries are nearly impossible to distinguish. The eastern margin of Channel 3 is relatively steep and stepped (Figs 7.2.8 and 7.2.9B) and is abruptly overlain by amalgamated, structureless sandstone (Facies 10). The easternmost part of Channel 3 is truncated by Channel 4 (UTM 449610 m E, 5825810 m N; Map 63, 081853 m E, 149016 m N). The lower bounding surface of Channel 4 has a steeply-dipping western margin and a relatively shallowly-dipping eastern margin. Strata in the western part of Channel 4 are predominantly amalgamated, structureless sandstone (Facies 10; Fig. 7.1.5) that lap out directly onto the margin. Strata in the eastern part of Channel 4 are westward-dipping, structureless sandstone (Facies 10; Fig. 7.1.5) units encased in shale-clast conglomerate (Facies 11) and are interpreted as LAPs (Fig 7.2.8). Channels 3 and 4 are asymmetrical in terms of facies and architecture.

Transfer to Stop 5

Stop 5 is located on the northern coast of Loop Head Peninsula, near the hamlet of Ross (Figs 7.1.2 and 7.2.10). Public parking is available in a car park west of Ross posted "Bridges of Ross" (Figs 7.1.2 and 7.2.10; UTM 441190 m E, 5827010 m N; Map 63, 073445 m E, 150334 m N). This stop can be accessed at any part of the tidal cycle, but should be

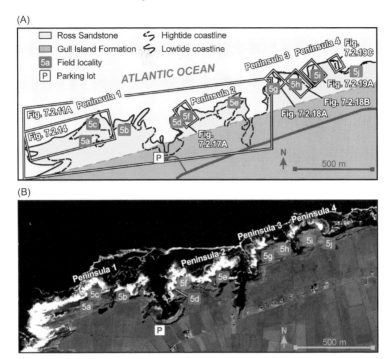

Fig. 7.2.10. A) Map of the Bridges of Ross exposure (Stop 5; see Figs 7.1.2 and 7.1.9 for location) documenting the location of field stops discussed in the text. B) Satellite photograph (courtesy GoogleEarth; Image ©DigitalGlobe) of the Bridges of Ross exposure (Stop 5) showing the localities discussed in the text. Modified after Pyles *et al.*, 2014 (courtesy of the Geological Society of America).

approached with caution as the rocks are very slippery when wet and waves are capable of overtopping the sea cliffs. This stop will require a full day to complete.

Stop 5: Bridges of Ross (UTM 441190 m E, 5827010 m N; Map 63, 073445 m E, 150334 m N)

This outcrop has the only 3D exposures of channels, lobes and slumps in the Ross Sandstone. The outcrop is ~2.5 km wide and the upper ~20 m of strata are the focus on this field stop (Fig. 7.2.10). Strata at this exposure predominantly dip gently northward although several small-scale anticlines and synclines are associated with locally vertically dipping beds, and dextral strike-slip faulting and pressure solution cleavage are located in the

(A)

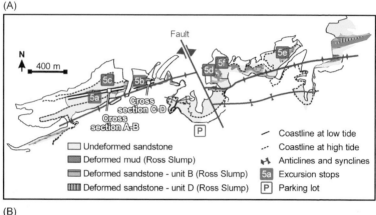

(B)

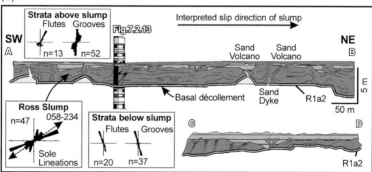

Fig. 7.2.11. A) Geologic map of the western part of the Bridges of Ross exposure highlighting the location of field stops and lithofacies associations. The location of the map is shown in Fig. 7.2.10. B) Cross-sections of the Ross Slump at Stop 1a and 1b. The locations of the cross-sections are documented in panel A. (Both modified after Strachan 2002a,b and Pyles *et al.*, 2014 (courtesy of the Geological Society of America).

southern part of the outcrop (Fig. 7.2.11A). Palaeocurrent directions measured from flutes, grooves, ripples and orientations of channel margins are bimodal to the north and east (Fig. 7.1.2).

Several authors document the stratigraphy at Bridges of Ross (Gill, 1979; Strachan, 2002a, b; Lien *et al.*, 2003; Pringle *et al.*, 2003; Pyles, 2004, 2008; Pyles *et al.*, 2014). The vast majority of published work focuses on deformed beds in the lower part of the succession and the channels that overlie them. Earlier studies interpreted the deformed strata as a slide but Wignall & Best (2000) and Strachan (2002a, b) documented internal deformational features to be more consistent with an origin as a slump and duly named these deformed beds the Ross Slump.

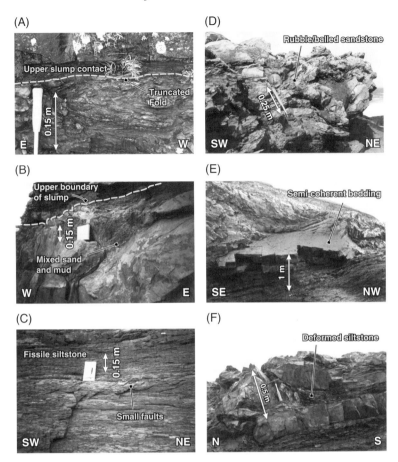

Fig. 7.2.12. Photographs of key features in the Ross Slump at Stop 1a (modified after Strachan, 2002a,b). A) Truncated folds against a fine-grained, undeformed horizon in the upper part of the Ross Slump. B) Mixed sand and mud units within the upper part of the Ross Slump. C) Small planar discontinuities that crosscut the primary foliation. D) Sandstone with a rubble/balled texture in the upper part of the Ross Slump. E) Deformed to semi-coherent sandstone raft in the Ross Slump. F) Deformed sandstone and siltstone in the upper part of the Ross Slump.

Strata above the slump are documented in earlier studies. Gill (1979) interpreted the erosional surface above the Ross Slump as slip scars related to slumping. More recently, Lien *et al.* (2003), Pringle *et al.* (2003), Pyles (2004) and Pyles *et al.* (2014) interpreted the erosional surfaces as the lower bounding surfaces of a channel. Herein, we document the strata above the Ross Slump to contain compensationally stacked lobes overlain by channels that cluster to form a channel complex. The upward succession is interpreted to reflect progradation of a distributary channel-lobe complex over the area (Pyles *et al.*, 2014).

The outcrop contains four peninsulas (Fig. 7.2.10) and this field stop is divided into 10 sections (Stops 5a through Stop 5j) reflecting a roughly eastward transect through the outcrop (Fig. 7.2.10).

Stops 5a through 5c focus on the strata north of the natural bridge known as the Bridge of Ross at Peninsula 1 (Figs 7.2.10 and 7.2.11; UTM 440840 m E, 5827155 m N; Map 63, 073097 m E, 150484 m N).

Stop 5a is focused on a shale-rich contorted unit (labelled '5a' in Figs 7.2.10A and 7.2.11A), near the north-eastern part of cross-section A-B in Fig. 7.2.11B (UTM 440780 m E, 5827180 m N; Map 63, 073037 m E, 150510 m N) and is interpreted as a slump, which Strachan (2002a, b) referred to as the Ross Slump.

The basal bounding surface of the Ross Slump is a décollement horizon coinciding with the $R_{1a}2$ condensed section (Fig. 7.2.11, and see also Figs 7.2.13 and 7.2.14 later; Gill 1979) that is interpreted to have acted as the dominant shear plane, along which most of the strain is concentrated. Although mainly planar, the basal décollement locally deviates from this trajectory and ramps down and up section where it: (1) locally erodes and entrains underlying beds (Bakken, 1987; Martinsen & Bakken, 1990), or (2) is deflected upward over dome and ridge deformational sandstone bodies (Strachan, 2002b). Strata underlying the slump are exposed in the embayment underlying the natural bridge.

The Ross Slump contains internal deformation structures including folds, faults, shear zones, boudins and highly deformed and sheared mudstone, that terminate against the upper bounding surface of the slump (Figs 7.2.12, 7.2.13 and 7.2.14). The upper bounding surface is planar and is conformably overlain by undeformed tabular sandstone.

Stop 5b is located east of Stop 5a (Figs 7.2.10 and 7.2.11; UTM 440960 m E, 5827200 m N; Map 63, 073218 m E, 150527 m N) and focuses on well-exposed deformational features in the Ross Slump documented in cross-section C to D (Fig. 7.2.11B).

Stop 5c is located north of Stop 5a and includes all strata above the Ross Slump (Figs 7.2.10, 7.2.11 and 7.2.13; UTM 440785 m E, 5827215 m N; Map 63, 073043 m E, 150545 m N), which are a mixed association of tabular and lenticular turbidites, interpreted as lobes and channels, respectively.

Tabular strata overlying the Ross Slump are divided into 3 units, and each is bounded by bedding-plane exposures (Fig. 7.2.14A). The units contain thickening-upward and coarsening-upward successions of strata composed of thin-bedded sandstone and mudstone (Facies 3 and 4; Fig. 7.1.5) overlain by thick-bedded, planar-laminated, and structureless sandstone (Facies 8 and 10; Fig. 7.1.5). The units are interpreted as lobes and are similar in character to those at Kilcloher Cliff (Stop 1; Figs 7.2.1

Lithology, grain size and structures

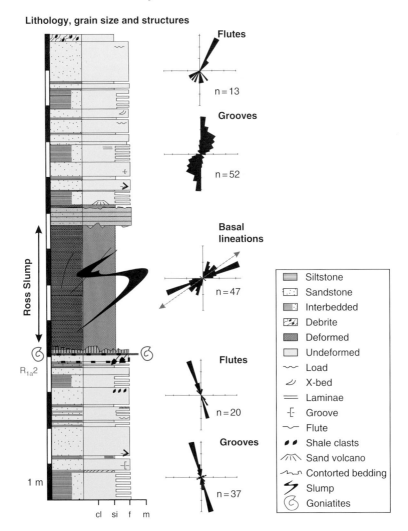

Fig. 7.2.13. Stratigraphic column of Bridges of Ross stratigraphy at Stop 5a (modified from Strachan, 2002a,b).

and 7.2.2), Rehy Hill (Stop 2; Fig. 7.2.4) and Kilbaha Bay (Stop 3; Figs 7.2.5, 7.2.6 and 7.2.7).

Lobes 1, 2 and 3 stack compensationally. The axis of Lobe 1 is located at the westernmost part of the exposure (Fig. 7.2.14B; UTM 440430 m E, 5827130 m N; Map 63, 072686 m E, 150464 m N) where it is >3 m thick and contains a thickening- and coarsening- upward succession of beds.

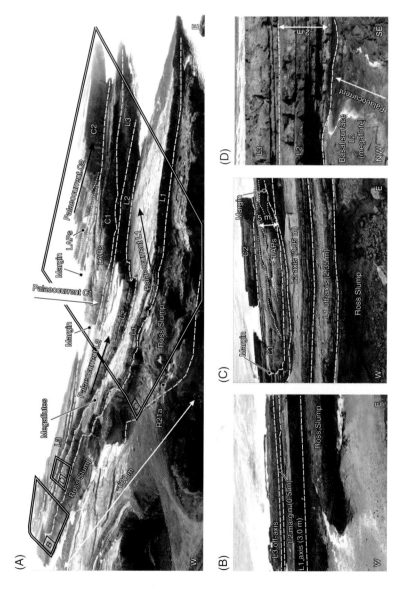

Fig. 7.2.14. A) Photo-panel of Stop 5c at Bridges of Ross (see Fig. 7.2.10 for location). B) Photograph of the western part of Stop 5c. C) Photograph of the eastern part of Stop 5c showing Channel 1. D) Photograph of the off-axis position of Lobe 2.

The lobe thins to the east and the beds de-amalgamate and become finer grained. Palaeocurrents measured from flutes and ripples in Lobe 1 are to the north (Fig. 7.2.14A). The axis of Lobe 2 is located ~500 m east of the axis of Lobe 1 (UTM 440748 m E, 5827210 m N; Map 63, 073006 m E, 150540 m N) and the lower bounding surface is mantled by megaflutes in both the axis and off-axis positions (UTM 440740 m E, 5827200 m N; Map 63, 072997 m E, 150530 m N). The locally abundant megaflutes erode up to 1 m into underlying amalgamated sandstone beds of Lobe 1 in the central part of the cliff (Figs 7.2.14A and D). Strata in the axis of the lobe thicken and coarsen upward and beds de-amalgamate and become thinner away from the axis. Palaeocurrents measured from megaflutes, flutes, ripples and dunes in Lobe 2 are north-east (Fig. 7.2.14A). Critically, palaeocurrent directions measured from megaflutes on the lower bounding surface of Lobe 2 are aligned with palaeocurrents measured from within the lobe. The megaflutes are also most abundant where Lobe 2 is thickest, indicating a genetic linkage between megaflutes and the overlying thickening-upward strata. Lobe 3 is poorly exposed and is not discussed herein.

Lenticular sandstone strata overlying the lobes (Fig. 7.2.14A) are interpreted as channels. Flutes on the basal bounding surfaces of sandstone beds and foreset orientations from cross beds, together with the orientation of the channel margins, indicate that the channel is oriented toward the south-east. The western margin of Channel 1 (UTM 440755 m E, 5827230 m N, Map 63, 073013 m E, 150560 m N; Fig. 7.2.14C) is relatively steep (>30 degrees), whereas the eastern margin (UTM 440800 m E, 5827260 m N; Map 63, 073058 m E, 150589 m N; Fig 7.2.14C) is relatively shallow and stepped. Strata in the north-eastern side of the channel are laterally-dipping sandstone units encased in shale-clast conglomerate and are interpreted as LAPs (Fig. 7.2.14C). Strata in the south-western part of the channel are amalgamated sandstone (Fig. 7.2.14C).

The lower bounding surface of Channel 2 is well exposed on its southern margin (UTM 440770 m E, 5827260 m N; Map 63, 073028 m E, 150590 m N) and is overlain by north-dipping sandstone units encased in shale-conglomerate interpreted as LAPs (Fig. 7.2.14A). Bedding-plane exposures of the sandstone units in the LAPs contain abundant dunes, from which an eastward palaeocurrent direction is evident. The axis of Channel 2 contains predominantly structureless sandstone (Facies 10; Fig. 7.1.5). From this location, one can look in the palaeocurrent direction (east) toward three other peninsulas. Each peninsula contains the same stratigraphy as this locality, with the most prominent, sand-rich strata at each peninsula being that of Channel 2.

Stops 5d through 5f focus on the strata exposed at Peninsula 2 (Fig. 7.2.10; UTM 442370 m E, 5827300 m N; Map 63, 074630 m E, 150608 m N). Collectively, these stops document an upward transect through the stratigraphy of the outcrop.

Stop 5d is focused on the Ross Slump on the western side of Peninsula 2 (Figs 7.2.10 and 26; UTM 441315 m E, 5827260 m N; Map 63, 073574 m E, 150582 m N). At this location, the Ross Slump has three distinct remobilization structures (Fig. 7.2.15, Strachan, 2002a, b): (1) deformational sandstone bodies, (2) sandstone dykes and (3) sandstone volcanoes. At this location several ~3 m tall and up to 30 m long dome-shaped and ridge-shaped deformed sandstone units are located directly beneath the slump. The Ross Slump thins significantly above them. Sandstone dykes several meters tall emanate from the underlying deformed sandstone units, extending upward in close proximity to sandstone volcanoes, which ornament the bedding plane above the Ross Slump (Fig. 7.2.15). Sandstone

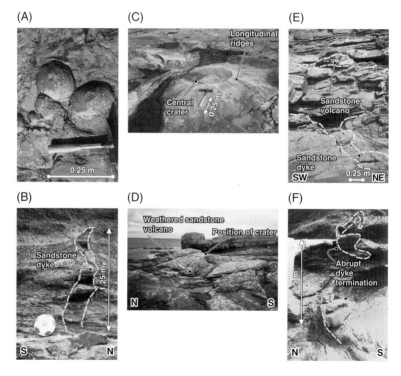

Fig. 7.2.15. Photographs of the Ross Slump at Stop 5d (see Figs 7.2.10 and 7.2.11 for location; modified after Strachan, 2002a,b and Pyles *et al.*, 2014 (courtesy of the Geological Society of America)). A) Sandstone volcanoes located on top of the Ross Slump. B) Sandstone dyke, which tapers upward and crosscuts two fabrics. C) Sandstone volcano with a well-formed central crater and ridges resulting from sedimentation along its flanks. D) Sandstone volcano. E) Sandstone dyke that crosscuts the Ross Slump; a sand volcano is located on top of the Ross Slump at this site. F) Sandstone dyke that terminates abruptly before reappearing and might represent a stepped geometry.

volcanoes above the Ross Slump are abundant and range from 0.05 to 3 m in diameter and 0.4 m in height. The flanks of the sandstone volcanoes are straight to slightly concave upward, commonly containing longitudinal ridges interpreted to result from sand flows (Gill & Kuenen, 1958; Fig. 7.2.15). The deformed sandstones, sandstone dykes and sandstone volcanoes are interpreted to result from fluidization and liquefaction associated with shear and loading coeval to emplacement of the slump (Strachan, 2002b; Pyles *et al.*, 2014).

Stop 5e is focused on the Ross Slump on the eastern side of Peninsula 2 (Figs 7.2.10 and 7.2.16; UTM 441640 m E, 5827320 m N; Map 63, 073900 m E, 150638 m N), where there are some of the best exposed deformed textures. The lower bounding surface (décollement) of the Ross Slump is well exposed at the NE end of the outcrop where lineations, interpreted to reflect local incision of the slump into underlying strata, are oriented ENE-WSW (Fig. 7.2.16). This trend is aligned with other deformation structures including folds, faults and boudinage (Bakken, 1987; Collinson *et al.*, 1991; Strachan, 2002b) and, together with fold vergence, imply movement toward the ENE (Wignall & Best, 2000; Strachan, 2002a, b). There are some deviations from this general transport trend, perhaps

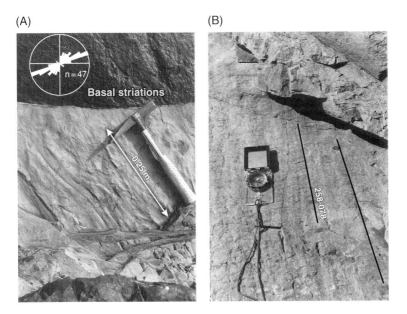

Fig. 7.2.16. Photographs of (A) striations at the base of the Ross Slump at Stop 5e, (B) lineations within the Ross Slump at Stop 5e (modified after Strachan, 2002a,b and Pyles *et al.*, 2014 (courtesy of the Geological Society of America)).

caused by flow perturbations as the unit interacted with local topography and expanded laterally. Gill (1979), in his summary of soft-sediment deformation structures in the area, proposed an alternative SE transport direction for the Ross Slump, but this direction is not supported by any subsequent studies (Bakken, 1987; Martinsen & Bakken, 1990; Collinson *et al.*, 1991; Strachan 2002a, b).

Stop 5f is located north-east of Stop 5d on the west side of the prominent sandstone outcrop (Figs 7.2.10 and 7.2.17; UTM 441320 m E, 5827300 m N; Map 63, 073579 m E, 150622 m N). At this location, Lobes 1 and 2 are 0.85 and 0.95 m thick, respectively, which is thinner than at Stop 5c (Fig. 7.2.17A and B). The lobes contain thickening-upward and coarsening-upward beds but their upper parts lack thick, amalgamated, structureless sandstone beds that are common to these lobes at Stop 5c. Due to the thinning of Lobes 1 and 2, and presumably Lobe 3, the bases of Channels 1 and 2 are only 2 m above the top of the Ross Slump, in contrast with >5 m at Stop 5c. Channel 1 crops out in the north-western part of the outcrop and can only be accessed at a low spring tide. The lower bounding surface of Channel 2 is locally flat in the axis and steps upward to the north where it erodes into Channel 1 and forms a well-exposed bedding plane from which the orientation of Channel 2 can be documented to trend to the north-east (Fig. 7.2.17A; UTM 441310 m E, 5827340 m N; Map 63, 073570 m E, 150662 m N). The southern part of the channel is tightly folded and is poorly exposed. The northern margin of Channel 2 is abruptly overlain by inclined structureless sandstone (Facies 10; Fig. 7.1.5) units encased in shale-clast conglomerate (Facies 11; Fig. 7.1.5) that are interpreted as LAPs (Figs 7.2.17A and B). These LAPs are located on the opposite margin of the channel than at Stop 5c and are onlapped by younger strata that fill Channel 2 (Fig. 7.2.17A and B). Strata overlying the LAPs on the northern margin, comprising the vast majority of the strata in the axis of the channel, are predominantly structureless sandstone (Facies 10) and to a lesser degree large-scale cross-stratified sandstone (Facies 7; Fig. 7.2.17C; UTM 441355 m E, 5827305 m N; Map 63, 073614 m E, 150626 m N).

Stops 5g and 5h focus on strata exposed at Peninsula 3 (Fig. 7.2.10; UTM 441930 m E, 5827500 m N; Map 63, 074192 m E, 150814 m N).

Stop 5g is located on the western side of Peninsula 3 (Figs 7.2.10 and 7.2.18A; UTM 441930 m E, 5827500 m N; Map 63, 074192 m E, 150814 m N). At this location, the Ross Slump is well exposed in three-dimensions and is ornamented by abundant sandstone volcanoes. Due to wave erosion, some of the sandstone volcanoes are eroded in such a manner that the internal stratigraphy of the sandstone volcanoes and feeder pipes are exposed. The feeder pipes weather out of the contorted siltstone

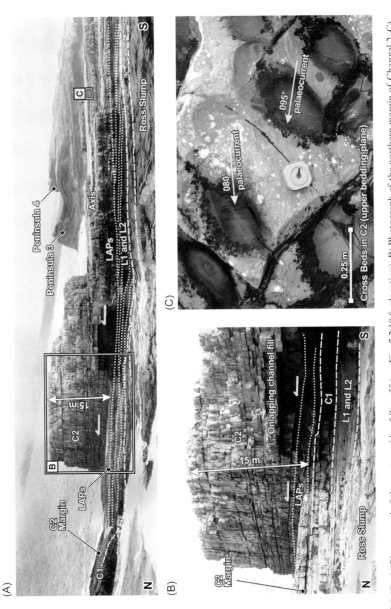

Fig. 7.2.17. A) Photo-panel of the western side of Stop 5f (see Fig. 7.2.10 for location). B) Photograph of the northern margin of Channel 2. C) Photograph of the upper surface of a large-scale cross-stratified sandstone bed that contains oval-shaped indentations that reveal large-scale foresets. Modified after Pyles *et al.*, 2014 (courtesy of the Geological Society of America).

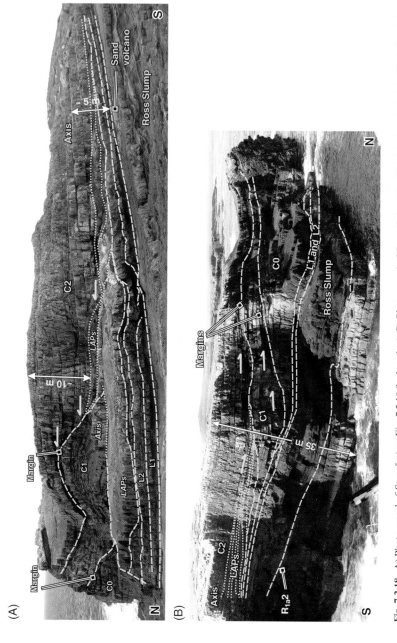

Fig. 7.2.18. A) Photo-panel of Stop 5g (see Fig. 7.2.10 for location). B) Photo-panel of Stop 5h (see Fig. 7.2.10 for location). Modified after Pyles *et al.*, 2014 (courtesy of the Geological Society of America).

as dome-shaped, very fine-grained sandstone knobs that are ~0.10 to 0.50 m in diameter. The Ross Slump is abruptly overlain by Lobes 1 and 2 (Fig. 7.2.18A) that are 0.5 and 0.7 m thick, respectively, which is thinner than at Stop 5f. Although the lobes are thin, they continue to form thickening-upward and coarsening-upward successions of beds (Fig. 7.2.18A). Three channels overlie Lobes 1 and 2 (Fig. 7.2.18A): Channels 0, 1 and 2. Channels 1 and 2 correlate to those described at Stops 5c and 5f (Figs 7.2.14 and 7.2.17), but Channel 0 is not exposed at these previous locations. Channels 0 and 1 are poorly exposed at this location and are not described herein. At this location, Channel 2 and its north-western margin are well exposed (Fig. 7.2.18A; UTM 441980 m E, 5827550 m N; Map 63, 074243 m E, 150863 m N). The lower bounding surface of Channel 2 is locally flat at the axis and it climbs upward to the north. Strata overlying the lower bounding surface in the northern part of the channel are laterally inclined structureless sandstone (Facies 10; Fig. 7.1.5) encased in shale-clast conglomerate (Facies 11) interpreted as LAPs (Fig. 7.2.18A). Strata overlying the LAPs are horizontal, structure-less and cross-bedded sandstone beds. The palaeocurrent directions measured from flutes, cross beds, flutes and orientation of the channel margin are eastward. The southern margin of Channel 2 is not exposed as it is covered by glacial till and grass.

Stop 5h is located on the east side of Peninsula 3 (Figs 7.2.10 and 7.2.18B; UTM 442100 m E, 5827560 m N; Map 63, 074363 m E, 150871 m N) but is best viewed from the opposite side of the embayment on Peninsula 4 (UTM 442210 m E, 5827605 m N; Map 63, 074474 m E, 150915 m N). The axis of Channel 2 is well exposed here, where the lowermost part of the channel is composed of structureless sandstone beds (Facies 10; Fig. 7.1.5) encased in shale-clast conglomerate (Facies 11; Fig. 7.1.5), interpreted as LAPs (Fig. 7.2.18B). The LAPs are overlain by structureless (Facies 10) and cross-bedded sandstone (Facies 7). From this location, there is an exceptional view of coevally deposited strata at Stop 5i (Fig. 7.2.19).

Stops 5i and 5j focus on strata exposed at Peninsula 4 (Fig. 7.2.10; UTM 442300 m E, 5827610 m N; Map 63, 074564 m E, 150919 m N).

Stop 5i is located on the western side of Peninsula 4 (Figs 7.2.10 and 7.2.19A) and an upward transect through the Ross Slump, Lobes 1 and 2 and Channel 2 is accessible. The Ross Slump (UTM 442230 m E, 5827563 m N; Map 63, 074493 m E, 150873 m N) is particularly well exposed here and the $R_{1a}2$ condensed section that forms the lower décollement of the Ross Slump is accessible. At this location, the condensed section is repeated in a thrust duplex (Fig. 7.1.3D). Lobes 1 and 2 overlie the Ross Slump and are 0.2 and 0.6 m thick, respectively. This is the most distal accessible exposure of these lobes and they continue

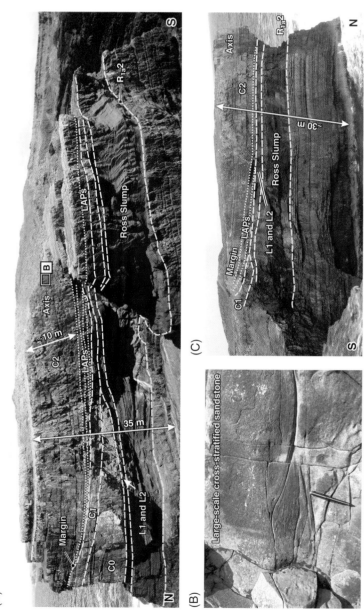

Fig. 7.2.19. A) Photo-panel of Stop 5i (see Fig. 7.2.10 for location). B) Photograph of large-scale, cross-stratified, medium-grained sandstone from the axis of Channel 2. C) Photo-panel of Stop 5j (see Fig. 7.2.10 for location). Modified after Pyles *et al.*, 2014 (courtesy of the Geological Society of America).

to contain thickening-upward and coarsening-upward beds. Although Channels 0, 1 and 2 are exposed at this locality, only the axis of Channel 2 is accessible (Fig. 7.2.19A). The north-western margin of the channel is exposed on the cliff on the northern side of the peninsula (Fig. 7.2.19A) and is overlain by inclined sandstone beds encased in shale-clast conglomerate that are interpreted as LAPs. The LAPs are accessible at a tower-shaped exposure (UTM 442185 m E, 5827575 m E; Map 63, 074448 m E, 150885 m N) and are overlain by planar strata consisting of structureless sandstone (Facies 10; Fig. 7.1.5) and large-scale cross-stratified sandstone (Facies 7; Fig. 7.1.5). The axis of the channel is accessible where some of the best examples of cross-stratified sandstone in the Ross Sandstone are exposed (Fig. 7.2.19B; UTM 442210 m E, 5827590 m N; Map 63, 074474 m E, 150890 m N). The upper surface of the channel is mantled with megaflutes from an overlying lobe that is locally exposed (UTM 442200 m E, 5827630 m N; Map 63, 074464 m E, 150940 m N). The northern margin of Channel 2 is covered by glacial till at this location, although it is easily viewed at Stop 5j.

Stop 5j is located on the eastern part of Peninsula 4 (Fig. 7.2.10; UTM 442350 m E, 5827635 m N; Map 63, 074614 m E, 150943 m N). The best view of the exposure is 100 m east along the cliff face (UTM 442530 m E, 5827640 m N; Map 63, 74794 m E, 150946 m N). The Ross Slump is exposed on the uppermost part of the cliff and can be correlated eastward for 4 km to the Toorkeal embayment (Figs 7.1.2 and 7.1.9). At Stop 5j, a remnant of Channel 1 is exposed, as is the southern margin of Channel 2 (Fig. 7.2.19C; UTM 442360 m E, 5827615 m N; Map 63, 074624 m E, 150923 m N). The southern margin of Channel 1 is overlain by LAPs, which dip to the north and are onlapped by structureless sandstone (Facies 10; Fig. 7.1.5) and large-scale cross-stratified sandstone (Facies 7; Figs 7.1.5 and 7.2.19C).

In summary, the upward succession at the Bridges of Ross (Stop 5) contains a slump (Ross Slump) overlain by compensationally stacked lobes, which are in turn overlain by channels that build a channel complex. This upward transect is interpreted to record progradation of a distributary channel-lobe complex (Pyles *et al.*, 2014).

Basin margin

Basin-margin strata are exposed in outcrops north of Ballybunnion and at Inishcorker Island (Pyles & Jennette, 2009; Figs 7.1.1 and 7.1.4). The Ballybunnion outcrop is detailed herein and contains the best and most complete view of strata interpreted herein as basin-margin in the Ross Sandstone (although see Chapter 6 for a different interpretation).

Chapter 7

Transfer to Stop 6

Stop 6 is located on the west coast of county Kerry, north of Ballybunnion, near Bromore East (Fig. 7.1.2). To reach the south side of the Shannon Estuary, a car ferry can be taken from Killimer on the north bank to Tarbert on the south bank (Fig. 1.2.1). Ballybunnion is then reached by driving westwards on the R551, via the village of Ballylongford. As you approach Ballybunnion, the main road takes a sharp turn to the south; at this point, turn north towards Bromore that is located approximately 1 km north of this road junction (Fig. 7.1.2). Parking is available on the coastal road near Bromore East adjacent to a bridge that crosses a small stream that leads to the coast (UTM 455570 m E, 5821950 m N; Map 63, 087762 m E, 145072 m N), or with permission at the farm at Bromore East (UTM 455190 m E, 5821955 m N; Map 63, 087382 m E, 145082 m N). Proceed toward the coast while walking parallel to the stream where it meets the coast as a waterfall. Access to the beach is available just south of the waterfall on a ledged cliff (UTM 454410 m E, 5821630 m N; Map 63, 086597 m E, 144768 m N). **This stop can only be accessed during a low spring tide, where access is limited to less than four hours per tidal cycle. The rocks on this beach are very slippery when wet and the cliffs are unstable.** This stop will require less than a half day to complete.

Stop 6: Ballybunnion (UTM 454410 m E, 5821630 m N; Map 63, 086597 m E, 144768 m N)

This outcrop is 500 m wide, 30 m high and has one of the best exposures of basin margin strata in the Ross Sandstone. The beds at this locality dip northward ~20°, such that a total of 150 m of strata are exposed (Fig. 7.2.20A). There is a Variscan fold near the top of the section where the dip increases locally to ~60°.

Several authors have documented the stratigraphy at the exposure (Collinson *et al.*, 1991; Lien *et al.*, 2003; Haughton *et al.*, 2009; Pyles & Jennette, 2009). Collinson *et al.* (1991) measured a stratigraphic column through the entire Ross Sandstone on the southern shore of the Shannon Estuary, and the lower ~150 m of their section corresponds to the Ballybunnion outcrop. Collinson *et al.* (1991) interpreted the overall coarsening upwards to record progradation of the Ross Sandstone, whereby the lowermost strata in the Ross Sandstone were deposited on the distal basin floor. Lien *et al.* (2003) published a more detailed section through the entire Ross Sandstone along the southern shore of the Shannon Estuary and described this outcrop as lacking any upward bed-thickness and palaeocurrent trends. Haughton *et al.* (2009) documented facies in the exposure and noted the presence of hybrid beds. Haughton *et al.* (2009)

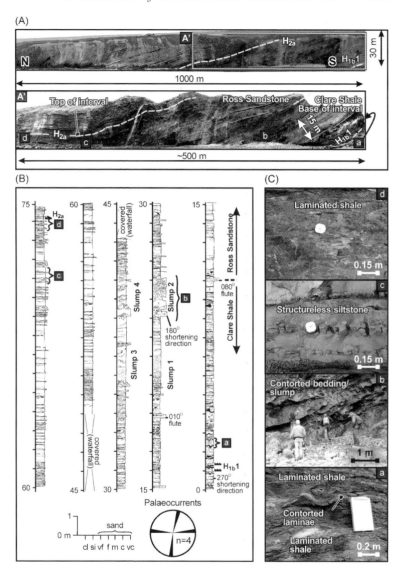

Fig. 7.2.20. A) Photo-panel of the Ballybunnion exposure highlighting Stop 6a (see Figs 7.1.2 and 7.1.9 for location). B) Stratigraphic column of the lower unit (locality 1) of the Ballybunnion exposure. Key for stratigraphic column is given in Fig. 7.2.22D. C) Photographs of key stratigraphic features of the unit. All modified from Pyles & Jennette (2009).

Chapter 7

interpreted the upward transect to record progradation with the lower-most strata, reflecting the distal most strata in the axis of the basin. Pyles & Jennette (2009) used detailed stratigraphic columns and regional correlations to constrain the location of the exposure on regional cross-sections and interpreted the exposure to contain strata deposited on the eastern margin of the basin, away from the main axis of deposition for the lower Ross Sandstone (Fig. 7.1.4). Pyles & Jennette (2009) further interpreted the co-genetic debrite-turbidite beds located in the middle of the outcrop to correlate laterally to the lobes at Kilcloher Cliff, whereas the lobes at Ballybunnion correlate laterally to the channels at Rehy Hill. Pyles & Jennette (2009) interpreted this upward succession to reflect outward expansion of the submarine fan through time, recording the progressive infill of the bowl-shaped basin. The co-genetic debrite-turbidite beds were interpreted to be located on the fringe of the submarine fan (Pyles & Jennette, 2009).

This stop is divided into three sections representing an upward transect through the lower Ross Sandstone (Figs 7.2.20, 7.2.21 and 7.2.22): (1) Stop 6a, the lowermost part of the outcrop where the transition between the Clare Shale and the Ross Sandstone is exposed, (2) Stop 6b where the middle part of the succession, which contains sandy siltstone, is exposed; and (3) Stop 6c where the Ross Sandstone transitions predominantly to sandstone.

Stop 6a focuses on the lower 75 m of strata in the succession, beginning at the peninsula that protrudes into the bay (UTM 454605 m E, 5821300 m N; Map 63, 086787 m E, 144435 m N). The upper part of Stop 6a is the base of the thick sandstone beds that form the overhanging ledge of the waterfall (Fig. 7.2.20A; UTM 454650 m E, 5821410 m N; Map 63, 086834 m E, 144545 m N). The lower strata are laminated shale beds (Facies 2) with chert and pyrite concretions (Fig. 7.2.20C; UTM 454605 m E, 5821300 m N; Map 63, 086787 m E, 144435 m N). These strata are locally folded into 0.02 to 0.2 m amplitude folds encased in undeformed, laminated shale beds. Fold axes strike north-south indicating locally east-west contraction, which is orthogonal to the regional tectonic deformation and are thus interpreted to result from synsedimentary movement. Overlying these strata is a 0.20 m thick interval of black, organic-rich shale with goniatites (Facies 1) and that is interpreted to be the horizon interpreted by Collinson *et al.* (1991) as the $H_{lb}1$ horizon, although that surface was not mapped in that study. The strata overlying this interval form a subtle thickening-upward and coarsening-upward succession to the top of the outcrop. The fine-grained strata above this condensed section gradually transitions upward from laminated shale to structureless, thin-bedded siltstone and eventually to thick-bedded coarse silt beds with flute structures on their base (Fig. 7.2.20B).

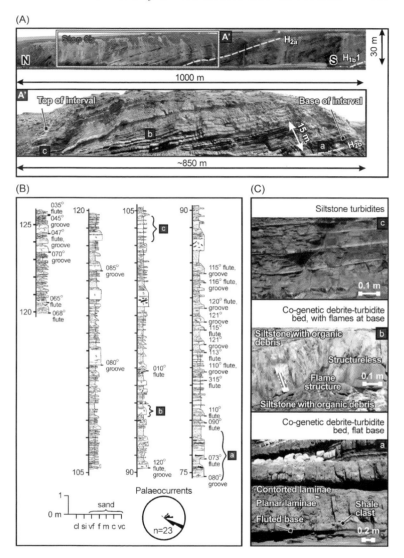

Fig. 7.2.21. A) Photo-panel of the Ballybunnion exposure highlighting Stop 6b (see Figs 7.1.2 and 7.1.9 for location). B) Stratigraphic column of the middle unit (locality 2) of the Ballybunnion exposure. Key for stratigraphic column is given in Fig. 7.2.22D. C) Photographs of key stratigraphic features of the unit. All modified from Pyles & Jennette (2009).

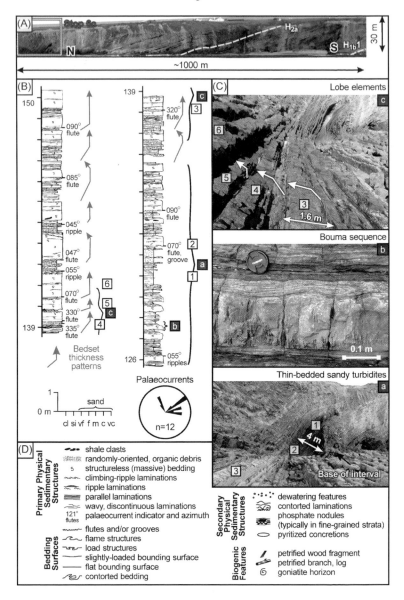

Fig. 7.2.22. A) Photo-panel of the Ballybunnion exposure highlighting Stop 6c
(see Figs 7.1.2 and 7.1.9 for location). Key for stratigraphic column shown in D.
B) Stratigraphic column of the upper unit (locality 3) of the Ballybunnion exposure.
C) Photographs of key stratigraphic features of the unit. D) Key for stratigraphic columns
shown in Figs 7.2.20 to 7.2.22. All modified from Pyles & Jennette (2009).

The first coarse siltstone bed is interpreted as the base of the Ross Sandstone (Fig. 7.2.20). The majority of strata in this unit are interbedded structureless siltstone and laminated shale (Facies 2; Fig. 7.1.5). Three slumps are located in this interval (Fig. 7.2.20). Slump 2 is most recognisable as it contains beds of coarse siltstone that locally protrude from the outcrop (Fig. 7.2.20C, photo C; UTM 454655 m E, 5821340 m N; Map 63, 086838 m E, 144474 m N). This slump thins to the south side of the peninsula where it transitions into bedded siltstone. A fold-fault pair in the slump crops out at the interface between the slump and the beach, indicating ~north-south synsedimentary contraction. This deformation is interpreted as synsedimentary as it is encased in undeformed strata. Outcrops of the other slumps are more subtly exposed as they are composed of fine-grained strata that are similar to subjacent and superjacent strata. The H_{2a} condensed section crops out immediately below the overhanging ledge of coarse siltstone that marks the base of Stop 6b (Fig. 7.2.20A; Pyles & Jennette, 2009), although Lien *et al.* (2003) inaccurately placed the interval 15 m higher in the stratigraphy.

Stop 6b focuses on the middle part of the succession. The lower part of Stop 6b is the base of the first thick-bedded coarse siltstone bed that forms the lip of the waterfall and, due to structural dip, these strata are accessible at the eastern edge of the outcrop (UTM 454565 m E, 5821450 m N; Map 63, 086749 m E, 144586 m N). The top of the succession is located just south of the boulder field at the base of the stepped entrance into the outcrop (UTM 454440 m E, 5821605 m N; Map 63, 086627 m E, 144743 m N). The interval is 51 m thick.

This interval mostly contains beds that contain lower bounding surfaces ornamented by flutes and grooves and is overlain by the following upward succession: (1) structureless coarse siltstone and very-fine grained sandstone with shale clasts; (2) planar-laminated, coarse siltstone and very-fine grained sandstone; (3) coarse siltstone and very-fine grained sandstone with contorted laminae; (4) normally graded bedding (coarse siltstone to siltstone) with dewatering features; and (5) structureless siltstone with organic debris and fluidization features. The only variation documented in the different beds of this unit are that bed thickness varies from 0.30 to 1.5 m and, in six of the 56 beds, the lower bounding surface is deformed by flame structures. The flames contain injections of structureless siltstone with organic debris from the upper division of the underlying beds. Each of the beds is interpreted to be the product of one sediment gravity flow, with each division representing a distinct flow behaviour that is interpreted as a co-genetic debrite-turbidite (Pyles & Jennette, 2009; *sensu* Talling *et al.*, 2004). Palaeocurrents in this interval are predominantly eastward, which is roughly perpendicular to the modal palaeocurrent directions recorded for coevally deposited strata on the basin-floor strata at Loop Head (Figs 7.1.2 and 7.1.10). This interval has no obvious bed-thickness trends (Lien *et al.*, 2003; Pyles & Jennette, 2009).

Stop 6c focuses on the upper 24 m of the succession. The base of this unit is located just south of the base of the stepped access point (Fig. 7.2.21; UTM 454440 m E, 5821605 m N; Map 63, 086627 m E, 144743 m N). The top of the interval is at the top of the stepped cliff (UTM 454390 m E, 5821640 m N; Map 63, 086577 m E, 144778 m N).

The lower part of this interval contains complete Bouma sequences that do not have any obvious patterns in bed-thickness (Fig. 7.2.21). A thin, recessive interval of laminated shale at 130 m (Fig. 7.2.22; UTM 454445 m E, 5821625 m N; Map 63, 086632 m E, 144762 m N) has the appearance of a condensed section but no goniatites were found in this interval. This could be an alternate location for the H_{2b} zone interpreted to be at the base of the lower and middle unit. The strata above this recessive interval contain units that thicken and coarsen upward from laminated shale and thin-bedded sandstone beds (Facies 3 and 4) to amalgamated, structureless sandstone beds (Facies 10). The thickening-upward and coarsening-upward units have a planar lower bounding surface and were interpreted by Pyles & Jennette (2009) as lobes, similar to those at Kilcloher Cliff (Stop 1; Figs 7.2.1 and 7.2.2), Rehy Hill (Stop 2; Fig. 7.2.4), Kilbaha Bay (Stop 3; Figs 7.2.5, 7.2.6 and 7.2.7) and the Bridges of Ross (Stop 5; Figs 7.2.10 to 7.2.19). The exposure at Stop 6c is narrower than the depositional limits of any one lobe and therefore axis-to-margin changes are not defined.

7.3 Synthesis

The localities described herein provide representative samples of the lower, middle and upper parts of the Ross Sandstone that were deposited in the axis of the basin and of strata in the lower part of the formation that were deposited on the basin margin (Figs 7.1.1 and 7.1.4). Key features are summarised below.

Axis of the basin

Lower Ross Sandstone outcrops described in this chapter are at Kilcloher Cliffs (Stop 1; Figs 7.2.1 and 7.2.2) and Rehy Hill (Stop 2; Figs 7.2.3 and 7.2.4). The middle Ross Sandstone outcrops at Kilbaha Bay (Stop 3; Figs 7.2.5, 7.2.6 and 7.2.7). Collectively, these outcrops document that the lower and middle parts of the Ross Sandstone are composed predominantly of compensationally stacked lobes and to a lesser amount channels and mudstone sheets. There are no channel complexes or laterally extensive slumps in the lower Ross Sandstone.

Outcrops of the upper Ross Sandstone documented in this chapter are at Rinevella Point (Stop 4; Figs 7.2.8 and 7.2.9) and Bridges of Ross (Stop 5; Figs 7.2.10 to 7.2.19). Collectively, these outcrops document that the upper Ross Sandstone is composed predominantly of lobes and to a lesser amount channels, channel complexes, slumps and mudstone sheets.

Margin of the basin

Basin-margin strata crop out at Ballybunnion (Stop 6; Figs 7.2.20 to 7.2.22), which contains strata from the lower Ross Sandstone. This outcrop contains mudstone sheets and slumps that are overlain by co-genetic debrite-turbidite beds, which are in turn overlain by lobes.

Palaeogeographic interpretation

The palaeogeographic map shown in Fig. 7.3.1A comprehensively combines all the observations described in this chapter. The map represents a time slice through the upper Ross Sandstone. Key attributes are:

1 The map illustrates a submarine fan that is oval-shaped in plan-view, the shape being constrained by the regional cross-sections from Pyles (2008) and Fig. 7.1.4. The stratigraphic architecture of the western and southern parts of the map are extrapolated.

2 The sediment was sourced from the south (Figs 7.1.1 and 7.1.2) and is presented as a point source, although there could have been multiple entry points from the south. The radially dispersive orientation of channels is constrained by local palaeocurrent measurements at Loop Head and Ballybunnion (Fig. 7.1.2) and Inishcorker (Fig. 7.1.1).

3 The axial position of the palaeogeographic map (Fig. 7.3.1A) contains a distributary channel-lobe system with spatially varying stratigraphic architecture. The axial-most position contains lobes, distributary channels and trunk channels. These bathymetric features are constrained by all the upper Ross Sandstone outcrops. Lobes are the dominant stratigraphic feature in the upper Ross Sandstone (e.g. Cloonconeen; Fig. 7.1.10C) and occupy the greatest area in this part of the map (Fig. 7.3.1A). Channel complexes (e.g. Rinevella and Bridges of Ross; Figs 7.2.8 and 7.2.9) are interpreted as trunk channels, whereas individual channels adjacent to lobes (e.g. Cloonconeen; Fig. 7.1.10c) are interpreted as distributary channels. Laterally extensive slumps are common in the upper Ross Sandstone but are not presented on this map as they would mask the architectural features described above.

(A)

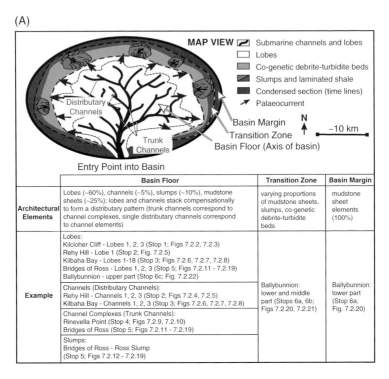

	Basin Floor	**Transition Zone**	**Basin Margin**
Architectural Elements	Lobes (~60%), channels (~5%), slumps (~10%), mudstone sheets (~25%); lobes and channels stack compensationally to form a distributary pattern (trunk channels correspond to channel complexes, single distributary channels correspond to channel elements)	varying proportions of mudstone sheets, slumps, co-genetic debrite-turbidite beds	mudstone sheet elements (100%)
Example	Lobes: Kilcloher Cliff - Lobes 1, 2, 3 (Stop 1; Figs 7.2.2, 7.2.3) Rehy Hill - Lobe 1 (Stop 2; Fig. 7.2.5) Kilbaha Bay - Lobes 1-18 (Stop 3; Figs 7.2.6, 7.2.7, 7.2.8) Bridges of Ross - Lobes 1, 2, 3 (Stop 5; Figs 7.2.11 - 7.2.19) Ballybunnion - upper part (Stop 6c; Fig. 7.2.22)	Ballybunnion: lower and middle part (Stops 6a, 6b; Figs 7.2.20, 7.2.21)	Ballybunnion: lower part (Stop 6a; Fig. 7.2.20)
	Channels (Distributary Channels): Rehy Hill - Channels 1, 2, 3 (Stop 2; Figs 7.2.4, 7.2.5) Kilbaha Bay - Channels 1, 2, 3 (Stop 3; Figs 7.2.6, 7.2.7, 7.2.8)		
	Channel Complexes (Trunk Channels): Rinevella Point (Stop 4; Figs 7.2.9, 7.2.10) Bridges of Ross (Stop 5; Figs 7.2.11 - 7.2.19)		
	Slumps: Bridges of Ross - Ross Slump (Stop 5; Figs 7.2.12 - 7.2.19)		

(B)

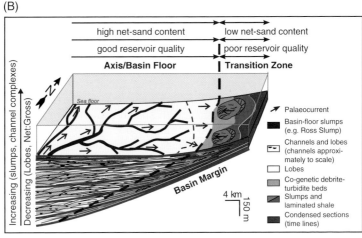

Fig. 7.3.1. A) Palaeogeographic map for the Ross Sandstone based on a comprehensive analysis of regional patterns and observations from each of the field stops (modified from Pyles & Jennette, 2009). Key features of the fan are listed for three distinctive parts of the system. B) Diagram of the interpreted stratigraphic pattern of the Ross Sandstone based on regional patterns, characteristics from the field stops and quantitatively constrained upward patterns from Fig. 7.1.11 (modified from Pyles & Jennette, 2009).

4 Sand-rich strata in the axis of the basin are illustrated to transition later-ally to co-genetic debrite-turbidite beds that are sourced from the same area. These beds onlap locally derived slumps and fine-grained strata, which in turn transition to condensed shale on the basin margin. This lateral association of facies and sedimentary architecture is constrained by the upward transect at Ballybunnion (Stop 6; Figs 7.2.20 to 7.2.22).

Palaeoenvironmental reconstructions of the lower and middle Ross Sandstone would be similar to that illustrated in Fig. 7.3.1A, except that the area of the maps would be significantly smaller (Fig. 7.1.4) and lack large trunk channels (channel complexes in the axial reaches of the basin).

The schematic block diagram shown in Fig. 7.3.1B synthesizes all observations described in this chapter and accounts for upward changes documented between the lower, middle and upper Ross Sandstone exposures. Key attributes are the following:

1 Condensed sections are extensive across the basin, as constrained by the cross-sections in Fig. 7.1.4.

2 The angle of the basin margin is approximately 0.75°, which is constrained by the rate of thinning of time-stratigraphic units (condensed section bounded cycles) from Loop Head to Inishcorker (Fig. 7.1.4) and also corresponds to the angle of the Namurian-Viséan boundary on a cross-section with a datum (Fig. 7.1.4).

3 Sandy turbidite strata of the Ross Sandstone laterally transition into the Clare Shale as constrained by all published cross-sections (Fig. 7.1.4).

4 An upward transect through the basin margin consists of mudstone sheets overlain by laterally restricted slumps and co-genetic debrite-turbidite beds, overlain by lobes. This upward transect is constrained by the patterns documented at Ballybunnion (Stop 6; Figs 7.2.20 to 7.2.22).

5 Upward patterns in stratigraphy at the centre of the basin are constrained by observations from Stop 1 through Stop 5 and the quantitative data presented in Fig. 7.1.11 where: (1) lobes are the dominant architectural elements at all stratigraphic positions; (2) laterally extensive slumps (e.g. Ross Slump) and channel complexes are restricted to the upper Ross Sandstone; and (3) net-sand content decreases upward.

6 The depositional area of the fan increased over time and, as a result, the point of onlap moved progressively outward onto the angled basin margin. This pattern is constrained by regional cross-sections from Pyles (2008; Fig. 7.1.4), who interpreted this trend to result from sediment supply exceeding the rate at which accommodation space was generated through differential subsidence in the basin.

7 The transition zone, which contains co-genetic debrite-turbidite beds and slumps, is located between the sand-rich basin-floor strata and the mud-rich strata deposited on the basin margin. This pattern is defined from the Ballybunnion outcrop.

Chapter 8
Evolving Depocentre and Slope: The Gull Island Formation

LORNA J. STRACHAN & DAVID R. PYLES

8.1 Introduction

The Gull Island Formation is the uppermost lithostratigraphic unit of the Namurian Shannon Group (Rider, 1974; Heckel & Clayton, 2006) and is best exposed on the Atlantic coast of southern and northern County Clare (Fig. 8.1.1). This chapter and excursion summarises and builds upon published observations and interpretations of the Gull Island Formation, with an emphasis on the stratigraphic architecture at several key outcrops, including those at Gull Island, Fisherstreet Bay and the Cliffs of Moher (Fig. 8.1.1). The chapter describes outcrops in the south (Stop 1) and north (Stops 2 and 3) of County Clare, where the varied nature of the Gull Island Formation can be viewed in a series of spectacular cliff sections. The transit time between these two areas is approximately 1 hour.

Strata of the Gull Island Formation are generally finer grained than the underlying Ross Sandstone Formation and consist of interbedded sandstone and mudstone (<30% sandstone) (Hodson, 1954a,b; Gill & Kuenen, 1958; Hodson & Lewarne, 1961; Rider, 1974; Gill, 1979; Martinsen, 1989; Collinson et al., 1991; Martinsen et al., 2000; Wignall & Best, 2000; Martinsen & Collinson, 2002; Martinsen et al., 2003; Pyles, 2008; Strachan & Alsop, 2006). Soft-sediment deformation is abundant and locally comprises up to 75% of the Formation (Hodson, 1954a; Hodson & Lewarne, 1961; Rider, 1974; Gill, 1979; Martinsen, 1989; Collinson et al., 1991). Several workers have used soft-sediment deformation features, including slumps, to reconstruct the geometry and orientation of the palaeoslope

A Field Guide to the Carboniferous Sediments of the Shannon Basin, Western Ireland, First Edition. Edited by James L. Best and Paul B. Wignall.
© 2016 International Association of Sedimentologists.
Published 2016 by John Wiley & Sons, Ltd.
Companion website: www.wiley.com/go/best/shannonbasin

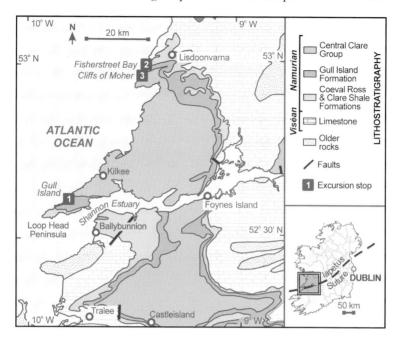

Fig. 8.1.1. Geological map of western Ireland. The Gull Island Formation and equivalent strata are shown in green, together with several key outcrop locations. Map is modified from Pyles (AAPG © 2008, used by permission of the AAPG whose permission is required for further use).

and to infer an unstable slope environment (Rider, 1974; Gill, 1979; Martinsen, 1989; Collinson *et al.*, 1991; Martinsen *et al.*, 2000; Strachan & Alsop, 2006).

The Gull Island Formation defined

The Gull Island Formation was formally defined by Rider (1974), who documented its lower bounding surface to record: 1) an abrupt lithofacies change from shale to slumped sandstones in northern Co. Clare, and 2) a gradational grain-size contact with the underlying Ross Sandstone Formation in southern Co. Clare (Fig. 8.1.2). The observations of Rider (1974) are in accord with those of Hodson (1954a) and Hodson & Lewarne (1961, Fig. 8.1.3), who documented an abrupt upward change to deformed, sandstone-rich, slumped intervals in north and south Co. Clare. Hodson (1954b) named the southern Co. Clare units 'Lower Sandy Shales and Sandstones', whereas Hodson & Lewarne (1961) termed the northern Co. Clare succession the Fisherstreet Slumped Beds, Cronagort Sandstone

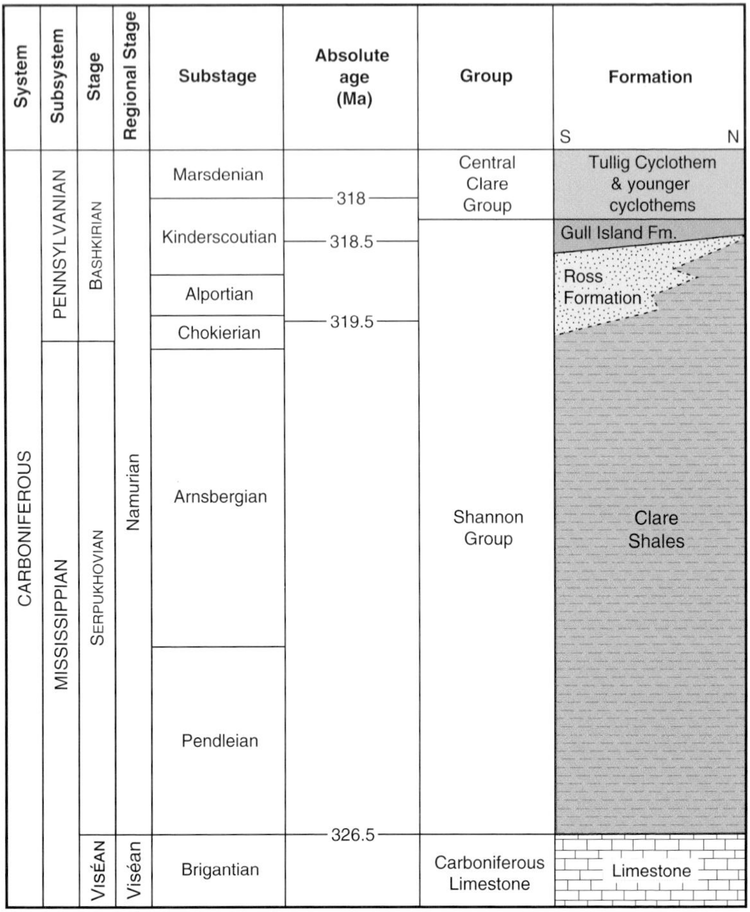

System	Subsystem	Stage	Regional Stage	Substage	Absolute age (Ma)	Group	Formation (S → N)
CARBONIFEROUS	PENNSYLVANIAN	BASHKIRIAN	Namurian	Marsdenian	318	Central Clare Group	Tullig Cyclothem & younger cyclothems
				Kinderscoutian	318.5		Gull Island Fm.
				Alportian			Ross Formation
				Chokierian	319.5		
	MISSISSIPPIAN	SERPUKHOVIAN		Arnsbergian		Shannon Group	Clare Shales
				Pendleian			
		VISÉAN	Viséan	Brigantian	326.5	Carboniferous Limestone	Limestone

Fig. 8.1.2. Regional stratigraphy of the Shannon Basin, adapted from Rider (1974), Heckel & Clayton (2006) and Pyles (2008). Figure from Pyles (AAPG © 2008) used by permission of the AAPG whose permission is required for further use.

Group and Doonagore Shales (Fig. 8.1.3). Rider (1974) replaced local names with the regionally-inclusive Gull Island Formation (Fig. 8.1.3).

Rider (1974) reports the Gull Island Formation contains "no fossils, falling between *Reticuloceras paucicrenulatum* fossil band" (R_{1a}3-5) "and the *Reticuloceras* aff. *stubblefieldi*" (R_{1b}3) fossil band (Fig. 8.1.3), and thus defining an exact biozone for the upper and lower boundaries of the formation is not possible. Furthermore, Rider (1974) placed the *R. stubblefieldi* (R_{1b}3) marine band at the top of the overlying Tullig Cyclothem

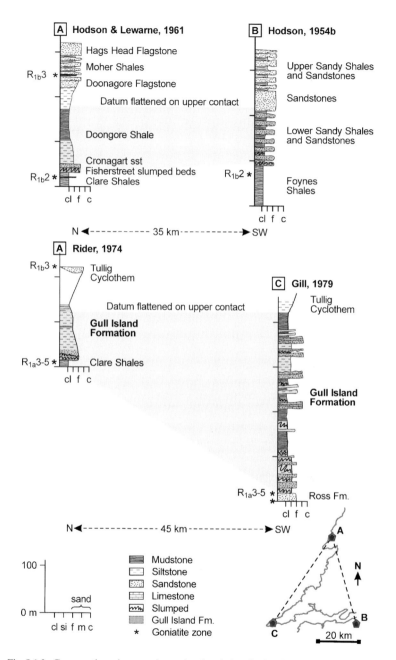

Fig. 8.1.3. Cross-sections documenting regional variations in the Clare Shale and Gull Island Formation. The logs have been redrawn from descriptions or modified from original figures. Cross-section A-B demonstrates the N-SSE development of Gull Island Formation strata from Hodson & Lewarne (1961) and Hodson (1954a) and thus uses the old nomenclature for stratigraphic units. Cross-section A-C is oriented N-SW with logs from Rider (1974) and Gill (1979; copyright Geological Society of Ireland) and show the regionally inclusive nomenclature now used widely.

(Fig. 8.1.3), effectively grouping the Gull Island Formation and Tullig Cyclothem between *R. paucicrenulatum* (R_{1a}3-5) and *R. stubblefieldi* (R_{1b}3). Although Rider (1974) and subsequent studies agreed with the interpretation that the Gull Island Formation is devoid of marine bands, his goniatite descriptions omitted the earlier pioneering work of Hodson (1954a, b) and Hodson & Lewarne (1961), who documented two significantly younger Clare Shale goniatites, *R. ornatum* (R_{1b}2) and *R. nodosum* (R_{1b}2), from northern and southern Co. Clare (Figs 8.1.3 & 8.1.4). Hodson & Lewarne (1961) placed *R. ornatum* (R_{1b}2) in the upper part of the Clare Shale, several metres below the Gull Island Formation (Rider, 1974; Figs 8.1.3 and 8.1.5). Therefore, the Gull Island Formation and overlying Tullig Cyclothem strata fall within two successive marine bands that record a mere *c.* 65 to 109 kyr, according to most estimates of zonal durations (e.g. Riley *et al.*, 1993; Menning *et al.*, 2000) (Fig. 8.1.5). Recent studies have also omitted *R. ornatum* (R_{1b}2) in biostratigraphic models, although Brennand (1965) documents a marine band that contains both R_{1b}1 and R_{1b}2 goniatite zones in Castleisland, Co. Kerry (Fig. 8.1.5).

Rider's (1974) designation of the lithostratigraphy of Namurian strata in western Ireland was embraced by subsequent studies, although many have modified the position of boundaries, both subtly and substantially (for example, Gill, 1979; Martinsen, 1989; Collinson *et al.*, 1991; Martinsen *et al.*, 2000; Wignall & Best, 2000; Pyles, 2008). Below, we document the utility of using the biostratigraphic information described above in an analysis of the Gull Island Formation.

While presenting a comprehensive description of soft-sediment deformation structures from the Shannon Basin, Gill (1979) provided a definitive lithostratigraphic upper Gull Island Formation boundary at the base of the Tullig Sandstone Member, where a coarsening-upward trend into delta-front sediments was documented (Fig. 8.1.3). This contact was well illustrated in a series of sedimentary logs from north and south Co. Clare (his fig. 4) that were sourced from Rider (1969).

Building on Gill's (1979) work, Martinsen (1989) described a range of mass transport deposits and water-escape structures in the Gull Island Formation. Whilst following the lithostratigraphic definition of Gill (1979), Martinsen (1989) reintroduced the *R. subreticulatum* (R_{1a}3) marine band, described previously by Hodson (1954a), and positioned it at the base of the Gull Island Formation in the southern part of the basin (Fig. 8.1.5). Placement of marine band R_{1a}3 was based on marine band observations of equivalent strata, known as the Cummer Formation, in north Co. Cork (Morton, 1965).

Martinsen (1989) divided the Gull Island Formation into lithostratigraphically distinct lower and upper units. The lower unit comprises mudstone slumps intercalated with sandstone and mudstone units, whereas the

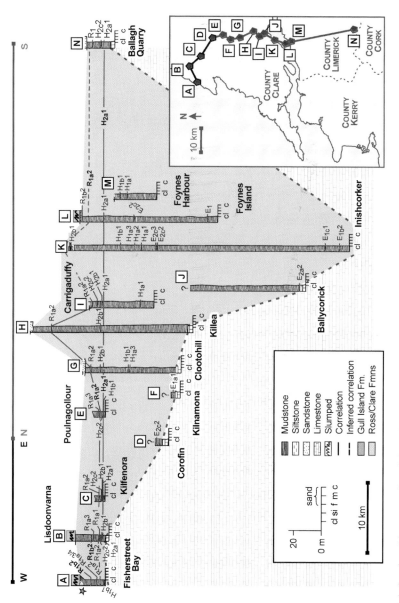

Fig. 8.1.4. Stratigraphic cross-section of the Gull Island Formation in eastern County Clare and County Limerick. Data from Hodson & Lewarne (1961), except for the ammonite in the Fisherstreet log adjacent to the star that comes from Hodson (1954a). The datum for the cross-section is the H$_{2a}$1 marine band (*Hudsonoceras proteum*). The location of the cross-section is shown in the inset map. The horizontal spacing of logs is to scale.

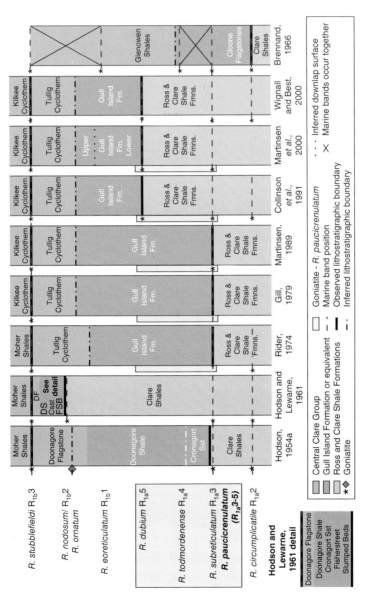

Fig. 8.1.5. Summary diagram showing biostratigraphic position only of the Gull Island Formation over time. Marine bands are indicated as reported from each case study. Note the consistency of the upper marine band *R. stubblefieldi* (R$_{1b}$3), which is a good biostratigraphic marker for the top of the Tullig Cyclothem and the range of ammonoid zones that *R. paucicrenulatum* (R$_{1a}$3 to 5) spans. The blue diamond goniatite in Hodson (1954a) was recognised but not put into stratigraphic context.

upper unit consists of undeformed mudstone and rare mudstone slumps and slides (Martinsen, 1989). Later work by Collinson *et al.* (1991) used detailed stratigraphic columns from Loop Head and Ballybunnion to present a sedimentological summary of the Shannon Group and Tullig Cyclothem. Collinson *et al.* (1991) document a previously unreported marine band, *Reticuloceras dubium* ($R_{1a}5$, Fig. 8.1.5), which was placed at the boundary between the Ross Sandstone and Gull Island formations (Fig. 8.1.5). The top contact of the Gull Island Formation was placed above the uppermost slump horizon (Collinson *et al.*, 1991).

Martinsen *et al.* (2000) expanded their earlier work, dividing the southern Co. Clare Gull Island Formation into a 420 m thick lower unit and a 200 m thick upper unit. The lower Gull Island Formation was interpreted as aggradational turbidites, whereas the upper Gull Island Formation was interpreted as a mudstone-dominated prograding slope succession (Martinsen *et al.*, 2000). This boundary between the lower and upper Gull Island Formation was interpreted as a regional downlap surface (Fig. 8.1.5). In northern Co. Clare, the base of the Gull Island Formation was placed at *R. paucicrenulatum* ($R_{1a}3$-5) and is overlain by ~50 m of lower Gull Island Formation strata and 220 m of Upper Gull Island Formation strata, that included the lowest part of the Tullig Cyclothem (Martinsen *et al.*, 2000). The biozone imprecision of *R. paucicrenulatum* ($R_{1a}3$-5) may allude to diachroneity of the base of the Formation, with Gull Island deposition being initiated first in northern Co. Clare and later in southern Co. Clare (Martinsen *et al.*, 2000).

In a reassessment of the basin, Wignall & Best (2000) adhered to the Gull Island stratigraphy of Collinson *et al.* (1991) in southern Co. Clare, but in northern Co. Clare placed a thin (~30 m) intervening Ross Sandstone Formation interval between the Clare Shale and Gull Island Formations (Fig. 8.1.5). Thus, the base of the Gull Island Formation was positioned in the shales overlying a thin interval of slumped and channelised sandstones (Fig. 8.1.5). In a more recent paper, Wignall & Best (2004) updated their northern Co. Clare model and reassigned the Ross Sandstone Formation interval to the Gull Island Formation (their fig. 2).

In summary, there is a lack of consensus with regard to lithostratigraphic definition and Gull Island Formation bounding marine bands (Fig. 8.1.5), largely due to the fact that submarine slumping, diagnostic of the base of the Formation (Rider, 1974), initiated and ended at different times across the basin, thus rendering lithostratigraphic and biostratigraphic approaches incompatible. Figure 8.1.6 contains a regional stratigraphic cross-section of the Gull Island Formation and documents that slumping initiated at $H_{2c}2$ times in the south, at $R_{1a}2$ times in the centre and at $R_{1b}2$ times in the north. Cessation of soft-sediment deformation varies, probably due to the limited aerial extent of slumps and

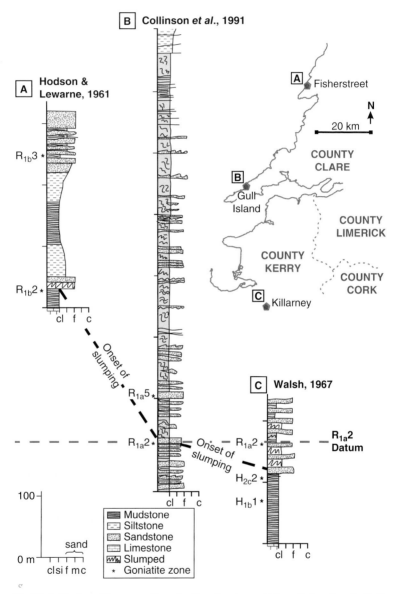

Fig. 8.1.6. Stratigraphic cross-section of sedimentary logs, which are flattened on the $R_{1a}2$ marine band. Slumping initiated in the southern part of the basin study area after marine band $H_{2c}2$ and moved northward through time. Slumping in the northern part of the study area initiated after marine band $R_{1b}2$.

slides, limited transport distances and the different number of events preserved at each outcrop (Fig. 8.1.6). These difficulties, combined with the lack of inclusion of the biostratigraphic work of Hodson & Lewarne (1961; Fig. 8.1.4), has made regional correlations difficult. As a result, attempts to correlate stratigraphic subdivisions of the formation are at best speculative.

The vast majority of published work in the Gull Island Formation describes the lower contact to underlie slumps and slides. As such, herein we define the base of the Gull Island Formation as a diachronous boundary between undeformed turbidite and hemipelagic strata of the coeval Ross Sandstone and Clare Shale formations and the deformed slump, slide, debrite and turbidite strata of the overlying Gull Island Formation (Fig. 8.1.7). The diachronous nature of the base of the Gull Island Formation means that its biostratigraphic position varies spatially and the onset of slumping is therefore time transgressive. For example, the base of the Gull Island Formation is positioned above marine band $H_{2c}2$ close to Killarney, Co. Kerry, above $R_{1a}2$ at Castleisland, Co. Kerry and above $R_{1b}2$ at Lisdoonvarna, Co. Clare (Fig. 8.1.7). Previous authors position the boundary between the Ross Sandstone and Gull Island Formation at the $R_{1a}5$ marine band at Loop Head Peninsula, Co. Clare. We follow this convention herein for this region, although recognise that slumping initiated before this time in several parts of the basin (Fig. 8.1.7). In this chapter, we place the top of the Gull Island Formation where strata begin to coarsen upward to the Tullig Cyclothem (Gill, 1979; Figs 8.1.3 and 8.1.7).

Sedimentology of the Gull Island Formation

The comprehensive documentation of the sedimentology of the Gull Island Formation provided below is based on data from numerous field studies from counties Clare, Kerry and Cork (Fig. 8.1.8). Using lithofacies and biostratigraphic descriptions, we identified strata that resemble Gull Island Formation sediments from neighbouring counties and refer to these strata as Gull Island Formation equivalents rather than their local names. Local names are provided in Fig. 8.1.8.

The thickness of the Gull Island Formation varies spatially, with a maximum thickness of 550 m on the west coast of the Loop Head Peninsula, Co. Clare (Rider, 1974; Martinsen, 1989) and a minimum of 12 m in north Co. Cork close to Buttevant (Hudson & Philcox, 1966, Philcox, 1961; Figs 8.1.8 and 8.1.9). A north-south transect documents a gradual thickening of strata from north to central Co. Clare, which thins slightly through northern and central Co. Kerry, finally thinning abruptly in south Co. Kerry (Fig. 8.1.7). Fig. 8.1.9 includes an isopach map of the Gull Island

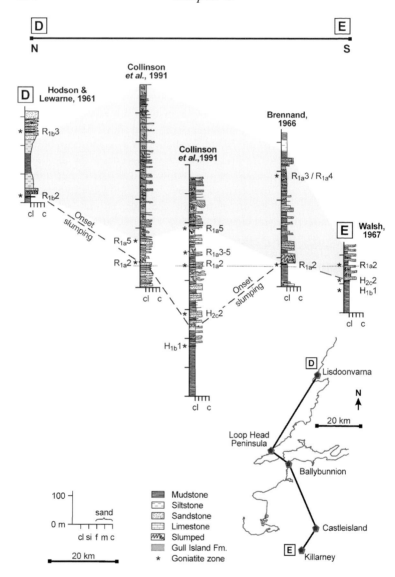

Fig. 8.1.7. North-south (D to E) stratigraphic cross-section through the Gull Island Formation in the Shannon Basin. Logs have been aligned on the $R_{1a}2$ marine band where present. Gull Island Formation sediments as defined herein are highlighted, together with the onset of slumping.

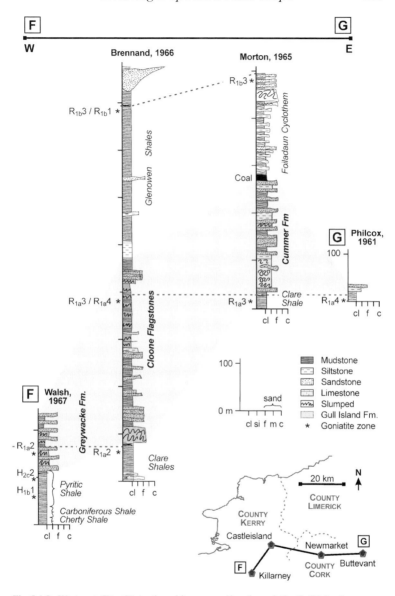

Fig. 8.1.8. West-east (F to G) stratigraphic cross-section through the Gull Island Formation in the Shannon Basin, showing a compilation of redrawn Gull Island Formation equivalent sedimentary logs from counties Kerry and Cork. The logs are aligned on marine bands $R_{1a}2$ and $R_{1a}3/R_{1a}4$ marine bands where present. Local formation names are indicated in bold italics. Note the diachronous base of the Gull Island Formation, which is defined by a change to a slump-dominated succession.

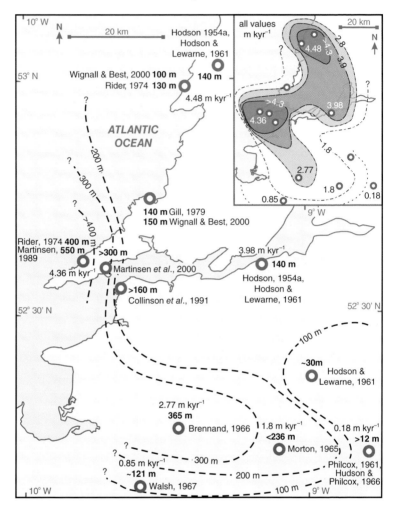

Fig. 8.1.9. Isopach map of the Gull Island Formation (contour interval is 100 m). Circles represent outcrop locations and are given together with citations for the source of data. Sedimentation rates are calculated by dividing the local thickness (in bold) by the duration of deposition using the time scale published by Riley *et al.* (1993). The Gull Island Formation is thickest at Loop Head and thins radially away from this area. Inset contour map of inferred sedimentation rates shows two sites of enhanced deposition.

Formation and reveals thinning towards the east and extreme thinning towards the south-east, showing that the distribution of the Gull Island Formation is more extensive than previously described.

The onset of slumping is determined using marine bands. Fig. 8.1.10 documents a systematic north-eastward migration of slumping through

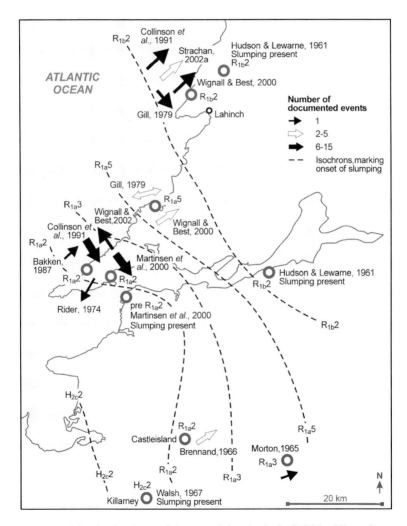

Fig. 8.1.10. Map showing the age of the onset of slumping in the Gull Island Formation with transport directions. The size of the arrows is proportional to the number of documented slumps. Contours are isochrons showing the onset of slumping as constrained by marine band biostratigraphy. Original authors of slump data are shown.

time that initiated in the SW after $H_{2c}2$ and $R_{1b}2$ in northern and eastern Co. Clare. The temporal variation in slumping is interpreted to reflect a systematic evolution of (slope) instability that migrated from the south-west to the north and north-east through time (Figs 8.1.10 and 8.1.11).

Gull Island Formation sedimentation rates are calculated using published sediment thickness and biostratigraphic data, adopting a marine band interval of 65 kyr (Riley *et al.*, 1993). The lack of Gull Island Formation marine bands, partial exposure of measured sections and the lithostratigraphic nature of the upper boundary, mean that some of these values probably have significant errors associated with them (Table 8.1.1). To overcome this difficulty, the calculated average sedimentation rates use marine bands that bound both the Gull Island Formation and Tullig Cyclothem (Table 8.1.1). In outcrops with complete Gull Island Formation successions and hence smaller errors, sedimentation rates range from 1.8 m kyr^{-1} at Newmarket, to 4.48 m kyr^{-1} at Fisherstreet (Fig. 8.1.9). These high rates confirm the calculations of Martinsen (1989) and Pyles (2008), who revealed that sedimentation rates were significantly higher in the Gull Island Formation than in the preceding Ross Sandstone and Clare Shale formations. Our calculations also document spatial variations in sedimentation rates over the basin, with significantly lower values in the south-east (Figs 8.1.9 and 8.1.11).

Figure 8.1.11 shows facies distributions and the location of depocentres at four distinct time periods (T1 to T4). The location of the depocentre and the distribution of slumped facies migrates to the north-east and east through time (Fig. 8.1.11). Enhanced deposition in the south and south-west part of the basin occurred during T1 to T3 (Fig. 8.1.11). During T4, the locus of slumping migrated to the northern and eastern parts of the basin (Fig. 8.1.11) although deposition continued in southern Co. Clare.

Lithofacies in the Gull Island Formation are turbidite sandstones and siltstones, hemipelagic mudstones and slumps and slides (Rider, 1974; Martinsen, 1989; Collinson *et al.*, 1991; Martinsen *et al.*, 2000; Wignall & Best, 2000; Martinsen *et al.*, 2003). Fine-grained to medium-grained sandstones and siltstones are thought to result from high and low concentration turbidity currents, generally deposited from waning flows within the deep marine Shannon Basin. A range of depositional elements are interpreted from Gull Island turbidites and include weakly and strongly confined channels, levees and lobes (Rider, 1974; Martinsen, 1989; Collinson *et al.*, 1991; Martinsen *et al.*, 2000; Wignall & Best, 2000; Martinsen *et al.*, 2003).

In northern Co. Clare, lithofacies in the basal part of the formation are primarily heterolithic slumps and lenticular sandstones, with subordinate tabular beds (Fig. 8.1.7). The overlying interval contains tabular bedded, laminated mudstone, siltstones and black shale, with rare lenticular

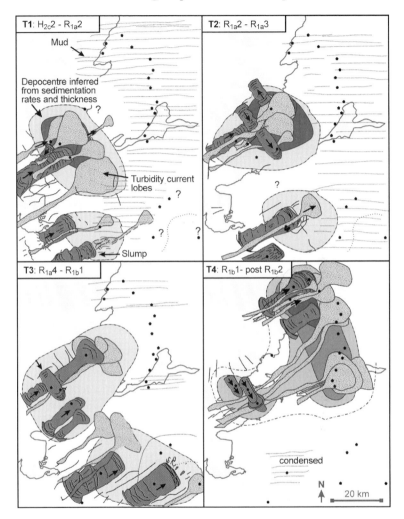

Fig. 8.1.11. Schematic maps showing the distribution of slumps, turbidites and mud deposition during $H_{2c}2$ to post $R_{1b}2$ (T1 to T4) times. Pentagons are locations of published sedimentary data that are sourced from Hodson (1954a,b), Hodson & Lewarne (1961), Morton (1965), Brennand (1965), Walsh (1967), Rider (1974), Gill (1979; copyright Geological Society of Ireland), Collinson *et al.* (1991), Martinsen *et al.* (2000), Wignall & Best (2000), Strachan (2002), Strachan & Alsop (2006) and Pyles (2008). Indicative transport directions are from published data. The distribution of slumps and slides appears to have gradually migrated north-eastward from $H_{2c}2$-$R_{1b}1$ (T1 to T3) times and then dramatically jumped northwards post $R_{1b}2$ (T4).

Table 8.1.1. Summary table showing calculated sedimentation rates for Gull Island Formation and Tullig Cyclothem sediments. Shaded rows represent incomplete sections. Authors and approximate locations of data are given. Authors are arranged in alphabetical order.

Author	Location	Thickness (m)	Basal marine band	Upper marine band	Formations included	Duration - kyrs (Riley et al., 1993)	Av sed. rate (m kyr⁻¹)
Brennand (1966)	Castle Island, Co. Kerry	365	R1a2	R1b4	Gull Island & Tullig	260	2.77
Collinson et al. (1991)	N. Co. Clare	260	R1a5	R1b3	Gull Island & Tullig	195	1.33
Collinson et al. (1991)	S. Co. Clare	850	R1a5	R1b3	Gull Island & Tullig	195	4.36
Gill (1979)	S. Co. Clare	400	R1a3 to 5	R1b3	Gull Island & Tullig	195 to 325	1.14 to 2.05
Hodson (1954b)	Foynes Island, Co. Clare	259	R1b2	R1b3 (inferred)	Gull Island & Tullig	65	3.98
Hodson & Lewarne (1961)	N. Co. Clare	291	R1b2	R1b3	Gull Island & Tullig	65	4.47
Martinsen (1989)	S. Co. Clare	850	R1a3 to 5	R1b3	Gull Island & Tullig	195 to 325	2.41 to 4.36
Morton (1965)	Newmarket, Co. Cork	470	R1a3	R1b3	Gull Island & Tullig	260	1.8
Philcox (1961)	Buttevant, Co. Cork	12	R1a4	R1a5 (inferred)	Gull Island	65	>0.18
Rider (1974)	S. Co. Clare	701	R1a3 to 5	R1b3	Gull Island & Tullig	195 to 325	2.16 to 3.6
Rider (1974)	N. Co. Clare	134	R1a3 to 5	R1b3	Gull Island	195 to 325	>0.41 to 0.69
Tanner et al. (2011)	Co. Clare	550	R1a5	R1b3	Gull Island & Tullig	195–325	2
Walsh (1967)	Killarney, Co. Kerry	110	H2c2	R1a2	Gull Island	130	0.85
Wignall & Best (2000)	S. Co. Clare	610	R1a5	R1b3	Gull Island & Tullig	195	3.12
Wignall & Best (2004)	N. Co. Clare	145	R1a5	R1b3	Gull Island	195	>0.74

sandstones. Palaeocurrent data from sole structures reveal NE and E flow directions (Fig. 8.1.12). In contrast, transport directions collected from slumps have been measured and conflicting interpretations given with movement towards the SE and NE (Gill, 1979; Collinson *et al.*, 1991; Wignall & Best, 2000; Strachan 2002a,b; Strachan & Alsop, 2006; Fig. 8.1.10). In southern Co. Clare, the most abundant lithofacies are slumps, laminated silty mudstones and graded, ripple-topped turbidite

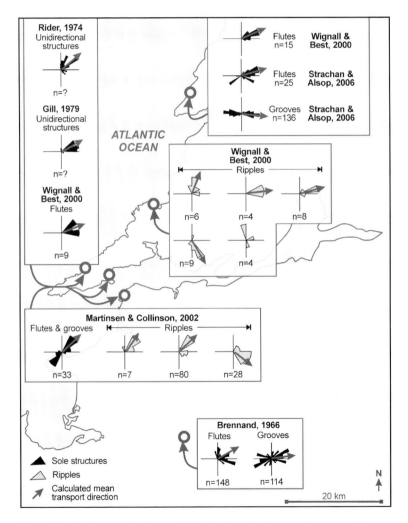

Fig. 8.1.12. Map documenting the spatial distribution of palaeocurrents measured from sole structures (black) and ripples (grey). n = number of measurements.

sandstones. Mudstones increase in abundance upward. Palaeocurrent data from sole structures are consistently to the NE, whereas palaeocurrents from ripple cross-laminae are more variable, to the NE, E and SW (Fig. 8.1.12). Slump transport directions, inferred from the orientations of fold axes and faults, were to the SE, NW and SW (Fig. 8.1.10). In counties Kerry and Cork, Gull Island Formation equivalent strata contain predominantly massive or graded turbidite sandstones and mudstones and a high proportion of slumps (Philcox, 1961; Hodson & Lewarne, 1961; Morton, 1965; Brennand, 1966; Hudson & Philcox, 1966; Walsh, 1967). Turbidite palaeocurrent directions collected from flutes, grooves and ripples record a mean transport to the NE, with a spread in directions (Fig. 8.1.12). Palaeocurrent directions collected from slumps record transport to the NE and ENE (Fig. 8.1.10). A single slump transport measurement of Morton (1965) was interpreted by Martinsen (1989) as a SE direction. However, because the movement direction was given using the "British right hand rule" convention, Morton's (1965) measurement of "160° N" actually implies a ENE directed flow (Fig. 8.1.10).

A compilation of published grain-size measurements reveals that the Gull Island Formation is mudstone dominated, although with variations in grain-size (Fig. 8.1.13). Most outcrops contain a broad fining-upward trend from sandstone rich bases, regardless of total net-to-gross or completeness of the measured section. The relative proportion of slumped to non-slumped strata varies spatially: northern outcrops contain an upward decrease in the proportion of slumps, whereas southern outcrops contain an upward increase in the proportion of slumps (Fig. 8.1.13). Furthermore, at Ballybunnion, slumps increase in proportion upward as the Formation becomes finer grained; however, at Fisherstreet the proportion of slumps decreases upwards as the interval fines. We interpret the ubiquitous slumping of the Gull Island Formation to record instability associated with relatively high sedimentation rates (Table 8.1.1), which are two orders of magnitude higher than the underlying Ross Sandstone and Clare Shale Formations.

The range of interpretations in slump transport directions has led to rigorous debate concerning the physiography of the basin (for example Martinsen & Collinson, 2002; Wignall & Best, 2002; Chapter 3). Deformational features in slumps can be used to reconstruct palaeo-slump and palaeoslope directions (Woodcock, 1979; Strachan & Alsop, 2006). However, a comprehensive analysis of slump data is required, together with an assessment of the significance of the slope; for example a local versus a regional palaeoslope (Woodcock, 1979; Strachan & Alsop, 2006; Debacker & De Meester, 2009). Deciphering the correct flow direction from slumps is difficult but like other palaeocurrent data, the greater the number of measurements the more robust the interpretation. Strachan &

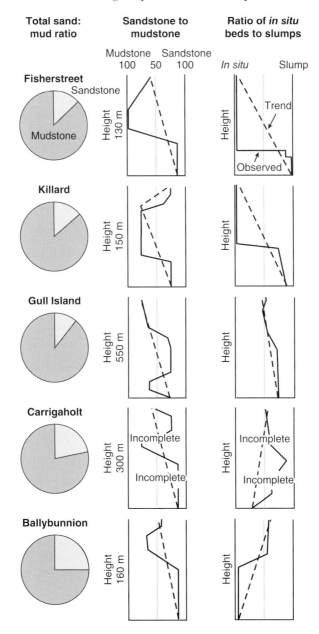

Fig. 8.1.13. Plots showing sandstone-to-mudstone ratios of Gull Island Formation sediments (left), upward patterns in sandstone-to-mudstone ratios (centre) and upward patterns in the proportion of slumping to *in situ* non-deformed units (right). Data from Collinson *et al.* (1991), Wignall & Best (2000) and Martinsen *et al.* (2000, 2003).

Alsop (2006) used 570 structural measurements (folds and faults) from a single slump sheet at Fisherstreet to determine palaeoflow direction. Using multiple transport determination techniques, they document that the Fisherstreet Slump moved ENE (Fig. 8.1.14). This direction is consistent with turbidite palaeocurrent directions collected from overlying units and therefore interpreted to represent the downslope orientation in northern Co. Clare. Other studies report slump transport directions using less rigorous analyses, either by using small numbers of measurements or by grouping measurements from multiple events onto single plots. Such approaches reduce the confidence level of the resulting interpretation. Additionally, slump folds and faults may form complex patterns, with dominant orientations anywhere between downslope to alongslope (Jones, 1939; Hansen, 1971; Lewis, 1971; Woodcock, 1979). As a result, multiple interpretations can result from the same data. For example, it is logically valid to interpret the raw slump-fold data of Strachan & Alsop (2006) to result from transport to the ENE, WSW, or SE direction (Fig. 8.1.14). However, their use of multiple techniques, in concert with data derived from fold vergence, allowed Strachan & Alsop (2006) to reject all other possible flow directions and robustly show an ENE flow direction. The raw structural data presented by Martinsen & Collinson (2002, their fig. 5), which group together measurements from several slumps, can similarly be interpreted in a number of different ways, with flows moving either towards the SE, NW and NE. For their data, Martinsen & Collinson (2002) opt for a SE flow interpretation.

In summary, a comprehensive analysis of the literature shows that the timing of slumping in the basin was diachronous, migrating northeastward through time. Thus the base of the Gull Island Formation is time transgressive. A complex spatial distribution of sediment thicknesses is interpreted to record the geometry of the basin. Calculated sedimentation rates are high in some areas suggesting more than one depocentre was present at a time. The Gull Island Formation is dominated by mudstone and this, together with high sedimentation rates, is interpreted to have promoted slope instability. The absence of marine bands prevents an accurate characterisation of the time at which slumping ended. To date, no provenance data have been collected to determine if the Gull Island Formation is a finer-grained equivalent of the Ross Sandstone Formation, or if it was from a different source area altogether.

Depositional environment of the Gull Island Formation

Since Rider (1974) named the Gull Island Formation, a variety of interpretations of the depositional environment have been made, as summarised below.

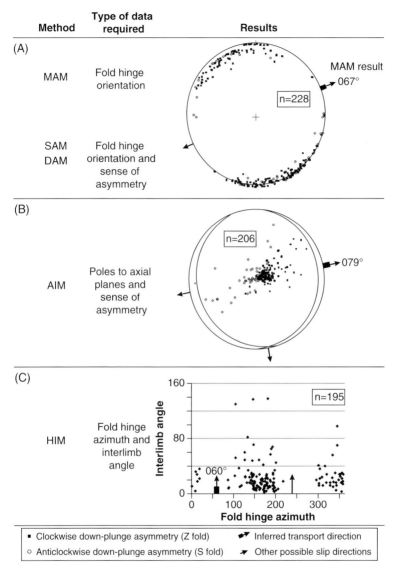

Fig. 8.1.14. Summary of techniques used to determine transport directions from slumps at Fisherstreet. All plots show several possible slip directions together with the inferred direction. n = number of data. (A) Stereonet showing data for fold hinge orientation and sense of asymmetry. The Mean Axis Method (MAM) suggests a transport direction of 067°. The Separation Arc Method (SAM) and Downslope Average axis Method (DAM) are inconclusive. (B) Stereonet of axial plane data and sense of asymmetry. If the Axial planar Intersection Method (AIM) is applied to the data, the transport direction is inferred to be 079°. (C) Fold Hinge azimuth and Interlimb angle Method (HIM) shows a preferred slope direction of 060°.

Rider (1974) interprets the vertical succession of the Shannon Group to record a shallowing-upward succession with the Clare Shale and Ross Sandstone Formations recording deposition in the basin overlain by prograding slope strata of the Gull Island Formation. Rider (1974) applied Walther's Law to the Shannon Group, interpreting the succession to be broadly progradational to the east. Gill (1979) broadly concurred with this model and highlighted a range of complex palaeogeographies, with slumps moving in multiple directions thought to record both local and regional slopes. Gill (1979) suggested that Gull Island Formation slumps moved towards the south-east, down "tectonically-generated slopes", but that the formation as a whole prograded eastwards.

Martinsen (1989) proposed a submarine delta slope environment for the Gull Island Formation, where the main locus of deposition was thought to be a narrow central trough, with its axis positioned close to the present day Shannon Estuary. Building upon this study, Collinson *et al.* (1991) interpreted the Gull Island Formation as sediments derived from a delta that collapsed down an ESE-SE palaeoslope and that periodically underwent avulsion. The delta was thought to have been positioned to the west during early Gull Island Formation deposition and moved towards the north-west with time.

Martinsen *et al.* (2000) developed a modified model for the environment of deposition, reflecting their division of Gull Island Formation stratigraphy into lower and upper units. Lower Gull Island sediments were interpreted as a mud-rich turbidite system recording aggradation of axially (ENE) transported basin floor turbidites and muddy slumps moving SE down palaeoslope. The source of the slumps was thought to be a degrading northern slope (Martinsen *et al.*, 2000). Lower Gull Island Formation sediments were thought to sidelap and onlap the narrow basin margin. Upper Gull Island Formation sediments were argued to result from progradation of the northern basin slope, which was deposited on top and over older basin floor sediments.

Wignall & Best (2000) reassessed the basin and envisaged a different model whereby the system was dominated by a NE facing submarine slope, which prograded to the NE with time. Like Rider (1974), their model evokes the vertical succession to record progradation of a linked shelf-slope-basin floor system and the Gull Island Formation sediments were interpreted as the direct record of progradation of that unstable NE slope.

Martinsen & Collinson (2002) questioned the interpretations by Wignall & Best (2000), with debate centring upon the geometry of the basin, the site of maximum subsidence and location of the depocentre. Martinsen & Collinson (2002) favour a basin axis positioned close to the present day Shannon Estuary, whereas Wignall & Best (2000, 2002) envisage the deepest most distal part of basin to be located in the NE. One of the lynch-pins behind this multi-faceted debate is the slump transport directions of the Gull Island Formation inferred by Martinsen and colleagues, which were refuted by Wignall & Best (2000, 2002). Strachan & Alsop (2006) used the north Co. Clare Fisherstreet Slump to robustly show a deep marine ENE dipping trough or sub-basin.

Despite the different interpretations summarised above, all authors agree the Gull Island Formation represents deposits from encroaching, unstable, submarine slopes associated with a dramatic increase in sedimentation rate. However, there remains a number of observational inconsistencies and questions concerning the Gull Island Formation that are currently only partly addressed. These include the orientation and shape of the Shannon Basin or sub-basins at the time of deposition, the orientation of submarine slopes, the location of sediment source areas, the internal stratigraphic evolution of the Gull Island Formation and the dominant control on instability migration from south to north through time.

8.2 Gull Island Formation Outcrops

Three outcrops of the Gull Island Formation are described below: Stop 1, Gull Island; Stop 2, Fisherstreet and Stop 3, Cliffs of Moher (Fig. 8.1.1). Collectively, these field stops document the regional variation in the Formation.

Directions to Stop 1 (UTM 445710 m E, 5829200 m N; Map 63, 077998 m E, 152462 m N)

Stop 1 is located on the north coast of the Loop Head Peninsula, ~2.5 km north-west of the village of Cross (Fig. 8.2.1). From Cross, travel west on the R487 for ~1.5 km, and turn north on the dirt road labelled 'Trusklieve'. Proceed north for ~1 km, and then turn west on another dirt road that ends at a quarry (Fig. 8.2.1). Note that after heavy rain, the surface of these roads can be difficult for normal vehicles. Park at the east side of the quarry. This stop is best exposed at low tide, although it is accessible at all times. **For safety, stay >5 m from the edges of the cliffs at all times. Care should be taken as the rocks are slippery when wet and strong wind gusts are frequent.** The stop takes one half to a full day to complete.

Stop 1 (Gull Island)

The Gull Island outcrop at Stop 1 is the type section of the Gull Island Formation (Rider, 1974; Figs 8.2.1 to 8.2.3). The contact between the Gull Island Formation and underlying Ross Sandstone is placed at the $R_{1a}3$ marine band, which occurs just below the first slumping in this area and is located at the southern part of the outcrop where strata dip ~70 degrees

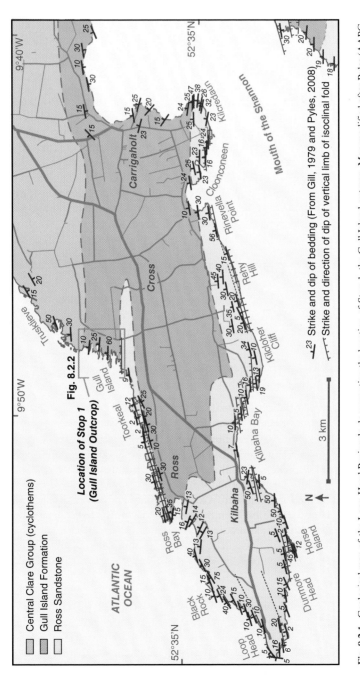

Fig. 8.2.1. Geological map of the Loop Head Peninsula documenting the location of Stop 1, the Gull Island outcrop. Map modified after Pyles (AAPG © 2008, used by permission of the AAPG whose permission is required for further use).

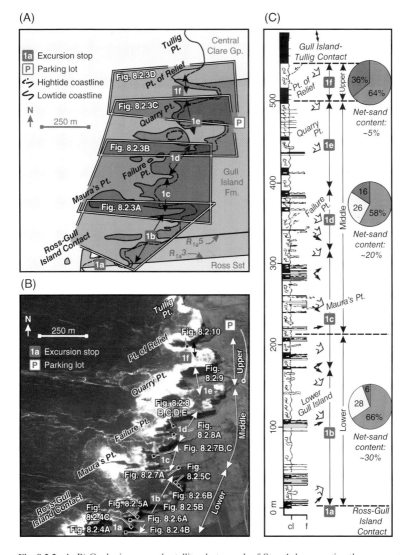

Fig. 8.2.2. A, B) Geologic map and satellite photograph of Stop 1 documenting the locations of the six field localities (image courtesy GoogleEarth; © DigitalGlobe). Note that the Ross Sandstone Formation - Gull Island contact is drawn on the $R_{1a}3$ (personal communication Trevor Elliott and Andrew Pulham) condensed section, which is a surface that records the onset of slumping in the area. C) Stratigraphic column through the Gull Island Formation at Stop 1 (modified from Collinson *et al.* (1991)). The base of the column coincides with the $R_{1a}5$ condensed section (see inset map for location). Wide arrows indicate palaeoslope measurements whereas smaller bold arrows indicate palaeocurrent measurements. Pie charts document proportions (by thickness) of architectural elements (blue = slumps; yellow = turbidite channels and lobes, grey = mudstone sheets).

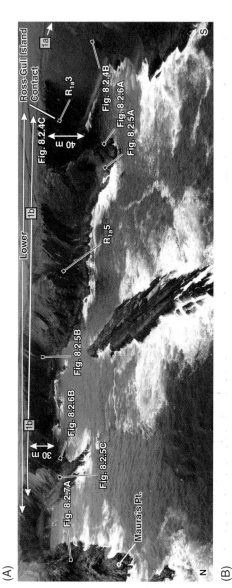

(A)

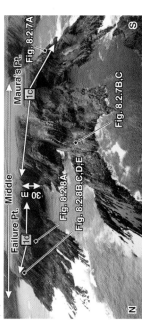

(B)

Fig. 8.2.3. Aerial photographs of Stop 1, the Gull Island Outcrop, documenting locations of the six field localities. The locations of photographs are shown in Fig. 8.2.2.

(C)

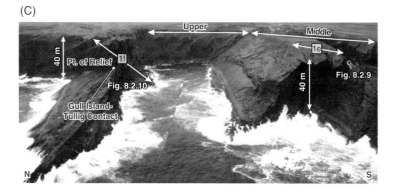

(D)

Fig. 8.2.3. (*Cont'd*)

north (Figs 8.2.2. and 8.2.3A), although Gill (1979) and Collinson *et al.* (1991) suggest that the $R_{1a}5$ condensed section be used as the contact. The contact between the Gull Island Formation and Central Clare group is located at the northern part of the outcrop where strata dip ~15 degrees north (Pyles, 2008; Figs 8.2.2 and 8.2.3). This contact occurs at the base of the coarsening-upward succession that leads to the Tullig Sandstone at Tullig Point.

This outcrop is extensively documented by Rider (1974), Martinsen (1989), Martinsen and Bakken (1990), Collinson *et al.* (1991), Martinsen *et al.* (2000, 2003), Wignall and Best (2000, 2002) and Martinsen and Collinson (2002). Palaeocurrent and palaeoslope measurements published by Collinson *et al.* (1991) are shown adjacent to the stratigraphic column in Fig. 8.2.2C. Martinsen *et al.* (2000; 2003) document sediment transport directions from flutes and ripples to the east and palaeoslope measurements from extensional faults and fold axes to the south-east. In contrast, Wignall & Best (2000, 2002) document palaeocurrent and palaeoslope measurements to the north and north-west.

Four lithofacies associations are recognised at this exposure and are interpreted to represent architectural elements:

1 Contorted mudstone and sandstone beds associated with shear planes, extensional faults and thrust faults are interpreted as slumps. Slumps are defined as displaced sediment bounded by a lower shear plane where internal strata are contorted and rotated (Stow, 1986).

2 Laminated mudstone beds are interpreted as mudstone sheets. Mudstone sheets are fine-grained sediment bounded by planar upper and lower bounding surfaces. They are laterally persistent and composed primarily of laminated mudstone (Pyles & Jennette, 2009).

3 Tabular turbidite beds with internal thickening-upward trends are interpreted as turbidite lobes. Turbidite lobes have planar lower bounding surfaces and broadly convex-upward upper bounding surfaces. They are thickest and sandstone rich in their axis and strata become thinner and finer-grained toward their lateral and distal margins (Pyles, 2007; Pyles & Jennette, 2009; Pyles & Strachan, Chapter 7).

4 Lenticular turbidite beds associated with erosional features that are >2 m in relief are interpreted as turbidite channels. Channels have convex-upward lower bounding surfaces and planar upper bounding surfaces except when truncated by younger elements. Channels are thickest and sand-rich in their axes and strata thin and become finer-grained toward their lateral margins (Pyles, 2008).

The stratigraphy at this outcrop is informally divided into three units – lower, middle and upper – representing an upward transect through the formation. Each unit is distinctive in terms of the physical observable characteristics of strata.

Lower Gull Island Formation

The lower Gull Island Formation is described in Stops 1a and 1b below (Figs 8.2.2, 8.2.3A and 8.2.4 to 8.2.6).

Stop 1a is located in the southern part of the outcrop at the contact between the Ross Sandstone and Gull Island formations (Figs 8.2.2 & 8.2.3). The lower Gull Island Formation contact is exposed where the contact dips into the ocean at the western part of the outcrop (Fig. 8.2.4A; UTM 444980 m E, 5828425 m N; Map 63, 077257 m E, 151697 m N) and adjacent to a steeply dipping dipslope in the upper Ross Sandstone (Figs 8.2.3A, 8.2.4B and C; UTM 445045 m E, 5828417 m N; Map 63, 077321 m E, 151688 m N; and UTM 445120 m E, 5828420 m N; Map 63, 077397 m E, 151690 m N). The contact between the formations is locally

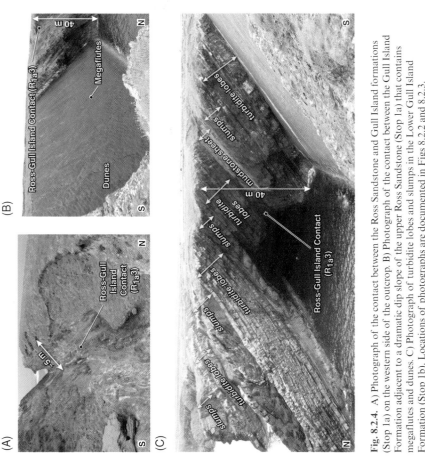

Fig. 8.2.4. A) Photograph of the contact between the Ross Sandstone and Gull Island formations (Stop 1a) on the western side of the outcrop. B) Photograph of the contact between the Gull Island Formation adjacent to a dramatic dip slope of the upper Ross Sandstone (Stop 1a) that contains megaflutes and dunes. C) Photograph of turbidite lobes and slumps in the Lower Gull Island Formation (Stop 1b). Locations of photographs are documented in Figs 8.2.2 and 8.2.3.

(A)

(B)

(C)

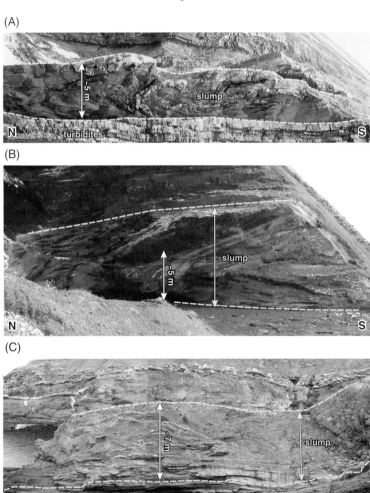

Fig. 8.2.5. Photographs of slumps in the lower Gull Island Formation (Stop 1b). The locations of photographs are documented in Figs 8.2.2 and 8.2.3.

the $R_{1a}5$ condensed section (Rider, 1974; Collinson *et al.*, 1991; Pyles, 2008) and is accessible on the western-most exposure (UTM 444980 m E, 5828425 m N; Map 63, 077257 m E, 151697 m N). At this locality the condensed section is a ~1 m thick interval of black shale with goniatites, which is encased in turbidites.

Fig. 8.2.6. A) Photograph of turbidite lobes in the lower Gull Island Formation (Stop 1b). B) Photograph of turbidite channels in erosional contact with a turbidite lobe in the lower Gull Island Formation (Stop 1b). The locations of photographs are shown in Figs 8.2.2 and 8.2.3.

Stop 1b comprises the lower 220 m of the Gull Island Formation (Fig. 8.2.2). The stop extends from the lower contact of the Gull Island Formation (UTM 444980 m E, 5828425 m N; Map 63, 077257 m E, 151697 m N) northward to the southern boundary of Maura's Point (Figs 8.2.2, 8.2.3A and 8.2.4C; UTM 445460 m E, 5828750 m N; Map 63, 077741 m E, 152015 m N). The lower Gull Island Formation contains sandstone-mudstone couplets that are tens of metres in thickness (Fig. 8.2.4C). The sandstone-rich parts of the couplets are composed of turbidite lobes and to a lesser extent turbidite channels, whereas mudstone-rich parts of the couplets are composed of slumps and to a lesser extent mudstone sheets (Figs 8.2.4 to 8.2.6). The sandstone-rich component of the couplets decreases in thickness upward from one couplet to the next, resulting in an upward decrease in net-sand content within the lower Gull Island Formation (Fig. 8.2.3A). The average net-sand content for the unit is ~30%, although the majority is located in the lower 100 m. This vertical stacking is similar to that of the upper Ross Sandstone at the Bridges of Ross (Pyles, 2008; Pyles & Strachan, Chapter 7, Stop 5).

Slumps are the most abundant architectural element in the lower Gull Island Formation and comprise ~66% of exposed strata (Fig. 8.2.2C). Three slumps are particularly well exposed at this locality (Fig. 8.2.5) and the characteristics common to all are: (1) their lower bounding surfaces are shear planes; (2) internally they contain contorted and rotated strata such as folds and thrust faults; and (3) the upper bounding surfaces of the slumps are flat erosional surfaces (Fig. 8.2.5). Distinctions between slumps in the lower Gull Island Formation are: (1) the amount of sandstone within them decreases from one to the next upward through the transect; (2) strata in sandstone-rich slumps are contorted and rotated and have thrust faults (Fig. 8.2.5A and B), whereas strata in mudstone-rich slumps are contorted, have thrusts and fluidized features such as sandstone balls and internally-deformed clasts in a mud-rich matrix (Fig. 8.2.5C); and (3) the slumps increase in thickness from one to the next through the upward transect.

Turbidite lobes, similar to those in the Ross Sandstone, are the second most abundant architectural element in the lower Gull Island Formation and comprise ~24% of exposed strata at this location (Figs 8.2.2C and 8.2.6A). Their characteristics are: (1) they are located in the sand-rich parts of sandstone-mud couplets in intervals 3 to 15 m in thickness (e.g. Fig. 8.2.6A); (2) internally they contain units with upwardly thickening and coarsening beds that are 1 to 5 m in thickness, with each individual upward coarsening unit being interpreted as a lobe; and (3) bedding is laterally persistent and there is little evidence of erosion between and within thickening-upward units. Lobes and associated sandstone-rich units decrease in abundance upward through the lower Gull Island Formation.

Mudstone sheets comprise ~6% of exposed strata in the lower Gull Island Formation at this locality (Figs 8.2.2C and 8.2.4C). They contain laminated and fissile, dark grey shale interbedded with structureless mudstone beds. Mudstone sheets increase in abundance upward through the lower Gull Island Formation and the lower shear planes of slumps commonly detach into mudstone sheets.

Turbidite channels are the least abundant architectural element in the lower Gull Island Formation and comprise ~3% of exposed strata (Figs 8.2.2C and 8.2.6B). They have erosional bases that are overlain by a lenticular mudstone layer, which is in turn onlapped by sandstone turbidites. Critically, the amount of erosion scales to the thickness of the interval. Channels are in erosional contact with underlying lobes and are ~5 m in thickness, this coinciding with the mean value measured for those in the Ross Sandstone (Pyles, 2007).

In summary, the lower Gull Island Formation is ~220 m thick and is composed of ~66% slumps, ~25% turbidite lobes, ~6% mudstone sheets and ~3% turbidite channels (Fig. 8.2.2). The average net-sand content of the unit is ~30%. The lower Gull Island Formation contains several sandstone-mudstone couplets. The sandstone-rich parts of the couplets are interpreted as turbidite lobes and minor channels whereas the mudstone-rich parts of the couplets are slumps and to a lesser degree mudstone sheets. Upward changes in stratal characteristics include: (1) a decrease in net-sandstone content; (2) a decrease in the abundance of turbidite lobes and channels; (3) a decrease in the abundance of sandstone within slumps; and (4) an increase in the abundance of mudstone sheets.

Middle Gull Island Formation

The middle Gull Island Formation is described in Stops 1c, 1d and 1e below (Figs 8.2.2B, 8.2.3, and 8.2.7 to 8.2.9).

Stop 1c is located north of Stop 1b and comprises all of the strata exposed on Maura's Point, the largest peninsula in the outcrop (Figs 8.2.2, 8.2.3B and 8.2.7). The lowermost strata at Maura's Point are well exposed on a >250 m wide outcrop on the southern part of the peninsula (Fig. 8.2.7A; UTM 445365 m E, 5828756 m N; Map 63, 077646 m E, 152023 m N). Strata at this location comprise interbedded slumps, turbidite lobes and mudstone sheets. Turbidite lobes are <5 m in thickness whereas the slumps are up to 15 m thick. These strata are accessible at the western-most part of the peninsula during low tide (UTM 445223 m E, 5828748 m N; Map 63, 077504 m E, 152017 m N).

Strata in the northern part of Maura's Point are described by Martinsen & Bakken (1990) (Fig. 8.2.7C; UTM 445255 m E, 5828820 m N; Map 63, 077537 m E, 152088 m N) and contain listric faults that detach into a lower

(A)

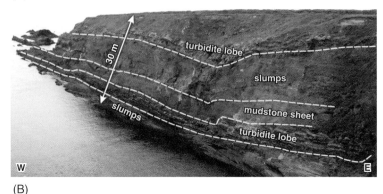

(B)

(C)

Fig. 8.2.7. A) Photograph of a laterally-continuous exposure in the middle Gull Island Formation at Maura's Point (Stop 1c). Turbidite lobes are thinner and lobes have a lower net-sand content at this locality than in the lower Gull Island Formation (Stop 1b). B, C) Line drawing modified from Martinsen & Bakken (1990) and photograph of a deformed unit at Maura's Point. The locations of photograph and line drawing are shown in Figs 8.2.2 and 8.2.3. Reproduced with permission from the Geological Society of London.

shale-rich interval. Shear sense indicators on the faults indicate dip-slip movement down to the east. Martinsen & Bakken (1990) documented wedge-shaped sandstone beds on the hangingwall of the faults that thicken toward the faults (see strata on hangingwall of Fault 2 in Fig. 8.2.7B) and interpreted them to record syn-sedimentary growth. The amount of offset on the faults ranges from several cm to several m. Through the use of cross-cutting relationships, Martinsen & Bakken (1990) documented the faults to become successively younger to the east and interpreted this pattern to record downslope propagation of the head region of a slide named the Maura's Point slide. However, as the unit is internally-deformed, we interpret this feature as a slump (*sensu* Stow, 1986). Maura's Point slump is significant as it is the oldest stratigraphic feature in the Gull Island Formation that contains extensional features – all of the underlying slumps contain predominantly contractional features (Figs 8.2.4C, 8.2.5 and 8.2.7A) and the upper bounding surface of the Maura's Point slump unit is faulted, unlike those in the lower Gull Island Formation. The irregular topography on the upper surface served to locally accommodate lenticular strata, such as unit 'B' in Fig. 8.2.7B.

Stop 1d is located north of Stop 1c and comprises all of the strata exposed at Failure Point (Figs 8.2.2, 8.2.3B and 8.2.8; UTM 445325 m E, 5828965 m N; Map 63, 077609 m E, 152232 m N). The lowermost strata at Failure Point are well-exposed on the western part of the peninsula and are predominantly slumps and mudstone sheets (Fig. 8.2.8A; UTM 445332 m E, 5828959 m N; Map 63, 077616 m E, 152226 m N). Although sandstone units are not abundant, when present they are lenticular. The upper strata at Failure Point are described by Martinsen & Bakken (1990) (Figs 8.2.8B, C, D and E; UTM 445360 m E, 5828990 m N; Map 63, 077645 m E, 152257 m N). The lower part of the unit is shale-rich and con-tains thrust faults that converge to a basal detachment surface (Figs 8.2.8B and C). The unit is interpreted as a slump. The upper surface of the slump is irregular and overlain by a lenticular sandstone unit that is thickest where the underlying slump is thinnest and contains load structures on its lower bounding surface (Figs 8.2.8B, C and D). The sandstone unit is interpreted as a turbidite that was deposited on the topographically irregular slump. The lenticular sandstone unit is not interpreted as a channel as the lower bounding surface is conformable with underlying strata and shows load structures.

Stop 1e is located north of Stop 1d and comprises all strata exposed at Quarry Point (Figs 8.2.2, 8.2.3C and 8.2.9). The lowermost strata at Quarry Point are well exposed on a >250 m wide outcrop on the southern part of the peninsula (Fig. 8.2.9A; UTM 445530 m E, 5829202 m N; Map 63, 077818 m E, 152467 m N). Strata at this location comprise interbedded slumps and mudstone sheets. Sandstones only occur as deformed clasts

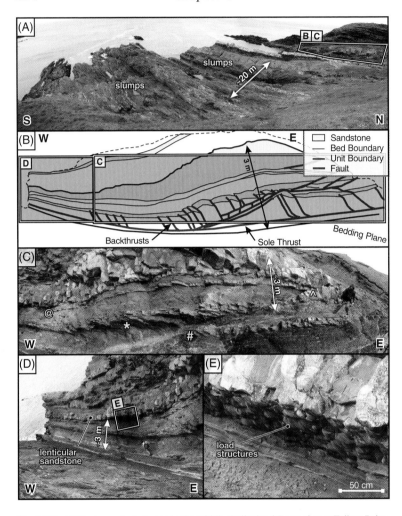

Fig. 8.2.8. A) Photograph of slumps in the middle Gull Island Formation at Failure Point (Stop 1d). The net-sand content of lobes is low. B, C, D, E) Line drawing modified from Martinsen & Bakken (1990) and photographs of a deformed unit at Failure Point. The locations of photographs and line drawings are shown in Figs 8.2.2 and 8.2.3. Reproduced with permission from the Geological Society of London.

within slumps (Fig. 8.2.9C). The most distinctive feature at this exposure is a large erosional surface that is ~20 m in relief. This surface truncates underlying mudstone sheets and slumps and the maximum depth of erosion is located in the centre of the exposure (Fig. 8.2.9A). The deepest part of the erosional surface is overlain by deformed sandstone clasts set in a mud

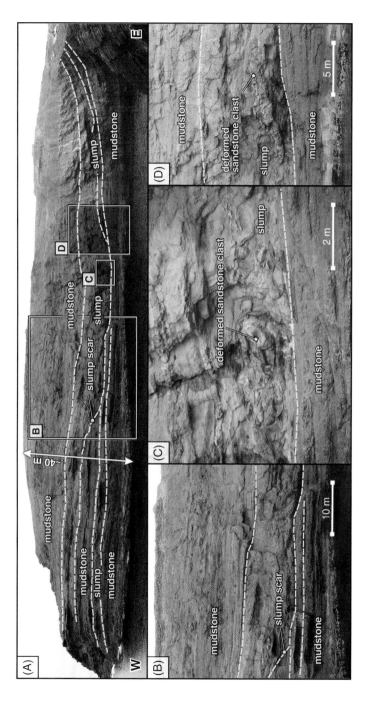

Fig. 8.2.9. Photographs of slumps and mudstone sheets in the middle Gull Island Formation at Quarry Point. This exposure contains a large-scale slump scar, slumps and mudstone sheets (Stop 1e). The slump scar is overlain by deformed sandstone clasts and contorted mudstone. The locations of photographs are documented in Figs 8.2.2 and 8.2.3.

matrix (Figs 8.2.9C, and D). The erosional surface is interpreted as a slump scar overlain by slumps, which are in turn overlain by mudstone sheets.

In summary, the middle Gull Island Formation includes the middle 300 m of the formation and is composed of ~58% slumps, ~26% turbidite lobes and ~16% mudstone sheets (Fig. 8.2.2). The average net-sand content of the unit is ~20%. Several stratal characteristics change upward through the unit: (1) net-sandstone content decreases; (2) mudstone sheets increase in abundance; (3) extensional features in slumps such as growth faults and slump scars are more abundant; (4) topography on the upper bounding surfaces of slumps are irregular; and (5) in the lower part of the succession, sandstone is laterally persistent and located in turbidite lobes, in the middle part of the succession sandstone is lenticular and located in topographically-confined areas above slumps, whereas in the upper part of the succession sandstone is located within slumps. The middle Gull Island Formation is distinct in the following ways: (1) it has a lower net-sandstone content than the lower Gull Island Formation; (2) slumps are predominantly composed of mudstone; (3) there are more abundant slumps with extensional features; (4) there is a higher proportion of mudstone sheets; and (5) slumps have irregular upper bounding surfaces.

Upper Gull Island Formation (UTM 445470 m E, 5829407 m N; Map 63, 077760 m E, 152673 m N)

Stop 1f is located north of Stop 1e and comprises ~100 m of strata exposed at the Point of Relief (Figs 8.2.2, 8.2.3D and 8.2.10), and contains the uppermost strata of the Gull Island Formation and the lowermost strata of the Tullig Cyclothem. The Point of Relief is best viewed from the promontory to the south (Quarry Point; UTM 445530 m E, 5829240 m N; Map 63, 077818 m E, 152505 m N).

The most prominent feature at the Point of Relief is a unit containing deformed strata exposed on the western half of the peninsula (Fig. 8.2.10) that Martinsen & Bakken (1990) describe in detail (Fig. 8.2.10B). The unit is ~35 m thick and contains three westerly-dipping growth faults (labelled 1 to 3 in Fig. 8.2.10B). Fault 1 is the eastern limit of the deformed unit. Shear sense indicators including drag folds are located on the hangingwall of the faults indicating extensional, normal dip-slip motion (Martinsen & Bakken, 1990). The deformed unit contains nine subunits, each of which is internally deformed and interpreted as slumps (Fig. 8.2.10B). Collectively, the deformed unit is interpreted as a slump complex, although Martinsen & Bakken (1990) interpret it as a slide. There is a general increase in the amount of internal deformation within subunits from subunits b through i. Fault terminations indicate that fault 3 is the oldest and fault 1 is the youngest (Fig. 8.2.10) and that they were active during deposition (they

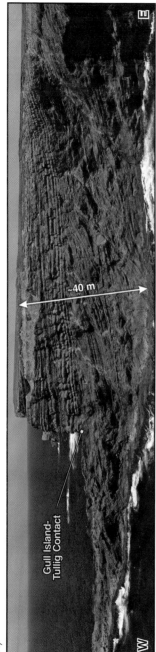

(A)

W

Gull Island-
Tullig Contact

~40 m

E

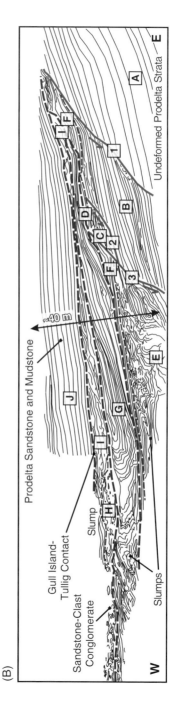

(B)

W

Prodelta Sandstone and Mudstone

Gull Island-
Tullig Contact

Slump

Sandstone-Clast
Conglomerate

Slumps

~40 m

J

I

H

G

E

F

C

D

B

A

Undeformed Prodelta Strata

1

2

3

E

Fig. 8.2.10. Photograph and line drawing of the deformed unit at the Point of Relief (Stop 1f). The upper surface of the Gull Island Formation is at the top of the deformed unit. The locations of photographs and line drawing are documented in Figs 8.2.2 and 8.3.3. Red lines are growth faults and record offset to the west-north-west. Blue lines are boundaries between the stratigraphic units described in the text. See online *GigaPan* image.

are growth faults) and probably created local accommodation space for the slumps to be deposited. The faults indicate extension to the west-north-west (Wignall & Best, 2000) although Collinson *et al.* (1991) report transport direction to the south-east. The upper bounding surface of the deformed unit is the contact between the Gull Island Formation and the Tullig Cyclothem (Fig. 8.2.10). Strata below this surface contain deformed mudstones whereas strata above this surface become thicker-bedded and coarser-grained upward and are part of the Tullig Cyclothem. This surface is planar, erodes into underlying strata and is conformably overlain by planar bedded strata and was therefore flat when the overlying strata were deposited (Fig. 8.2.10). The composition of strata in the lower most Tullig Cyclothem (unit J) is interbedded, thin-bedded sandstone and siltstone with rare *Zoophycos* and *Paleodictyon* trace fossils interpreted as prodelta deposits. The strata in the underlying slump complex are similar in composition, although deformed.

In summary, the upper Gull Island Formation is ~100 m thick and comprises 36% mudstone sheets and 64% slumps (Fig. 8.2.11). The average net sand content of the unit is ~5%. The primary distinction between the upper and middle parts of the Gull Island Formation is: (1) the upper Gull Island Formation has a lower net-sandstone content; (2) the upper Gull Island Formation contains a higher proportion of mudstone sheets; and (3) the deformed unit in the Upper Gull Island Formation is the most complex of those in the entire formation in South Co. Clare.

South County Clare Summary

The Gull Island outcrop (Stop 1) is the type section of the Gull Island Formation and contains a complete vertical transect through the formation. At this locality the formation is informally divided into three units on the basis of physical, observable stratal characteristics (Fig. 8.2.11): (1) lower, (2) middle; and (3) upper. Six characteristics decrease upward through the transect (Fig. 8.2.11): (1) net-sand content, (2) proportion of turbidite lobes and channels, (3) net-sand content in slumps, (4) amount of contractional features within slumps, (5) lateral continuity of sandstone units; and (6) the thickness of sandstone units. Five characteristics increase upward through the transect (Fig. 8.2.11): (1) proportion of slumps, (2) proportion of mudstone sheets, (3) amount and size of extensional features in slumps, (4) amount of topography on upper surface of slumps; and (5) lenticularity of sandstone units. The upward changes described herein are interpreted to reflect a temporal change from ponded (lower Gull Island) submarine fan strata, similar in character to those in the upper Ross Sandstone (Pyles, 2008), to a prograding slope in the middle and upper Gull Island Formation, although the direction of progradation is debated.

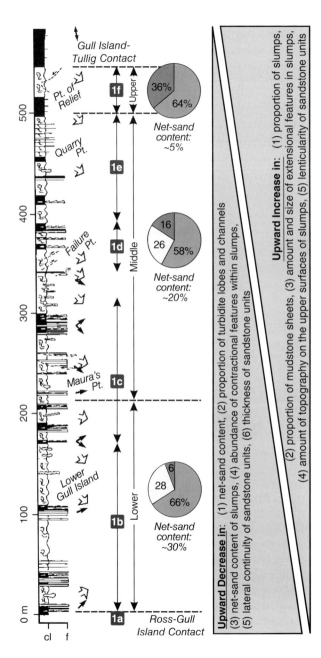

Fig. 8.2.11. Stratigraphic column and pie charts documenting upward patterns in the Gull Island Formation at the Gull Island outcrop (Stop 1). Labels 1a – 1f denote locality numbers. Stratigraphic column modified from Collinson *et al.* (1991).

Upward Decrease in: (1) net-sand content, (2) proportion of turbidite lobes and channels
(3) net-sand content of slumps, (4) abundance of contractional features within slumps,
(5) lateral continuity of sandstone units, (6) thickness of sandstone units

Upward Increase in: (1) proportion of slumps, (2) proportion of mudstone sheets, (3) amount and size of extensional features in slumps, (4) amount of topography on the upper surfaces of slumps, (5) lenticularity of sandstone units

Northern County Clare

Two stops are detailed from which the Gull Island Formation can be observed (Fig. 8.2.12):

Stop 2: Fisherstreet Bay, where the Clare Shale and Gull Island Formation contact is marked by a spectacular slump horizon (Gill, 1979; Strachan & Alsop, 2006).

Stop 3: The Cliffs of Moher, where a complete Gull Island Formation succession is observable.

Transfer to Stop 2

From Stop 1, head back to the R487 and then drive northwards to Kilkee. From here, drive north along the N67 coast road and through Milltown Malbay to Lahinch, where you should take a left turn just past the town

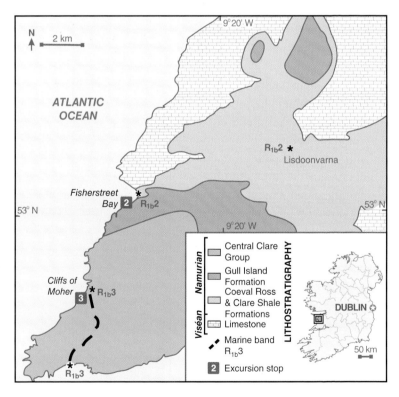

Fig. 8.2.12. Geologic map of northern County Clare documenting the location of field stops and goniatites. Map based on Hodson (1954a; reproduced with permission from the Geological Society of London) and Gill (1979; copyright Geological Society of Ireland).

centre and head to Liscannor on the R478. Drive through Liscannor and continue on this road and past the Cliffs of Moher, until you turn left on the R479 towards the village of Doolin. Turn left at the junction at the bottom of the hill, and drive along the R459 into the small village of Fisherstreet, where parking is available. The drive between Stops 1 and 2 will take approximately 1 hour. Once you have parked your car, walk southwestwards and up the hill out of the village along River Vale Road. The Fisherstreet outcrops are easy to access from a track that leaves River Vale Road (UTM 472997 m E, 5872508 m N; Map 51, 105900 m E, 195411 m N). Walk south-west along the track, leaving it after ~2.5 km to walk westward. This stop takes one half to a full day to complete.

Stop 2: Fisherstreet Bay

The outcrop is 2.5 km long, ~30 m thick and comprises rock platforms and cliffs. The outcrop readily divides into 3 sections, separated by impassable cliffs. Only the southernmost section is suitable for larger groups, although it is easily accessed (by following a stream at UTM 472469 m E, 5872154 m N; Map 51, 105367 m E, 195064 m N) and affords exceptional three-dimensional exposures of the Fisherstreet Slump and overlying non-slumped units (Gill, 1979; Strachan & Alsop, 2006; Figs 8.2.13 and 8.2.14). The basal contact of the Fisherstreet Slump is also exposed within an inaccessible cliff close to Doolin (UTM 473722 m E, 5873284 m N; Map 51, 106636 m E, 196177 m N) where a planar, concordant surface separates underlying Clare Shales from the Gull Island Formation (Fig. 8.2.15; and see Chapter 6.2, Stop 2).

Seven lithofacies are recognised (Tables 8.2.1 and 8.2.2) and grouped into 3 distinct depositional elements:

1 A 20 m interval of intensely folded and faulted interbedded sandstones and mudstones bound by a sharp basal décollement and erosive upper contact (lithofacies 7). This interval is interpreted as a slump (*sensu* Stow, 1986), herein referred to as the Fisherstreet Slump (Strachan & Alsop, 2006). The contrast between the extensively deformed Fisherstreet Slump and overlying flat-lying strata (lithofacies 1 to 6) exhibits the penecontemporaneous nature of slump deformation that predated, and was eroded by, the overlying interval.

2 Tabular mudstone-dominated sheets that display chaotic folding and faulting, deformed sandstone clasts, anastamosing cleavage, sharp flat-lying bases and faulted top bedding planes (lithofacies 3) are interpreted as debrites (Middleton & Hampton, 1973).

3 Non-slumped interbedded sandstones and mudstones (lithofacies 1, 2 and 4 to 6) reveal a range of tabular and lenticular bed geometries that infill compensationally the topography of the Fisherstreet Slump. Formative flows frequently avulsed and shallowly incised (<1 m) into the substrate and

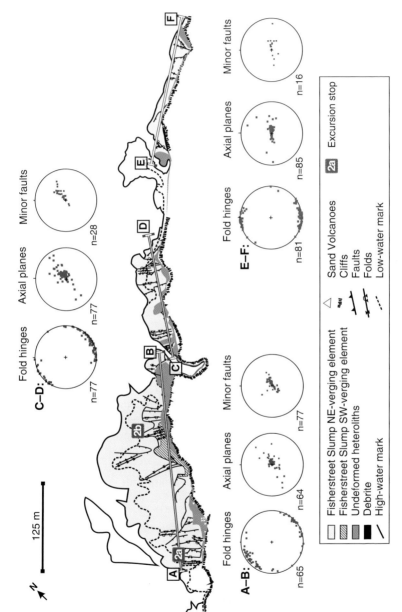

Fig. 8.2.13. Detailed geological map of Fisherstreet showing slump folds and faults, together with stereonets showing fold and fault data. Figure modified from Strachan & Alsop (2006).

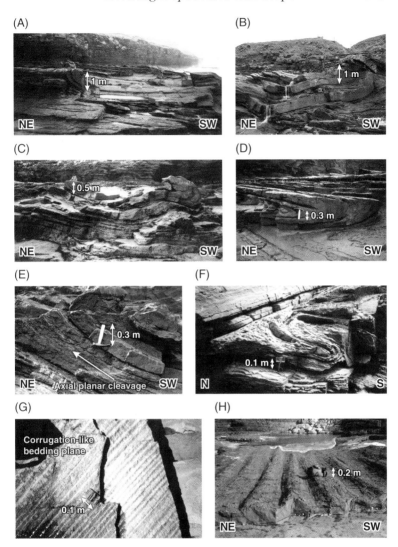

Fig. 8.2.14. Photographs of slump deformation structures from the southern end of the Fisherstreet outcrop (Fig. 8.2.13). A to F. Slump folds truncated and overlain by flat lying strata. G to H. Corrugate-like internal textures (upright folds).

the resultant deposits become increasingly amalgamated and thicken up-section. Sedimentary structures, palaeocurrents and bed geometries suggest deposition from unconfined flows in a turbidite channel-to-lobe transition (e.g. Prélat *et al.*, 2009; Eggenhuisen *et al.*, 2011; Macdonald *et al.*, 2011).

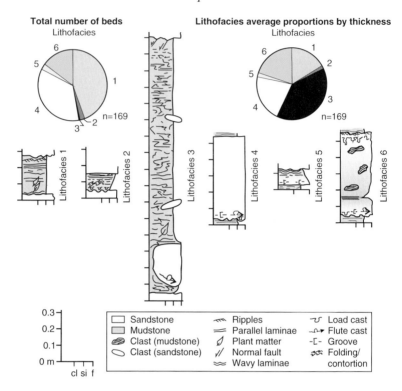

Fig. 8.2.15. Schematic sedimentary log of lithofacies 1 to 6, together with pie charts showing the total number of beds and lithofacies average proportions by thickness in the study area. n = number of measurements.

The 26 m thick stratigraphy at Fisherstreet Bay is divided into two units (Fig. 8.2.16) - the Fisherstreet Slump and the overlying non-slumped strata.

Fisherstreet Slump

The Fisherstreet Slump (lithofacies 7, Tables 8.2.1 and 8.2.2) is described at Stops 2a (UTM 472317 m E, 5872015 m N; Map 51, 105213 m E, 194927 m N) and 2b (UTM 472442 m E, 5872218 m N; Map 51, 105340 m E, 195128 m N; Fig. 8.2.13). At location 2a, a series of recumbent cylindrical slump folds are sharply truncated by flat-lying lithofacies 1 to 6 (Tables 8.2.1 and 8.2.2; Fig. 8.2.14). Folds and faults are concentrated into discrete areas separated by intervening zones of relatively low strain. Folds are typically tight (0 to 30° interlimb angle) although a small number of gentle (120 to 180°) examples are observed. These more open folds may refold the earlier tight folds, suggesting local polyphase deformation (Fig. 8.2.13). Attitudes are characterised by recumbent and gently inclined folds (Figs 8.2.13 and

Table 8.2.1. Summary table of lithofacies observations. Lithofacies are arranged into grainsize divisions.

Lithofacies name	Grain size	Bed thickness	Facies proportion	Number of beds	Sedimentary structures	Bed boundaries	Bed geometry
1. Undeformed tabular mudstone	Silt	0.01 to 0.24 m	16%	71	Massive, parallel laminae, low amplitude ripples, fissile, some thickness variations along section. May contain plant matter.	Sharp, sometimes undulating upper surface. Undulating base where sediment infills topography from underlying ripples or sandstone volcanoes.	Tabular
2. Upward-coarsening lenticular mudstone	Silt	0.02 to 0.1 m	2%	3	Inversely graded, massive, parallel laminae, loaded base, convolutions.	Sharp upper and lower surfaces.	Lenticular
3. Deformed tabular mudstone	Silt	1.3 to 3 m	39%	2	Highly disrupted mudstone. Folding, faulting of mudstone with sandstone clasts. Anastamosing cleavage, highly fissile and also has lenticular sandstones within base. Highly weathered.	Irregular, normally-faulted upper contact. Flat lying base. Sandstone bed above infills topography.	Tabular
4. Massive tabular sandstone	Fine sand	0.01 to 0.5 m	21%	56	Massive sandstones, erosive sole structures (grooves, flutes), occasionally loaded base, asymmetric rippled upper bedding plane.	Sharp, parallel sided.	Tabular
5. Fining-upwards tabular sandstone	Fine sand	0.04 to 0.09 m	2%	4	Normally graded, grooved base, parallel laminae, wavy laminae at the top, contortion near top.	Sharp base, eroded top, loaded top, sharp top.	Tabular
6. Massive lenticular sandstone	Fine sand	0.04 to 0.5 m	20%	24	Massive sandstone bed, sole structures (grooves, tool marks, flutes), ripples on top bedding plane, loaded bases, rounded mud clasts, convolution at the top, wavy lamination.	Sharp.	Lenticular. De-amalgamates over 10 m.
7. Intensely folded sandstones and mudstones	Silt and fine sand	20 m	N/A	1	Slump folds, faults, boudinage, shear zones, sand volcanoes.	Sharp.	Tabular

Table 8.2.2. Summary table of lithofacies interpretations. Lithofacies are arranged into grain-size divisions.

Lithofacies name	Interpreted Bouma divisions	Flow processes	Depositional element
1. Undeformed tabular mudstone	Td, Tde, Tcd	Dilute mud-rich turbidity currents. The presence of plant material implies proximity to a foliage-covered hinterland and suggests a direct link to river-flood effluent derived from hyperpycnal flows. Or alternatively plant material is far travelled out into the basin.	Turbidite channel-to-lobe transition.
2. Upward coarsening lenticular mudstone	Tdc, Tcd, Td, Te	Dilute turbidity currents, which pinch out laterally. Their lenticular geometry reflects flow size.	Turbidite channel-to-lobe transition.
3. Deformed tabular mudstone	N/A Debrite	Debris flow deposits. The inclusion of sandstone rafts in the lower part of the bed indicates highly erosive flows that were able to erode and entrain the seafloor substrate, although the planar lower contact of debris flows suggests they were not erosional in the study area. Preservation of the irregular upper contact suggests that the succeeding sandstone was deposited a short time after debrite deposition.	Debrites.
4. Massive tabular sandstone	Ta, Tac	Sandy turbidity currents that were erosive and deposited rapidly. Flows were unconfined. The absence of fining-upwards beds favours aggradation from a fast-moving steady flow, whereby the finer sediment fractions bypass downslope.	Turbidite channel-to-lobe transition.
5. Fining upwards tabular sandstone	Tbc	Rapidly-decelerating unconfined sandy turbidity currents.	Turbidite channel-to-lobe transition.
6. Massive lenticular sandstone	Ta, Tabac, Tac	Sandy turbidity currents, which were erosive leading to frequent amalgamation. Bed types suggest complex changes in flow velocities with time.	Turbidite channel-to-lobe transition.
7. Intensely folded sandstones and mudstones	N/A Slump	Siliciclastic slump horizon that remobilized a heterolithic assemblage of interbedded turbidite sands and muds.	Slump.

8.2.14), with a number of steeply inclined examples also being found within the uppermost part of the slump. The observed folds are rootless and display fold layer shapes ranging from class 1B (parallel) to class 2 (similar; Hudleston, 1973), have fold amplitudes ranging from 0.3 to 5 m and bed thicknesses that vary greatly from 0.005 to 4.1 m.

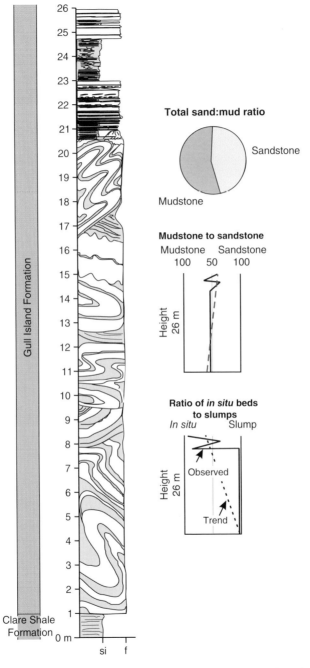

Fig. 8.2.16. Summary log showing entire interval at Fisherstreet, together with ratios of sand to mud and *in situ* beds to slumps.

Examination of the recumbent folds at Stop 2a reveals a dominance of synformal structures with south-west closing fold hinges (Fig. 8.2.14). These folds are cylindrical and may be traced for up to 30 m (Fig. 8.2.13). Each fold is juxtaposed against a slump thrust fault that terminates at the top of the slump. The dominance of synformal folds without their anti-formal pair implies that all antiformal folds have been thrust away from their synformal counterpart. The complete absence of antiformal structures further suggests that they have been removed via erosion from the upper part of the slump (Strachan, 2002a). The planar nature of the upper slump contact is spectacular (Figs 8.2.14A to 8.2.15). A thin sandstone horizon tops the slump, appears to have infilled the slump generated topography and is interpreted as a turbidite (Strachan & Alsop, 2006). Palaeocurrents from ripples on the upper bedding plane indicate a preferred south-westerly orientation of flow and are interpreted as reflected flows from the slump topography (Fig. 8.2.17). At the base of the cliff, spectacular sandstone volcanoes of 0.1 to 1 m in diameter can be viewed on the well-exposed bedding plane that marks the top of the slump (Fig. 8.2.18). Volcanoes are composed of very fine-grained sandstone that is lighter in colour than the surrounding strata. The volcano flank profiles may be convex or concave, with dips ranging from 1 to 19° (Gill & Kuenen, 1958), and the central craters are well-developed with preserved laminae on their flanks, suggesting long duration, unsteady, dewatering flows (Fig. 8.2.18). Sandstone volcanoes are positioned directly above slump faults and were formed via fluidization of sediments focused along slump faults, which acted as

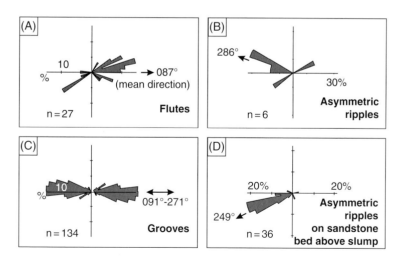

Fig. 8.2.17. Palaeocurrent measurements from strata overlying the Fisherstreet Slump grouped by type of sedimentary structure and relative location.

(A)

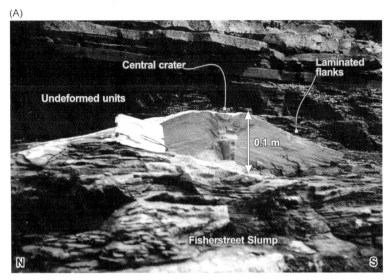

(B)

Fig. 8.2.18. Photographs showing the range of sand volcano types. A) Well-formed large isolated example with prominent central vent. B) Numerous small-scale sand volcanoes completely covering a bedding plane.

conduits (Gill & Kuenen, 1958; Strachan & Alsop, 2006). Development of sand volcanoes post-dates deposition of the thin ripple-topped sandstone upon which all sand volcanoes sit. This implies that dewatering occurred a short time after slump emplacement, with the trigger for dewatering being interpreted to be compaction and loading of the slump (Strachan, 2002b). The section between locations 2a and 2b reveals an array of fold structures of varying size. Fold hinges are typically sub-horizontal, with trends varying from NW-SE and axial planes that fan about the associated fold axes, resulting in a weak girdle on stereonet plots (Fig. 8.2.13). Rare, gently curvilinear, fold hinge-lines are locally observed. Fold asymmetry varies along the exposure with two distinct NE-verging and SW-verging packages (Fig. 8.2.13). The north-east-verging folds predominate, with southwest-verging folds located close to location 2b (UTM 472442 m E, 5872218 m N; Map 51, 105340 m E, 195128 m N) where they directly overlay north-east-verging structures (cross-section A to B, Fig. 8.2.13). Fold hinge data along the entire outcrop exhibits a change in trend from NW-SE (sections A to B and C to D) to more N-S (section E-F, Fig. 8.2.13). This change in fold hinge orientation by ~35°, without a corresponding change in fold style, is most readily explained as lateral spreading at a frontal- or lateral- lobe within the toe region of the slump (Strachan & Alsop, 2006; their fig. 13). Such locally-divergent flow patterns suggest that constant transport directions should not automatically be assumed for slump sheets even when folds appear to display consistent geometries along the exposure (Fig. 8.2.13). The systematic use of slump transport determination techniques, as conducted by Strachan & Alsop (2006), reveals a mean downslope transport direction of 067° (Fig. 8.2.17). This direction supports the findings presented by both Collinson *et al.* (1991) and Wignall & Best (2000), but is contrary to those of Gill (1979) who interpreted a SE transport direction. Gill (1979) observed a distinct internal fabric at location 2b, that he interpreted as groove-like mullions formed parallel to flow. Fig. 8.2.14H shows these so-called grooves but it is apparent that this bed has been folded and not scoured and thus was formed by layer-parallel compression and not transport-parallel grooving.

Non-slumped strata

Five sedimentary logs from non-slumped units positioned above the Fisherstreet Slump reveal a heterogeneous assemblage of lithofacies (1 to 6, Figs 8.2.19 and 8.2.20) and lenticular sandstones that amalgamate and de-amalgamate over 10s of metres (Tables 8.2.1 and 8.2.2; Fig. 8.2.21). Lithofacies 1, 2 and 4 to 6 are interpreted to have been deposited by weakly confined to unconfined, dilute and concentrated turbidity currents that exhibit a range of temporal behaviours including waxing, waning and steady flows (Kneller, 1995; Table 8.1.1, Figs 8.2.19 and 8.2.20). The total

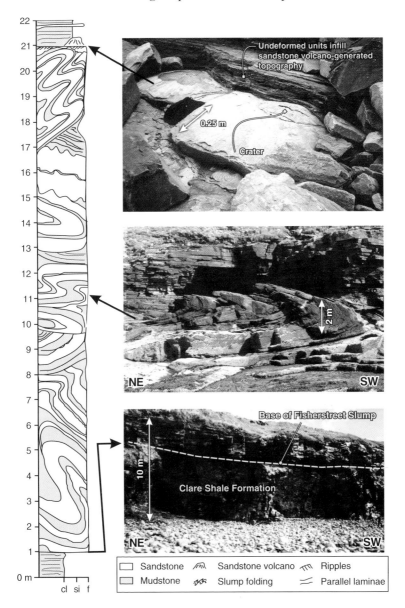

Fig. 8.2.19. Sedimentary log of lithofacies type 7 – the Fisherstreet Slump, together with representative photographs showing the basal surface, folding and the upper surface.

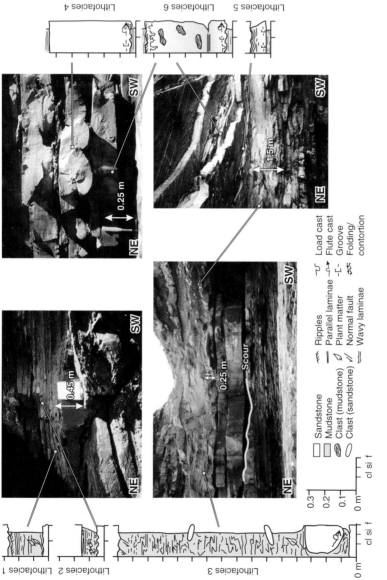

Fig. 8.2.20. Photographs and stratigraphic columns of lithofacies 1 to 6.

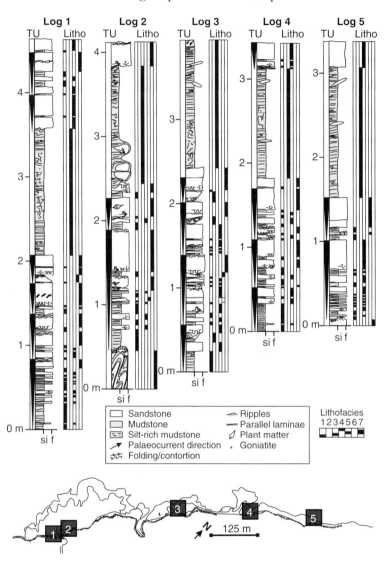

Fig. 8.2.21. Five measured sedimentary logs showing the vertical and lateral variation in strata above the Fisherstreet Slump. Location of logs is shown in the lower map.

thickness and number of beds varies at each logged site. Nevertheless, a consistent upward trend in lithofacies is observed, characterised by: 1. interbedded lithofacies 1 and 4, 0 to 1 m above the slump; 2. interbedded lithofacies 1 and 6, 1 to 2 m above the slump; and 3. a single, laterally extensive bed of lithofacies 3, 2 to 4 m above the slump (Tables 8.2.1 and 8.2.2). Sandstone beds reveal an array of thickening upward (TU) motifs and large shallow scours (1 to 2 m across, 0.2 m deep,) are observed on the base of several sandstone beds, which often truncate mudstones (Fig. 8.2.17). Erosive sole structures are well-exposed in the roof of a small cave positioned at Stop 2a (Figs 8.2.13 and 8.2.20), with grooves being the dominant form but flutes and prod marks are also present (Figs 8.2.17 and 8.2.20). The basal bedding planes exposed in the cave exhibit an exquisite array of structures and a range of palaeocurrent flow directions that display the scatter in flow directions of successive turbidity currents (Fig. 8.2.20).

Lateral correlation of beds shows that maximum bed thicknesses coincide with amalgamated beds (Fig. 8.2.22). Amalgamated beds typically have one thickened area, interpreted as the lobe or weakly confined channel axes, with beds thinning and pinching laterally. Bed aspect ratios are dominated by thin, wide beds generally less than 0.4 m thick and 0.9 to 2.6 km wide that are more typical of lobes (Fig. 8.2.22). Sedimentary log correlation reveals an irregular Fisherstreet Slump upper surface, characterised by a broad u-shaped conduit ~1 m deep and 2 km wide (Fig. 8.2.22). Overlying beds have compensationally infilled this topography, with thinner intervals positioned on relative highs (Fig. 8.2.22). Many lobe or channel apices are positioned close to log 3 (UTM 472683 m E, 5872381 m N; Map 51, 105584 m E, 195288 m N; Fig. 8.2.22), which was a site of preferential aggradation. The irregular slump surface is interpreted to result from significant erosion, which cut a broad, shallow conduit into the top of the slump. We speculate that a debris flow was the eroding agent that removed the upper parts of the Fisherstreet Slump, some of which may be preserved at the northern end of the outcrop, where a 3 m thick debrite (lithofacies 3) overlies the slump and is locally incorporated into it (Fig. 8.2.23). The fact that this unit is locally incorporated into the slump implies coeval formation with the Fisherstreet Slump and suggests that significant surface erosion and transformation occurred during translation (Fisher, 1983; Strachan, 2008; Fig. 8.2.23).

Palaeocurrent directions have been measured in sandstones (lithofacies 4 to 6) from flutes, grooves and ripple cross beds. Figure 8.2.17 shows a compilation of rose diagrams from Fisherstreet Bay, with groove casts being the dominant preserved structure revealing W-E trending flows. Flute casts reveal a polymodal distribution, with some transport towards the SE and WSW, with the statistical mean towards the ENE (Fig. 8.2.17). Only a small number of ripple cross-beds have been measured in the upper few centimetres of lithofacies 6 beds and reveal an oblique distribution with flows directed towards the NE, SW and WNW (Fig. 8.2.17). Flow directions from flute casts and cross-beds can vary within the same bed,

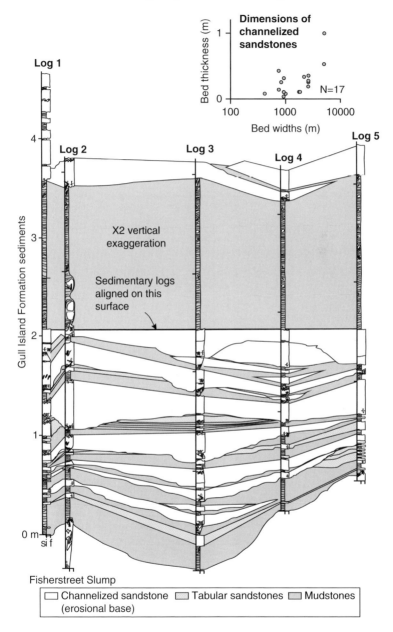

Fig. 8.2.22. Stratigraphic cross-section of the Fisherstreet Slump. Location of logs and cross-section is shown in Fig. 8.2.21. The plot above the cross-section documents the thicknesses and widths of lenticular, channelized, sandstone beds. Logs are aligned on the basal surface of the debrites.

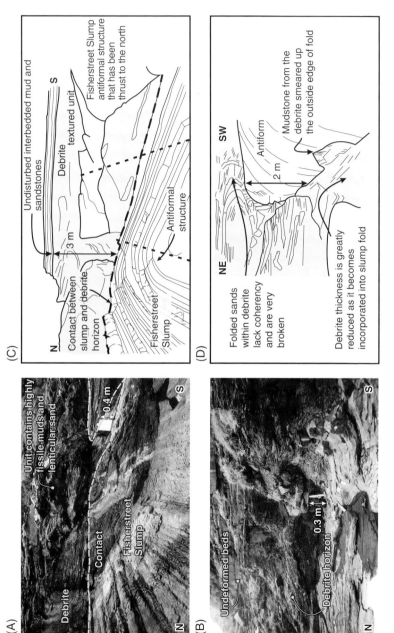

Fig. 8.2.23. Annotated photographs (A and B) and field sketches (C and D) documenting the relationship between the Fisherstreet Slump and overlying debrites. Grey shading represents the debrites.

for example flute casts on the base of a single bed suggest flow towards 093°, with overlying ripples implying a 235° flow direction. This suggests that such flows were deflected or reflected against the sea floor topography. Probable candidates for such barriers include local fault traces, slump topography and frontally-emergent slumps (Frey Martinez *et al.*, 2006). Dominant ENE and E transport directions are interpreted as the principal flow direction (Fig. 8.2.17). Several other transport directions are present towards the WSW and SE and may imply: 1) unconfined flows that radiate from a central feeder; 2) that the turbidity currents were channelised and deposited within meandering thalwegs; and 3) that turbidity currents were deflected and reflected by sea floor topography. Based on field observations, the range of transport directions is attributed to radiating unconfined flows and flow deflection and reflection against sea floor topography.

The environment of deposition was prone to large-scale sediment failure and deposition of unconfined turbidites, with instability probably driven by high sedimentation rates. Sea floor topography on the Fisherstreet Slump had a profound influence on ensuing turbidity currents that were focussed and compensationally stacked upon the surface. Turbidity current flows were erosive and often bypassed the study area, although the local sea floor topography reflected some flows. The slope is inferred to have faced towards the ENE because the Fisherstreet Slump moved towards 067° (Strachan & Alsop, 2006) and because the majority of turbidity currents flowed in this direction. Sandstone turbidite bed thickness distributions reveal an exponential distribution (Fig. 8.2.24), which are thought to be typical of proximal base-of-slope settings (Sinclair & Cowie, 2003). Based on all available data, we suggest the environment of deposition to be a submarine slope or base of slope of either the Shannon Basin or a local sub-basin. Non-slumped units were deposited in a proximal lobe position close to a channel-lobe transition. We suggest that the lobe location was relatively proximal (e.g. crevasse splay) given evidence of slope collapse and sediment bypass. Water depths are not known, but there is as an absence of storm-wave generated bedforms.

Transfer to Stop 3

From Fisherstreet, drive back to the R478 and head south-west to park at the Cliffs of Moher visitor centre (UTM 471134 m E, 5869244 m N; Map 51, 103990 m E, 192171 m N). Walk to the cliffs and follow the footpath northwards, 300 m past O'Brien's Tower towards Knockardakin (Fig. 8.2.25). By looking south at this location, visitors are able to view the outcrops below. Gull Island Formation sediments are exposed in the bottom half of the ~180 m high cliff. **The outcrop is dominated by shear vertical cliffs that are partially covered in rock fall. Take care.** This stop takes 2 hours to complete.

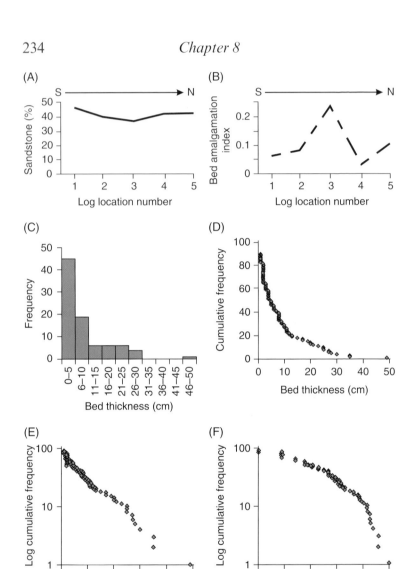

Fig. 8.2.24. Plots showing: A) spatial changes in percentage sandstone; B) spatial changes in bed amalgamation index; C) frequency histogram of bed thickness; and D to F) cumulative frequency of bed thickness. Location of logs is shown in Fig. 8.2.21.

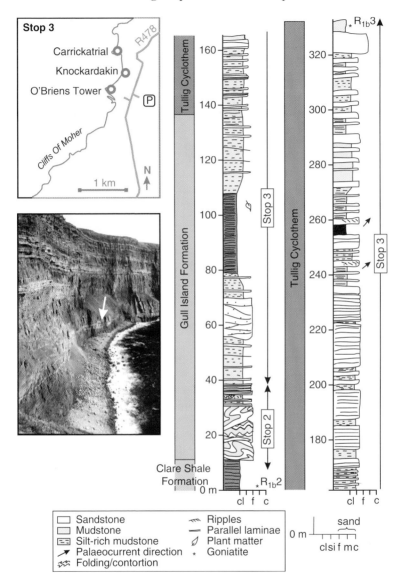

Fig. 8.2.25. Map of Stop 3, Cliffs of Moher and schematic composite stratigraphic column of the Gull Island Formation and Tullig cyclothem at Stops 2 and 3. The Cliffs of Moher portion of the log is redrawn from previous descriptions (Hodson, 1954a; Collinson *et al.*, 1991; Wignall & Best, 2000). White arrow on inset photograph points to a channel in the Gull Island Formation. Photograph courtesy of Paul Wignall.

Stop 3: Cliffs of Moher (UTM 471134 m E, 5869244 m N; Map 51, 103990 m E, 192171 m N)

Several authors document the Gull Island Formation at the Cliffs of Moher (Hodson, 1954a; Martinsen *et al*., 2000; Wignall & Best, 2000, 2004) and this data is used herein to describe the interval. The exposed Gull Island Formation is ~100 m thick (Martinsen *et al*., 2000; Wignall & Best, 2000; Martinsen *et al*., 2003; Wignall & Best, 2004) that, when added to the Fisherstreet Bay section, produces a total north Co. Clare Gull Island Formation thickness of *c.* 130 to 150 m (Rider, 1974; Gill, 1979; Martinsen *et al*., 2000; Wignall & Best, 2000).

The exposed stratigraphic interval at the Cliffs of Moher is younger than at Fisherstreet, containing younger strata from within the $R_{1b}2$- $R_{1b}3$ bio-stratigraphic cycle (Hodson & Lewarne, 1961; Figs 8.1.4 and 8.1.5). Marine band $R_{1b}3$ has been identified within a ~7 m thick black marine shale (Fig. 8.2.25) located in the highest part of the cliff, some 100 m above the Gull Island Formation (Hodson, 1954a). The top of the Gull Island Formation is placed at the base of a distinct coarsening-upwards trend that marks the transition to deltaic Tullig Cyclothem sediments (Gill, 1979; Chapter 10.2, Stops 2 and 3).

Gull Island Formation sediments contain abundant undeformed tabular silty mudstones with thin sandstone beds (Hodson, 1954a; Gill, 1979; Martinsen *et al*., 2000; Wignall & Best, 2000; Fig. 8.2.25). Within these units, an 8 to 15 m thick, 200 m wide lenticular sandstone unit is observed (Martinsen *et al.* 2000; Wignall & Best, 2000; Martinsen *et al*., 2003). Internally, the lenticular unit is composed of several turbidite sandstone beds that are speculated to be structureless with laminated tops (Martinsen *et al*., 2000). Laterally, these sandstone beds amalgamate and de-amalgamate (Martinsen *et al*., 2000; Wignall & Best, 2000; Martinsen *et al*., 2003). Martinsen *et al*., (2003) report that these turbidites flowed towards the east, although it is unclear how this direction was acquired given the precipitous and inaccessible nature of the outcrop. An upper extensive sheet sandstone extends out across lower lenticular sandstones (Martinsen *et al*., 2003). The undeformed tabular silty mudstones with thin sandstone beds fine upwards to black shales (Hodson, 1954a; Fig. 8.2.25). Rare plant matter has been found within fallen blocks of black shale at Carrickatrial and includes *Calamites* sp. and *Mariopteris muricata* (Hodson, 1954a).

The silty mudstone character of these Gull Island Formation sediments is interpreted to have resulted from hemipelagic deposition punctuated by rare turbidity currents that scoured slope channels transporting coarser grained sediment further downslope (Wignall & Best, 2000; Fig. 8.2.25). The overlying black shale part of the interval has not been described since the detailed biostratigraphic study of Hodson (1954a). The omission of this interval from more recent descriptions is unclear (for example,

Collinson *et al.*, 1991; Wignall & Best, 2000; Martinsen *et al.*, 2003) but may be because some of the cliff has been obscured by rock fall.

Few deformation structures are reported and as a consequence slumping and sliding are thought to be minimal at this locality (Hodson, 1954a; Gill, 1979; Collinson *et al.*, 1991; Wignall & Best, 2000; Martinsen *et al.*, 2003; Fig. 8.2.25). The lack of soft-sediment deformation has been attributed to relatively low sedimentation rates (Collinson *et al.*, 1991) and/or a basin floor location (Wignall & Best, 2000).

Several environmental models have been proposed to explain the North Co. Clare Gull Island Formation. In a summation of his regional observations, Rider (1974) suggested that these sediments represent a shelf and unstable continental slope, with north-eastward progradation of an overlying delta with time. In a reinterpretation of Rider's (1974) work, Collinson *et al.* (1991) suggested that these sediments were deposited in a pro-delta or slope setting linked to south-east progradation of a genetically linked (Tullig) delta. The apparent absence of turbidites was explained by ubiquitous sediment bypass, which delivered coarser-grained material further downslope via isolated erosive channels. Thus, this location was envisaged to be a relatively inactive northern slope or basin margin to a prograding delta, dominated by low sedimentation rates (Collinson *et al.*, 1991; Martinsen *et al.*, 2000). Martinsen *et al.* (2003) further suggested that northern Co. Clare only became a site of active deposition after southern and central Gull Island Formation accommodation space had become filled. Wignall & Best (2000) similarly envisaged low sediment flux at this location, attributing this to a distal basin floor position rather than an inactive slope. In their model, Wignall & Best (2000) interpreted the Gull Island Formation as representing slope progradation towards and down a NE-dipping slope.

The inaccessibility of the Gull Island Formation outcrops at the Cliffs of Moher has prevented detailed sedimentological quantification and has led to some observational inconsistencies and differences in interpretation. In addition, recent workers have discarded the biostratigraphic work of Hodson (1954a) and Hodson & Lewarne (1961), who identified marine band $R_{1b}2$ (*R. nodosum*) beneath the Fisherstreet Slump. The omission of this biostratigraphic marker in later studies is significant because sedimentation rates were tentatively estimated. By resurrecting the original north Co. Clare biostratigraphic work of Hodson (1954a) and Hodson & Lewarne (1961), who described ~291 m of strata between marine bands $R_{1b}2$ and $R_{1b}3$, an extraordinary average sedimentation rate of ~4.48 m kyr^{-1} is calculated (Fig. 8.1.9) that is not in accord with the low rates inferred by Martinsen *et al.* (2000). Although this is an average sedimentation rate for both the Gull Island Formation and Tullig Cyclothem, we suggest that equally high rates persisted for part of the Gull Island Formation depositional period. This necessitates a reinterpretation of the mode of deposition for the undeformed tabular silty mudstones and thin sandstone beds, which we speculate were the products of unconfined turbidity currents. These formative flows were deposited as turbidite lobes or as turbidite channel levee deposits, adjacent to small erosive channels on a submarine slope whereby finer grained sediments were sourced from higher in the flow. The erosive turbidite channel unit preserves evidence of infrequent small confined channelized flows (Rider, 1974; Collinson *et al.*, 1991; Martinsen *et al.*, 2000; Wignall & Best, 2000; Martinsen *et al.*, 2003). The aspect ratio of these lenticular sandstones is similar to the Ross Sandstone Formation (Fig. 8.2.25).

There is also some disagreement in the literature relating to the vertical succession of lithofacies types at the Cliffs of Moher. Hodson (1954a) reported that silty mudstones and thin sandstone beds fine upwards to significant black shales (~80 m thick), while subsequent workers described a coarsening-upwards motif (Rider, 1974; Collinson *et al.*, 1991; Martinsen *et al.*, 2000; Wignall & Best, 2000; Martinsen *et al.*, 2003). Today the ~80 m of black shale is not visible but for completeness we cautiously include the black shale interval of Hodson (1954a) and also note that coarsening-upwards demarcates the base of the Tullig Cyclothem here (Fig. 8.2.25).

Due to the precipitous nature of the Cliffs of Moher, a number of observational and interpretative questions remain. Nevertheless, an alternate model for the environment of deposition is offered, based on all available published data. Here, we suggest that the Gull Island Formation in the lower part of the Cliffs of Moher records an interval of rapid deposition via unconfined turbidity currents on a submarine slope or basin floor. Here, despite the high rates of deposition, low slope gradients prevented slope failure in the Gull Island Formation. The central part of the Cliffs of Moher is composed of the Tullig Cyclothem that is described in Chapter 10.2, Stops 2 and 3.

Northern County Clare summary

The outcrops exposed at Fisherstreet and the Cliffs of Moher reveal a series of slumps and turbidites deposited in submarine slope and base of slope environments, with extremely high aggradation rates (~4.48 m kyr^{-1}). We consider the northern region of Co. Clare became an active depocentre after $R_{1b}2$ times and therefore is one of the youngest sites of deposition in the Gull Island Formation. Sedimentation was initially dominated by slumping and instability, followed by turbidity currents that were not remobilized. Overall the degree of deformation is much lower than in southern Co. Clare.

8.3 Synthesis

The review and field excursions described herein document the range of superb Gull Island Formation outcrops from northern and southern Co. Clare and has unearthed some 'lost' data that has enabled refinement of our understanding of the Formation. The base of the Gull Island Formation is a diachronous boundary between the coeval Ross Sandstone and Clare Shales, which was initiated in the SW and migrated north and east with time. The Gull Island Formation is defined on lithostratigraphic grounds characterised by ubiquitous slumping, with the high proportion of slumps being interpreted as a response to high sedimentation rates. In our view, the spatial distribution of sediment thicknesses records complex and rapidly evolving basin geometry, with sedimentation rates suggesting more than one active depocentre at a time.

Chapter 9
The Tullig and Kilkee Cyclothems in Southern County Clare

JIM BEST, PAUL B. WIGNALL, ELEANOR J. STIRLING, ERIC OBROCK & ALEX BRYK

9.1 Introduction

The Tullig Cyclothem marks the first and thickest of the repeated cyclical pulses of sedimentation that filled the Shannon Basin after deposition of the Gull Island Formation, with a total of five such deltaic cycles being seen in County Clare. Unlike the thickness trends in the early part of the fill of the Shannon Basin, the Tullig Cyclothem shows only slight thinning to the north-west of County Clare, whilst the overlying Kilkee Cyclothem shows a slight thickening towards the NE (the reverse of the trend present in the underlying Ross Sandstone and Gull Island Formations). These cyclical sediments have been widely interpreted to be the product of repeated delta progradation into the basin, accompanied by periods of sea-level rise and fall (Rider, 1974; Pulham, 1989; Collinson *et al.*, 1991; Wignall & Best, 2000). This chapter provides details of excursions to view the principal localities of these fluvio-deltaic sediments in southern County Clare, with the exposures in northern County Clare being detailed in Chapter 10. Much of the exposure is along sea cliffs or low coastal outcrop **and great care should be taken on slippery surfaces, when walking near high, steep cliffs, and in poor weather with high winds.**

The erosive contacts at the base of some of the cyclothem sandstones, and the lateral change in facies, have been used to provide support for sequence stratigraphic explanations of the stratigraphy, including suggestions that the bases of the major fluvial sandstones in the cyclothems form Type 1 sequence boundaries due to base-level fall and basinward migration of the fluvial channels (Hampson *et al.*, 1997). The cyclothems broadly consist of small-scale, coarsening-upwards cycles/parasequences in which the individual sandstone

A Field Guide to the Carboniferous Sediments of the Shannon Basin, Western Ireland, First Edition. Edited by James L. Best and Paul B. Wignall.
© 2016 International Association of Sedimentologists.
Published 2016 by John Wiley & Sons, Ltd.
Companion website: www.wiley.com/go/best/shannonbasin

bodies generally become thicker upwards, although each cycle is associated with mouth-bar progradation and abandonment/retrogradation. As such, a wide range of lithologies and sedimentary environments are exposed in these cyclothems, from offshore muds to shoreface sands and from mouth-bar sediments to fluvial channels and interdistributary bay fills.

The lower part of each cyclothem shows mudstone- and siltstone-dominated prodelta facies, often with evidence of slope collapse (slumps, growth faults) and diapiric movement, whilst mouth-bar and distributary channel facies dominate the central parts of each cycle (Rider, 1974; Pulham, 1989). These mouth-bars and channels, in turn, are sharply/erosively overlain by fluvial sandbodies (e.g. the Tullig and Kilkee Sandstones) that may be capped by palaeosols. Other evidence for emergence includes coal horizons and fossil forests with *in situ* tree stumps. These fluvial sandbodies and terrestrial facies record the culmination of progradation within the cyclothems but do not mark the formational boundaries. Instead, marine bands were chosen as the markers to subdivide the stratigraphy by Hodson & Lewarne (1961), and this approach is still adhered to. Thus, the fluvial sandbodies are overlain by retrogradational (transgressive) successions that typically show a strong marine influence and can include shoreface sandstone bodies, with smaller parasequences within these transgressive sediments (Davies & Elliott, 1996; Wignall & Best, 2000; Obrock, 2011). In sequence stratigraphic parlance, transgressive systems tracts therefore form the upper part of the cyclothems and maximum flooding surfaces (marine bands) mark their lithostratigraphic boundaries.

There has been considerable debate as to the exact nature and direction of the deltaic, mouth-bar and fluvial progradation within the Shannon Basin (see Chapter 3) and the Tullig Cyclothem provides key exposures and palaeocurrents to assess the direction of sediment transport at this time in the basin evolution. Palaeocurrent data collected by a range of workers (Rider, 1969, 1974; Gill, 1979; Pulham, 1989; Williams & Soek, 1993; Wignall & Best, 2000, 2002; Stirling, 2003) has shown that the vast majority of palaeocurrents within the Tullig Cyclothem demonstrate flow towards the N/NE, suggesting a source to the S/SW. This has been interpreted either to represent the progradation direction of the major delta and its fluvial channels (Fig. 9.1.1A; Rider, 1974; Gill, 1979; Wignall & Best, 2000, 2002; Stirling, 2003) or a distributary channel that is running at a high angle to the principal south-easterly progradation direction (Fig. 9.1.1B; Pulham, 1989). Collection of palaeocurrent data, together with data that show the changing size and style of sedimentation within the Tullig Cyclothem, and in particular within the fluvial Tullig Sandstone, are clearly vital to resolve the differences in these models. The outcrops detailed in this chapter present the principal locations where data can be gained to address this key debate regarding the evolution of the basin.

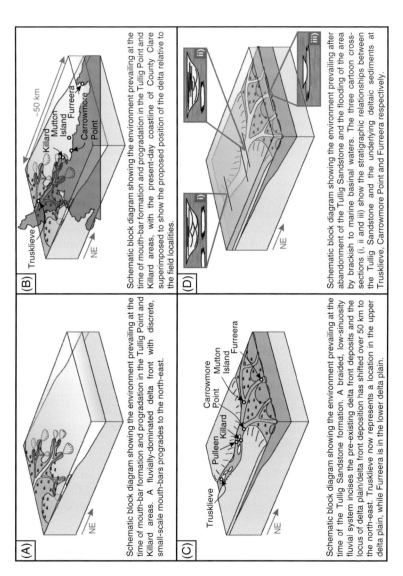

Fig. 9.1.1. Models of delta progradation and evolution during the Tullig Cyclothem (after Stirling, 2003). A) Principal flow direction to the NE. (*Figure continued on next page*).

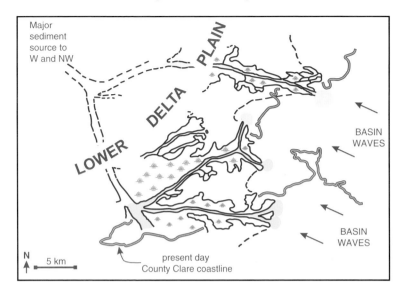

Fig. 9.1.1. (*Cont'd*) B) Dominant flow to the SE with a major distributary channel running to the NE (redrawn from Pulham, 1989; reproduced with permission from the Geological Society of London). The yellow shading shows sediment transported through the distributary channels and deposited as mouth-bars at the delta front, whilst the green vegetation symbols represent plants and subaerial exposure.

This chapter contains details of several excursions to key localities of the Tullig and Kilkee Cyclothems in southern County Clare, which display a wide range of features associated with progradation of these major fluvial channels and subsequent transgressions that occurred at the top of each cyclothem. The localities detailed are at Tullig Point, Trusklieve, the coast road south of Kilkee, Kilkee, Killard, Carrowmore Point and Mutton Island. Each locality is detailed as a separate excursion in order that the reader can plan their own itinerary, with there being several stops at all localities except Tullig Point.

9.2 Tullig Point (UTM 445490 m E, 5829400 m N; Map 63, 077780 m E, 152665 m N)

The transition between the Gull Island Formation (Strachan & Pyles, Chapter 8) and the Tullig Cyclothem is visible at Tullig Point, above a large growth fault complex in the upper Gull Island Formation, as described in Chapter 8 (see Fig. 8.2.10, Chapter 8, Stop 1f). To reach Tullig Point, travel west from the village of Cross on the R487 for ~1.5 km and then turn north on the unsurfaced road towards Trusklieve. Proceed north for ~1 km and then turn west on a dirt road that ends at a quarry, where you can park on the east side of the quarry. After a short walk to the north side of the quarry, the lowest Tullig Cyclothem strata are viewed in spectacular, but largely inaccessible, cliff sections. The Tullig Cyclothem has its maximum thickness (200 m) at its southernmost outcrop and type section at Tullig Point (UTM 445490 m E, 5829400 m N; Map 63, 077780 m E, 152665 m N; Fig. 9.2.1A). The outcrops described at this locality may be accessed by walking along the coast to the north of the Point of Relief to Tullig Point and for a short distance to the north of here. This location will take approximately 2 hours to visit.

Laminated silty shales dominate the lower part of the cyclothem (see Fig. 8.2.10, Chapter 8) but gradually coarsen-up into siltstones and fine sandstones. These sediments can be examined in the cliff top exposures at UTM 445610 m E, 5829365 m N (Map 63, 077900 m E, 152629 m N) and UTM 445570 m E, 5829540 m N (Map 63, 077862 m E, 152804 m N). The increasing frequency of thin, wave-rippled sandstones, with some surfaces showing rare thread-like trace fossils, records the progradation of a delta into a wave-dominated basin. A mouth-bar sandbody caps the coarsening-up trend that can be viewed in the cliffs at Tullig Point (Fig. 9.2.1A,B) and in which large cross-beds can be seen at the cliff top. Several further cycles are developed in the Tullig Cyclothem. Pulham (1989) clearly showed the extremely limited lateral extent of the cycle-top sandbodies (a few kilometres at most). This is considered a consequence of their point-sourced distributary and mouth-bar origin.

The style of sedimentation of the Tullig Cyclothem changes abruptly with the appearance of the Tullig Sandstone, a sharp-based, extensive sheet sandbody. The Tullig Sandstone reaches its maximum thickness of 60 m at Tullig Point (Figs 9.2.1 and 9.2.2) but can be traced along the entire length of the County Clare coastline. In this cliff section, the rather irregular internal geometry of the Tullig Sandstone suggests stacked channel sandbodies of a fluvial system. This sandbody will be examined in detail at Trusklieve (see 9.3 below). It represents the most proximal facies of a prolonged, continuous phase of basin fill that began with the basin-floor turbidites of the Ross Sandstone Formation and ended in the fluvial facies of Tullig Point.

Exposures to the north of Tullig Point show the initial flooding surface above the Tullig Sandstone (UTM 445435 m E, 5829719 m N; Map 63, 077730 m E, 152985 m N) that contains abundant *Zoophycos* traces and then is overlain by a series of well-laminated siltstones (Fig. 9.2.3) with each lamina showing a fining-upwards motif. The thin, dark laminae are between 1 and 10 mm thick and lighten in colour upwards, whereas the light-coloured laminae are between 0.01 and 0.1 m thick, fine-upwards

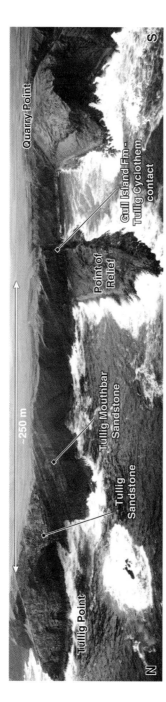

Fig. 9.2.1. A) Aerial photograph of the cliffs at the Point of Relief and Tullig Point, with the Tullig mouth-bar sandstone, Tullig Sandstone and Tullig Cyclothem – Gull Island Formation contact indicated (photo courtesy of David Pyles).

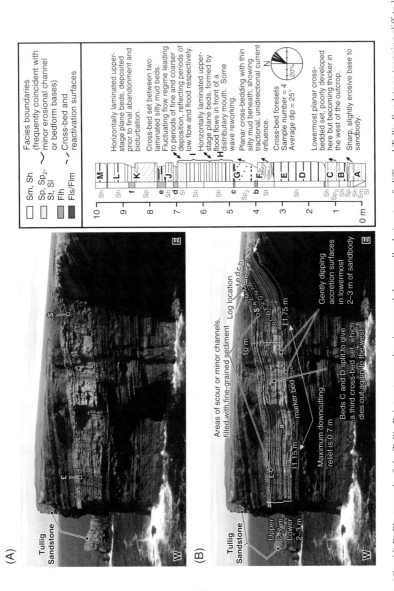

Fig. 9.2.1. (*Cont'd*) **B** Photograph of the Tullig Point outcrop, showing mouth-bar sandbody (centre of cliff) and Tullig Sandstone (capping cliff to left, labelled). Log shows succession from sharp base to top of mouth-bar sandbody. Arrows on log indicate palaeocurrent directions.

~17 m

Fig. 9.2.2. The Tullig Sandstone at its type locality at Tullig Point, where it is over 50m thick. The lower part of the sandbody shown here shows irregular bedding, suggesting stacked channels, whilst the upper part is better bedded and may represent a mouth-bar facies.

and contain occasional vertical burrows (Fig. 9.2.3). These siltstones probably represent the products of hypopycnal flows at the delta front, with sediment settling from suspension. The outcrop to the north is covered until the top of a sandstone bed that is exposed at UTM 445570 m E, 5829868 m N (Map 63, 077867 m E, 153132 m N), where a wave-rippled, bioturbated sandstone is overlain by *c.* 3 m of dark grey sideritic siltstone that is, in turn, capped by a marine band containing the goniatite *Reticuloceras* aff. *stubblefieldi*, representing the maximum flooding surface at the top of the Tullig Cyclothem. The stratigraphic distance between the top of the Tullig Sandstone and this marine band is *c.* 77 m at this locality, with the wave-rippled sandstones towards the top of this transgressive package demonstrating a shallowing-upwards sequence before the final transgression that produced the maximum flooding surface at the top of the Tullig Cyclothem (Wignall & Best, 2000; Obrock, 2011). The entire sequence within the Tullig Cyclothem can only be viewed partially at Tullig Point, due to the inaccessibility of the outcrop. However, other sections along the coast afford opportunities to examine all parts of the Tullig Cyclothem, and its lateral and vertical variability, in detail.

(A)

(B)

Fig. 9.2.3. Well-laminated siltstones within the Tullig Cyclothem at Tullig Point showing: A) distinct fining upwards within each lamina, representing deposition from suspension fallout from hypopycnal plumes; and B) occasional minor bioturbation.

Transfer to Trusklieve

From the Tullig Point outcrop, turn north on the gravel track and follow the road through the hamlet of Tullig and then down the hill where you will reach a crossroads. Turn left and follow the road towards Trusklieve, where after *c.* 2 km you will reach a tight right-handed hairpin in the road (UTM 447929 m E, 5831175 m N; Map 63, 080245 m E, 154407 m N), with a small track leading off to the left (west). Take this track for *c.* 0.75 km where it is possible to park off the track and around 100 m away from the large coastal inlet that forms part of the Trusklieve outcrop. Note that this last gravel track is small, uneven, the verges are very soft and care should be taken in driving along here.

9.3 Trusklieve (UTM 447929 m E, 5831175 m N; Map 63, 080245 m E, 154407 m N)

The coastal and cliff-top exposures of the Tullig Cyclothem near the small hamlet of Trusklieve (Fig. 9.3.1) allow viewing of, and access to, some of the most important and beautifully exposed Tullig sections. The sections near Trusklieve are exposed in a cliff-top amphitheatre, in vertical cliffs and in smaller cliff-top exposures, and a full day is required to fully examine the sediments exposed here. The sections described below are mainly to the south of a very prominent inlet in the coastline (Fig. 9.3.1; UTM 447193 m E, 5831220 m N; Map 63, 079509 m E, 154463 m N) that has sheer vertical sides of some 50 m in height. Great care should be exercised at all times at these localities as the loose edges, slippery surfaces and winds all call for extra safety vigilance.

Pulham (1989) provided a detailed study of the Tullig Cyclothem at Trusklieve and presented a sequence of logs and photographs that illustrate the sedimentology of a prograding mouth-bar and overlying fluvial channel sequence. Later work (Davies & Elliott, 1996) explored the sequence stratigraphic importance of these outcrops and discussed the erosive nature of the boundary between the Tullig Sandstone and the underlying sediments, interpreting this surface as a major sequence boundary that has regional relief purported to be ≤30 m. Stirling (2003) provides the most detailed account of the sedimentology of the Tullig Cyclothem fluvial facies and recognised a suite of sedimentary facies within this cyclothem (Table 9.3.1). These are used herein in many of the logs and photomontages from the Tullig Cyclothem. They reflect a range of processes operative in fluvial, shallow marine and deltaic environments.

Five localities at this site are detailed below:

i) Two spectacular cliff-top outlooks where the whole sequence exposed can be viewed and the Tullig Cyclothem and its broad grain-size trends can be placed in context.

ii) A superb exposure of the contact between the Tullig Sandstone and the underlying fluvial-deltaic deposits.

iii) An exposure at the top of the Tullig Sandstone that shows the initial flooding surface during the start of the transgression over these fluvial sands.

iv) A cliff that shows the transgressive sediments that overlie the Tullig Sandstone and record the transition to the Kilkee Cyclothem.

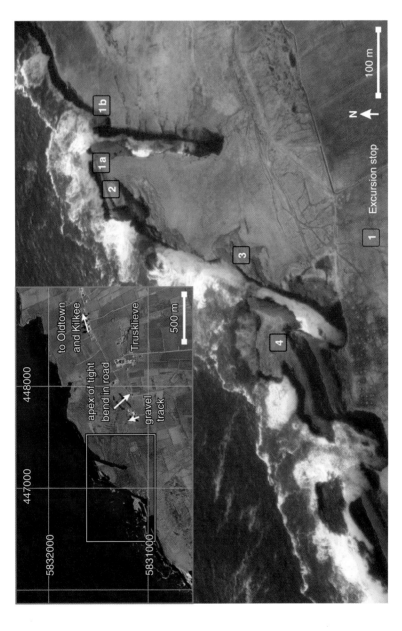

Fig. 9.3.1. Location map of the outcrops at Stops 1 to 4 at Trusklieve. Coordinates in inset map are UTM. Image courtesy GoogleEarth: © DigitalGlobe.

Table 9.3.1. Lithofacies descriptions and interpretations of facies in the Tullig Cyclothem. From Stirling (2003). Facies codes are from Miall (1996) where interpretations agree with Miall's fluvial facies scheme; other codes, not representing existing facies schemes, have been chosen specifically for facies seen in the outcrops studied.

Code	Lithofacies type	Description	Depositional process/interpretation	Facies association
Sm	Massive to faintly bedded sandstone	Well-sorted, clean, yellow, massive sand with occasional faint, patchy lamination.	Deposited by sediment gravity flows, for example in small channels due to bank collapse. Alternatively due to very rapid deposition and subsequent homogeneous dewatering.	Smc, Sh, Shc, Sl, Sp, St, Sc, Ccm, Sd, Sr (channel-fill facies)
Smc	Sm with conglomeratic lag and/or floating clasts	Well-sorted, clean, yellow, massive sand with occasional faint, patchy lamination. Clastic lag at base and rare floating clasts.	Deposited as for facies Sm, but with additional coarse material deposited at the base of the bed (as flow wanes) or higher in the bed (due to particle overpassing).	Sm, Sh, Shc, Sl, Sp, St, Sc, Ccm, Sd, Sr (channel-fill facies)
Sh	Horizontally-laminated sandstone	Well-sorted, clean, yellow sand, parallel laminated, with primary current lineation.	Bedforms swept out by high velocity unidirectional flow, forming upper-stage plane beds.	Sm, Smc, Shc, Sl, Sp, St, Sc, Ccm, Sr (channel-fill facies)
Shc	Sh with conglomeratic lag and/or floating clasts	Well-sorted, clean, yellow sand, parallel laminated, with primary current lineation. Clastic lag at base and rare floating clasts. Shows crude imbrication.	Upper-stage plane beds (as above) with deposition of coarse bedload fraction at base of bed.	Sm, Smc, Sh, Sl, Sp, St, Sc, Ccm, Sr (channel-fill facies)
Sl / Sl₂	Low-angle cross-bedded sandstone	Low-angle (<10°) cross-beds in well-sorted yellow sandstone. Sl₂ contains fine grained silty laminae.	Sl develops at the transition from dunes to upper-stage plane beds as bed forms are swept out by unidirectional flow. Sl₂ formed by deposition from suspension on gently dipping surfaces.	Sm, Sh, Shc, Sp, St, Sc, Ccm (channel fill facies) (plus Flh, Fls, Flm for Sl₂)
Sp/Sp₂	Planar cross-bedded sandstone	Varying scale of planar cross-bedding in well-sorted, clean yellow sandstone. Foresets are typically asymptotic towards the flat lower set surfaces, and in Sp₂ show mud drapes.	Migration of straight-crested (2-D) dune bedforms within unidirectional flow.	Sm, Sh, Shc, St, Sc, Sl, Ccm (channel-fill facies) (plus Sh₂, Sl₂)

(Continued)

Table 9.3.1. (Cont'd)

Code	Lithofacies type	Description	Depositional process/interpretation	Facies association
St	Trough cross-bedded sandstone	Scoop-shaped erosional bases to sets of trough cross-beds averaging 0.5 to 1.5 m thick. Foresets are asymptotic at base. Well-sorted, clean, yellow sands.	Migration of sinuous-crested (3-D) dunes within unidirectional flow.	Sm, Smc, Sh, Shc, Sp, Sc, Sl, Ccm (channel-fill facies)
Sc	Sp, St or Sl with clast lag	As for Sl, Sp and St, with a lag of coarse material (gravel to pebble sized) at the base, and rare "floating" clasts. Clasts predominantly mud/silt, angular, and frequently deformed.	Erosion of fine channel-fill or overbank material during channel reoccupation or splay channel breakout. Deposition of gravel/cobble fraction at base of new channel scour as peak flows wane.	Sm, Smc, Sh, Shc, Sp, St, Sl, Ccm (channel-fill facies)
Ccm	Clast-supported massive to crudely bedded conglomerate	Gravel to cobble-sized clasts, predominantly mud/silt. Poorly sorted, angular to rounded, including deformed clasts. Organic debris often present. Patchy crude imbrication and horizontal stratification. Matrix of fine sand to muddy silt showing compaction-related deformation.	Eroded overbank or channel abandonment material incorporated in coarse bedload. Gravel and cobble fraction deposited at base of newly formed channel scour as strongly erosive peak flows wane.	Sm, Smc, Sh, Shc, Sp, St, Sc, Sl (channel-fill facies)
Sd	Deformed sandstone	Clean, well-sorted yellow sandstone with convolute lamination, loading and other deformational features. Sometimes without any remaining signs of bedding.	Formed by soft-sediment deformation due to liquefaction and fluidization, possibly caused by a seismic shock.	Sm, Smc, Sh, Shc, Sp, St, Sc (channel-fill facies)
Sr	Ripple cross-laminated sandstone	Cross-lamination (5-40 mm high) in well-sorted but slightly silty sand. Occurs in sandstone beds or as discrete thin (5-30 cm) beds within finer sediments.	Develop at low to moderate flow velocities in a unidirectional current.	Sp, Flh, Fls, Flm (channel-fill, mouth-bar or bay-fill facies)
Src	Climbing ripple cross-laminated sandstone	Climbing cross-lamination in well-sorted but slightly silty sand. Stoss and lee sides of ripples preserved.	Develop during rapid deposition within a waning unidirectional current.	Sp, Flh, Fls, Flm (bay-fill facies)
Srw	Wave ripple cross-laminated sandstone	Symmetric ripples with continuous crests and bidirectional cross-lamination in well-sorted, slightly silty sand.	Develop during wave reworking of previously deposited sands.	Sp, Flh, Fls, Flm (mouth-bar and bay-fill facies)

Code	Facies	Description	Interpretation	Associated facies
Shcs	Hummocky cross-stratified sandstone	3-D cross beds in dome-shaped features in well-sorted clean yellow sandstone. Dome surfaces often preserved, with wave-reworked tops.	Formed by storm wave reworking of sediment, usually in shelfal conditions but also nearer shore.	Sp, St, Sr, Flh, Fls, Flm (mouth-bar facies)
Sh_2	Horizontally laminated sandstone (type 2)	Well-sorted, pale yellow, parallel-laminated sandstone with thin (<10 mm) silty laminae.	Formed by rapid deposition from suspension immediately in front of distributary mouth, producing laminated sands with silt layers at times of lower flow.	Sp, St, Sl_z, Sr, Flh, Fls, Flm (mouth-bar facies)
Flh	Interlaminated sand, silt and mud	Parallel to occasionally wavy-bedded interlaminated fine sands, silts and muds. Dark grey where muddier, through mid-grey silts to pale cream sands. Sand layers commonly show lenticular bedding (uni- and bi-directional).	Deposited from suspension with weak tractional currents and/or wave reworking producing ripple trains in sand and sandy silt. Interdistributary bay to general lower delta plain overbank environment.	Fls, Flm, Sh_z, Sr, Srw, Src, Shcs (mouth-bar or bay-fill facies)
Fls	Laminated silt	Parallel-laminated pale grey silt. Occasional bioturbation in some beds.	Fine material deposited from suspension in quiet water (marine, below wave base, or (where no bioturbation) overbank or abandoned channel environment.	Flh, Flm, Sh_z, Sr, Srw, Src, Shcs (mouth-bar or bay-fill facies)
Flm	Laminated mud	Parallel-laminated dark grey mud. Occasional bioturbation.	As Fls, but further from the sediment source.	As above (mouth-bar or bay-fill facies)

*Stops 1a and 1b: Overview of the sequence (UTM 447204 m E,
5831500 m N; Map 63, 079524 m E, 154743 m N)*

Walk towards the prominent inlet and then proceed on its western side to
a ledge (UTM 447204 m E, 5831500 m N; Map 63, 079524 m E, 154743 m N;
Fig. 9.3.1) where you can view across the inlet and to the cliff face to the
north. **Great care should be taken when accessing and walking on this
outcrop, especially when the surface is wet and the weather rainy and windy.**
There is no need to go very near the cliff edge at this locality. This viewpoint of the northern cliff (Figs 9.3.2 and 9.3.3) provides a superb overview of the section, showing an overall coarsening-upwards sequence
in the Tullig Cyclothem that is capped by a very thick, amalgamated
sandstone (the Tullig Sandstone). Within this overall coarsening-upwards
trend, several smaller parasequences can be discerned that Davies & Elliott
(1996) interpret as a transgressive systems tract (Fig. 9.3.3):

i) A lower coarsening-upward sequence (denoted as a large coarsening-
upwards trend, Fig. 9.3.2A) that starts at sea-level and possesses several
laterally continuous sandstones near its top, eventually terminating in a
prominent 2 to 3m thick sandstone that forms the large dipping bedding
surface on the small island at the mouth of the inlet. The lower 60% of this
cycle consists of siltstones and shales and the upper 40% then becomes
dominated by sandstones that can be traced along the cliff line. This
sequence has been interpreted as the product of progradation of a major
axial mouth-bar (the Tullig mouth-bar) of the Tullig delta (Pulham, 1989).
ii) Above this main parasequence, but still below the Tullig Sandstone, a
sequence of up to three thinner parasequences is present (denoted as smaller
coarsening-upwards trends, Fig. 9.3.2A), each commencing in a shale/silt-
stone, with Parasequences 2 and 3 being cut into by the prominent sandstone
in the middle of the outcrop (Figs 9.3.2 and 9.3.3). The vantage point from
which you view this cliff is on the top sandstone of the second parasequence.
The fourth parasequence, seen only in the cliff-face running into the inlet, consists solely of siltstones before being erosively truncated by the Tullig Sandstone.
The interpretation of the upper three parasequences has varied between them
representing shallow-water deltas during a transgressive period (Davies &
Elliott, 1996; Figs 9.3.3 and 9.3.4), to them representing levee (Parasequence 2)
and interdistributary bay fill sequences (Parasequences 3 and 4; Pulham, 1989;
Fig. 9.3.5). These interpretations will be returned to below.
iii) Cutting erosively downwards into these underlying parasequences is
the Tullig Sandstone: the erosive surface (Fig. 9.3.2A) can be clearly seen
to cut down towards the north (seawards) and removes the whole of
Parasequence 4 and approximately half of Parasequence 3 at the northern
end of the cliff (Figs 9.3.2 and 9.3.3). The depth of erosion at this locality
is approximately 6 to 7m. The Tullig Sandstone has a different character
from the sandstone of the underlying parasequences: it shows concave-up

Fig. 9.3.2. A) Photomontage of the Tullig Cyclothem from Stop 1a at Trusklieve. The four parasequences are labelled with arrows 1 to 4, showing coarsening-upwards trends in parasequences 1 to 3. Note also the erosive base to the Tullig Sandstone. See online *GigaPan* image. B) Close-up of cliff showing a mudstone-filled channel (CF); C) Close-up of cliff showing large-scale cross-stratification (CS) and erosive base of Tullig Sandstone.

Fig. 9.3.3. Tullig Cyclothem at Trusklieve viewed from Stop 1a, showing Tullig Sandstone and underlying systems tracts. The logs from Davies & Elliott (1996; redrawn with permission from the Geological Society of London) are superimposed on the photograph showing: i) Gamma ray total counts; and ii) Th/K ratio, with reference line at Th/K=6. These logs show the gamma correlation to the parasequence flooding surfaces and at the erosional unconformity. Note the 6 to 7 m of erosive relief at the base of the Tullig Sandstone. IFS=initial flooding surface; PS=parasequence boundary.

erosion surfaces, which are laterally discontinuous, with the sandbody comprising a series of amalgamated sandstones. A beautifully exposed cross-section through a shale-siltstone infilled channel is also present at the top of the outcrop (arrowed CF in Fig. 9.3.2B) and large-scale cross-stratification can be seen at several parts of the Tullig Sandstone (arrowed CS in Fig. 9.3.2B). The nature of the erosive contact at the base of the Tullig Sandstone has attracted considerable attention and has been inter-preted as either due to autocyclic scour by the major river channels (Wignall & Best, 2000) or a Type-1 sequence boundary with the Tullig Sandstone being the infill of an incised valley (Davies & Elliott, 1996; Hampson *et al.*, 1997; Figs 9.3.3 and 9.3.4). Davies & Elliott (1996) argue the underlying three small parasequences are deposits of a transgressive systems tract, with the primary parasequence representing a prograding lowstand wedge (mouth-bar; Fig. 9.3.3). These issues can be further addressed by examin-ing in detail the sediments in part of this sequence at Stop 2 (see below).

This location provides an overview of much of the Tullig Cyclothem. However, another view is obtained by walking around to the north side of the inlet and viewing back towards the south and the cliff top on which Stop 1a is located. This view from Stop 1b (UTM 447323 m E, 5831500 m N; Map 63, 079643 m E, 154741 m N) is illustrated in Pulham (1989, Fig. 9.3.5)

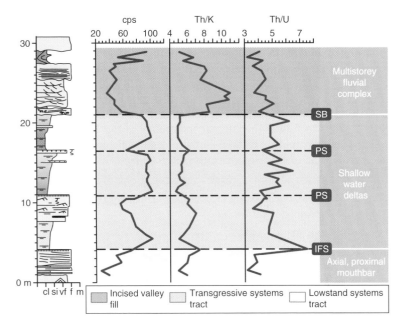

Fig. 9.3.4. Sedimentary log, gamma ray (cps), Th/K and Th/U profiles of the Tullig Sandstone and underlying systems tracts at Trusklieve (redrawn after Davies & Elliott (1996) with permission from the Geological Society of London). In the transgressive systems tract (4.5 to 21 m), there is a decreasing sand:silt ratio upwards that is characterised by a trend towards lower Th/K ratios relative to the successive parasequences. Also, note the low counts in the Tullig Sandstone, whose erosive base has an intraclast conglomerate. IFS=initial flooding surface; PS=parasequence boundary; SB=sequence boundary.

Fig. 9.3.5. Photograph of succession taken from Stop 1b at Trusklieve with the interpretation of palaeoenvironments taken from Pulham (1989) superimposed.

in his interpretation of the entire sequence as a mouth-bar with levee/
interdistributary bay fill sequences that are then overlain by the Tullig
Sandstone – a minor Tullig distributary channel in his interpretation
(Fig. 9.1.1B). The Stop 1b viewpoint also allows the erosive boundary of
the Tullig Sandstone to be viewed in the steep faces of the gully
(Fig. 9.3.6A), as well as the internal architecture of the Tullig Sandstone,
which displays numerous erosion surfaces, large-scale cross-stratification
and partially mud-/silt- filled channels (Fig. 9.3.6B).

Stop 2: Trusklieve amphitheatre (UTM 447145 m E, 5831475 m N; Map 63, 079465 m E, 154718 m N)

From Stop 1a, walk approximately 100 m to the south-west where you will
come to an extensive dipping sandstone bedding plane (Fig. 9.3.1), which
represents the top of the third parasequence seen at Stop 1a. From this
exposed surface, the outcrop in front of you forms a natural amphitheatre
that trends NNE-SSW and exposes a superb section through the upper
parasequences viewed at Stop 1 and within the Tullig Sandstone. **Please be
aware of the loose nature of rocks at the top of the amphitheatre outcrop and
avoid any areas that show signs of recent rockfalls. Care should also be taken
when walking on this bedding plane in wet weather.**

The amphitheatre exposure allows access both to the sediments in the
top parasequences and the strongly erosive base of the Tullig Sandstone;
and it is possible to climb and examine the sandstones and shales within
the Tullig Sandstone at the southern end of the outcrop. The thin, bioturbated sandstone that forms the main bedding surface shows symmetrical
oscillatory flow ripples on its top surface and some surfaces possess small
U-shaped *Arenicolites* burrows. The sediments immediately below the
erosive contact of the Tullig Sandstone show magnificent preservation of
a range of ripples, with a starved sand supply at this point, with both
flaser, wavy and lenticular bedding (Fig. 9.3.7) and the ripples showing
both oscillatory and longer-period reversing flows.

The erosive surface of the Tullig Sandstone is beautifully exposed in this
outcrop and can be seen to cut down by up to 2.5 m into the underlying
siltstones and sands. Numerous intraclasts of the underlying sediment, as
well as clasts of darker shales (a lithology not seen in the immediately
underlying strata) and large organic fragments and chunks of wood, can
be found in coarse lag deposits within the lowermost sands of the Tullig
Sandstone. Erosive flutes on the underside of the lowest Tullig sand show
north-easterly flow directions.

The Tullig Sandstone exposure at Trusklieve affords one of the best
opportunities to examine the detailed internal characteristics of this extensive sandbody, which has been interpreted widely as fluvial in origin (Rider,

(A)

(B)

(C)

Fig. 9.3.6. Photographs of the Tullig Cyclothem taken in the coastal inlet at Stop 1b.
A) The erosive base of the Tullig Sandstone (labelled 'e') and the underlying
parasequences; B) Detail of the internal geometry of the Tullig Sandstone showing
large-scale internal erosion surfaces (e), cross-stratification (cs) and silt-filled channels or
erosive remnants (s); C) Detail of the internal geometry of the Tullig Sandstone shown in
Fig. 9.3.6B, showing cross-stratification (cs) and silt-filled channels or erosive remnants (s).

Fig. 9.3.7. Interbedded siltstones and sandstones directly below the erosively-based Tullig Sandstone, Trusklieve ampitheatre, Stop 2. Note both the ripples with opposed current directions in this section, as well as the symmetric ripples formed under oscillatory flows.

1974; Pulham, 1989; Williams & Soek, 1993; Stirling, 2003). Note that there is also a plaque mounted halfway up the outcrop dedicated to the memory of Trevor Elliott.

The thickness of Tullig Sandstone here is approximately 22 m and Stirling (2003) identified five storeys, based on the erosional contacts between the beds and the intervening fine-grained sediments that have been interpreted as overbank or channel-fill facies (Davies & Elliott, 1996; Stirling, 2003). The thickness of the fines is seen to vary laterally across the outcrop, with one bed reaching up to 4.7 m thickness, although all these fine-grained beds are absent at the northern end of the amphitheatre outcrop. The five storeys can be examined in detail across and vertically upwards in the outcrop (Figs 9.3.8 to 9.3.10); and have the following principal characteristics:

1 Storey 1: reaches up to 8.5 m in thickness and has a pronounced basal erosion surface with a relief of up to 2.5 m. At the northern end, it shows well-developed lateral accretion surfaces which, when corrected for tectonic tilt, dip gently to the NNE and WSW, with slickensides showing evidence for some later structural movement along these surfaces. These beds fine upwards from a conglomeratic lag at their base to fine-grained to medium-grained sandstones with planar cross-stratification and horizontal laminae. Stirling (2003) has interpreted these beds as representing lateral accretion on a barform, with predominant flow to the NNE. Towards the centre of the outcrop, Storey 1 becomes more heterogeneous and is dominated by concave-upwards erosional surfaces with

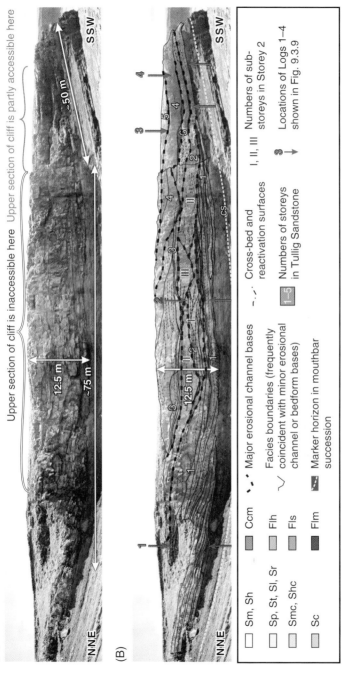

Fig. 9.3.8. A) Panorama of the Trusklieve amphitheatre at Stop 2 outcrop looking east-south-east; B) Facies interpretation of Fig. 9.3.8A, with the principal channel storeys indicated by numbers and the sub-storeys being denoted by Roman numerals and corresponding to those discussed in the text. Red arrows and numbers refer to the position of logs shown in Fig. 9.3.9. Facies descriptions are given in Table 9.3.1.

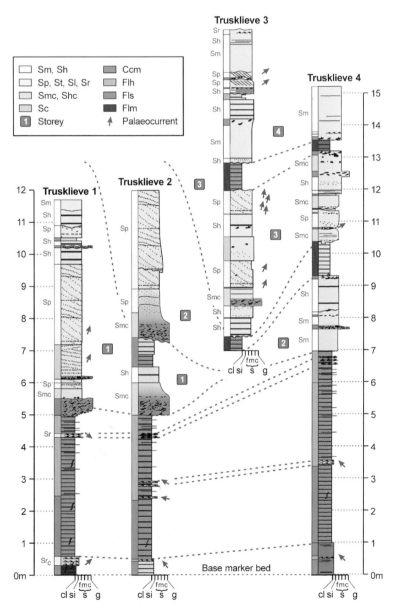

Fig. 9.3.9. Correlation panel of logs at Trusklieve amphitheatre, Stop 2, with logs 1 to 4 running from north (left) to south (right) – see localities on Fig. 9.3.8B. Note that none of these logs reach the top of the sandbody. Numbers refer to channel storeys as denoted on Fig. 9.3.8B and arrows indicate palaeocurrent directions. Facies descriptions are given in Table 9.3.1.

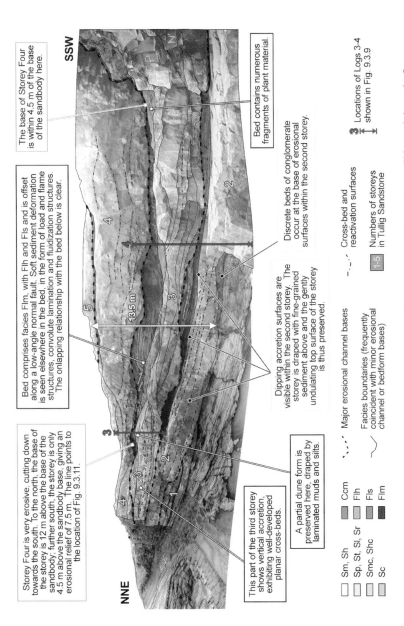

The base of Storey Four is within 4.5 m of the base of the sandbody here.

Bed contains numerous fragments of plant material.

Bed comprises facies Flm, with Flh and Fls and is offset along a low-angle normal fault. Soft sediment deformation is seen elsewhere in the bed, in the form of load and flame structures, convolute lamination and fluidization structures. The onlapping relationship with the bed below is clear.

Discrete beds of conglomerate occur at the base of erosional surfaces within the second storey.

Storey Four is very erosive, cutting down towards the south. To the north, the base of the storey is 12 m above the base of the sandbody; further south, the storey is only 4.5 m above the sandbody base, giving an erosional relief of 7.5 m. The line points to the location of Fig. 9.3.11.

Dipping accretion surfaces are visible within the second storey. The storey is draped with fine-grained sediment above and the gently undulating top surface of the storey is thus preserved.

This part of the third storey shows vertical accretion, exhibiting well-developed planar cross-beds.

A partial dune form is preserved here, draped by laminated muds and silts.

13.5 m

SSW

NNE

- Sm, Sh
- Sp, St, Sl, Sr
- Smc, Shc
- Sc
- Ccm
- Flh
- Fls
- Flm

- ·-·-· Major erosional channel bases
- Facies boundaries (frequently coincident with minor erosional channel or bedform bases)
- --·-- Cross-bed and reactivation surfaces
- 1-5 Numbers of storeys in Tullig Sandstone
- 3 Locations of Logs 3–4 shown in Fig. 9.3.9

Fig. 9.3.10. Photomontage of the southern end of the outcrop at Trusklieve (Stop 2) looking north-east. This panel shows the five storeys present, major erosional contacts and the principal facies. Facies descriptions are given in Table 9.3.1.

conglomeratic lags being common. Fine-grained sediments are interbedded with these sands and conglomerates but are discontinuous and extend laterally for between 4 and 28 m. One of these fine-grained pods of sediment has yielded the freshwater conchostracan *Hemicycloleaia* (Wignall & Best, 2000). The flow directions in the sandstones here are NE-ENE and these sediments have been interpreted as representing accretion of a bar form within low-sinuosity channels that were of the order of 6m deep (Stirling, 2003).

2 Storey 2: this erodes into Storey 1 progressively from north to south across the outcrop and at the southern end of the amphitheatre completely cuts out Storey 1, producing an erosional relief of *c.* 8.5 m. Storey 2 is between 2 and 10.5m thick and predominantly comprises poorly preserved planar cross-strata with occasional impersistent conglomeratic mudclast beds, with some evidence for larger-scale accretion surfaces also being present. The sandbody can be further subdivided, based on the erosional contacts, into three substoreys (Fig. 9.3.8), with Substorey III showing a beautiful concave-upwards erosional contact and the sandy fill of a small channel, whose fill is approximately 20m wide and 3.2m thick. This channel-fill probably represents a smaller, secondary channel cutting across the bar-tops of the underlying Substoreys I and II (Fig. 9.3.8), with the axis of this small channel being approximately normal to the outcrop face. However, these sediments are inaccessible and the absolute flow direction in this channel cannot be discerned. Storey 2 is draped by a thin, laminated siltstone and mudstone to the south (logs Trusklieve 3 and 4, Fig. 9.3.9), which contains occasional thin, sandy lenses; both are interpreted as overbank and abandonment facies.

3 Storey 3: this storey is only accessible (with great care) at the southern end of the outcrop, although it can be viewed across the entire amphitheatre and cuts down into the underlying fine-grained abandonment facies by up to 1.5 m locally (Fig. 9.3.10), with the total erosive relief across the outcrop being *c.* 5.5 m. Storey 3 is between 5.5 and 10.7m thick and the basal beds show occasional loading into the underlying muds. Cross-stratification produced by unidirectional sand dunes is more prevalent in this storey than in those below, with a predominant flow direction to the N-NE, but the internal geometry of the beds is quite variable. The lower parts of Storey 3 in the south of the outcrop possess conglomeratic lags, and several of the beds are low-angle and possibly represent accretion surfaces on the edges of bars. At the top of the sandstones of Storey 3, the upper surface of a partially preserved dune form is present (Fig. 9.3.10 and in foreground of Fig. 9.3.11) with planar cross-stratification 0.5m thick. Laterally, this dune pinches out and in its place the sandbody is capped by a bed with a very irregular top surface containing numerous large fragments of compressed woody material including large trunks of *Lepidodendron*.

Fig. 9.3.11. Photograph of the fine-grained channel fill at the top of Storey 3 (Facies Flm/Fls) and erosional base of Storey 4 at the Trusklieve amphitheatre (Stop 2). Facies descriptions are given in Table 9.3.1.

Above this irregular surface, a laminated silty mud package up to 2.9m thick and 35m across drapes onto and infills the irregular topography (Fig. 9.3.11). This infilling fine-grained bed also shows ripple cross-lamination, load structures and low-angle faults with small offsets and a slight rollover. This surface and these overlying sediments represent the predominantly fine-grained infill of an abandoned channel.

4 Storey 4: the base of this storey is again highly erosional into the underlying sediments (Figs 9.3.10 and 9.3.11), with up to 7.5 m of relief and a maximum thickness of 8.5 m. There are numerous sole structures on the base that show flow towards the WNW and ENE. The basal sandstones of Storey 4 locally contain lenses of conglomeratic material which may only extend a few metres laterally but possess intraclasts of silt and large fragments of woody material. The sandstones of this storey are predominantly parallel laminated, or contain low-angle bedding surfaces. Many beds are difficult to discern and are cut out by minor concave-upwards scour surfaces. The palaeocurrents obtained from foresets here are broadly to the NE, with the bed set boundaries dipping to the WNW and ENE at around 8° when corrected for regional dip. Adopting the methodology of Leclair & Bridge (2001) produces an estimated average flow depth of *c.* 4.6 m for Storey 4. This storey probably represents the filling of fluvial channels, firstly by a basal lag, then succeeded by deposition upon barforms with low-angle accretion surfaces and associated dunes.

5 Storey 5: this erodes down into a thin (<0.45 m) siltstone that lies at the top of Storey 4 and extends laterally for up to 16.5 m. This siltstone is partially removed by erosion at the base of Storey 5, with up to 2 m of relief on the erosional surface. Storey 5 sandstones reach up to 6.5 m in thickness and are dominated by trough cross-stratification with some parallel-bedded sandstones. Planar cross-stratification occurs mainly at the base of the storey and is replaced by trough cross-stratification, especially at the very top of this outcrop, where the most well-developed trough cross-stratification in the whole of the outcrop is found. These sediments represent deposition from sinuous-crested sand dunes (Williams & Soek, 1993) that have preserved sets up to 0.51 m thick, with a mean of 0.27 m. Using the methodology of Leclair & Bridge (2001) yields an average flow depth of *c.* 4.8 m.

The geometry of these fluvial deposits of the Tullig Sandstone at Trusklieve thus beautifully displays the internal character of a multi-storey sandbody, which comprises five distinct sandbodies that are laterally and vertically erosive into each other and are separated by finer-grained siltstones and mudstones deposited in overbank areas and abandoned channels. The depth of erosion at the base of these individual sandbodies in this amphitheatre exposure ranges from 2 to 7.5 m, and these sediments show a trend to a simpler, and less heterogeneous, internal sedimentary architecture in the upper storeys.

Stop 3: The top of the Tullig Sandstone and overlying transgressive sediments (UTM 446985 m E, 5831170 m N; Map 63, 079301 m E, 154415 m N)

After examining the uppermost storey of the Tullig Sandstone exposed in the amphitheatre, walk approximately 300 m to the south-west to a small, cliff-top exposure (Fig. 9.3.1) at the top of the Tullig Sandstone that

Fig. 9.3.12. The Tullig Sandstone viewed from near Stop 3. Note the sharp erosive basal contact of the Tullig Sandstone (yellow arrows), numerous internal erosion surfaces (white arrows) and shales within the sands (yellow arrow labelled 's').

displays the overlying mudstones and siltstones. This location also allows views of the Tullig Sandstone in the dramatic cliffs to the south. These cliff-top views show the complex internal geometry of the Tullig Sandstone and the erosional remnants of some of the fine-grained sediments that separate the individual storeys (Fig. 9.3.12). **Extreme care should be taken on these occasionally slippery surfaces, especially near the cliff edge.**

The top bedding surface of the Tullig Sandstone is well exposed at this locality, revealing it to be intensely bioturbated with many *Zoophycos* traces (Fig. 9.3.13), often preserved in siderite. *Skolithos* and *Arenicolites* burrows are also present and both fish scales and teeth also occur. Directly overlying this bioturbated surface are a few decimetres of dark, shaley mudstone that yield, approximately 0.1 to 0.2 m above the top of the sandstone, small, pyritized shells of bivalves and articulate brachiopods (productids and chonetids) along with gastropods and scaphopods.

The top of the Tullig Sandstone and the marine band thus bear witness to the transgression and initial flooding surface that occurred at the end of Tullig Sandstone times, with the top of the fluvial sandstone being extensively reworked at this locality but with little evidence of any intervening shelf or shallow marine sandstone deposition (but see Stop 4 below). The 15 m of sediments exposed above the marine band are characterised by interbedded siltstones and shales, with many possessing sideritic concretions and bands that suggest the influence of more brackish pore waters. The upper siltstones and fine sandstones at this locality occasionally yield finely preserved

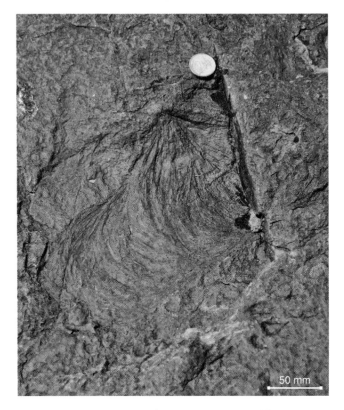

Fig. 9.3.13. *Zoophycos* exposed on top of Tullig Sandstone at Trusklieve, Stop 3.

Diplichnites tracks. This locality can thus be used to examine the nature of this initial transgression and the first marine band after the Tullig delta had prograded into this part of the basin. If the cliffs to the south are viewed from here, the sediments overlying the Tullig Sandstone can be seen, which show a general coarsening upwards to a very prominent sandstone ledge. This next outcrop forms the final stop for this part of the sequence.

Stop 4: The Tullig transgression and lower Kilkee Cyclothem (UTM 446830 m E, 5831100 m N; Map 63, 079145 m E, 154348 m N)

Proceed approximately 450 m to the south-west along the coast (Fig. 9.3.1) and around the inlet to the south-west of Stop 3, where fine views of the Tullig Sandstone and its erosive basal contact can be gained. Once around this inlet, it is possible in good weather to climb down a slope onto the top of the sandstone that was viewed at the end of the itinerary for Stop 3 (Figs 9.3.14 and 9.3.15). A climb up this steeply dipping sandstone bed, which dips at 40°

Fig. 9.3.14. The package of transgressive sediments that lies above the Tullig Sandstone at Trusklieve. This view is taken from the inlet to the south-west of Stop 3 and shows the sediments detailed at Stop 4.

to the SSE, allows careful access down the outcrop, eventually ending up on the top of the *Zoophycos* surface on top of the Tullig Sandstone once again (Fig. 9.3.15). From here, it is possible to examine the entire transgressive succession of the top Tullig Cyclothem to the marine band that separates the Tullig from the overlying Kilkee Cyclothem (Fig. 9.3.15).

The upper Tullig Sandstone at this locality possesses an undulatory upper surface, with the presence of hummocky-type cross-bedding (Fig. 9.3.16), showing that in the initial stages of the transgression, combined flows were capable of reworking the underlying Tullig sands into new bedforms. Above the *Zoophycos* surface there is again a thin, dark, mudstone with marine fossils, which then passes vertically into siltstones, with individual siderite concretions as well as thin, more continuous, beds of siderite. The overlying 20 m of interbedded shales and siltstones progressively coarsen upwards, with the bottom 13.5 m containing scattered siderite nodules and few interbedded thin sands (Fig. 9.3.15). Above this height in the upper half of the outcrop, thin, fine-grained sandstones exhibit the increasing occurrence of symmetrical oscillatory flow ripples, which bear witness to shallowing and an increasing influence of waves. The uppermost sandstones in the outcrop become thicker and internally consist of alternating sands and siltstones/mudstones that display ripples migrating in opposing flow directions (although the majority show migration towards the NE-ENE), symmetrical oscillation ripples at the top of the thin sandstones and well-preserved fine-scale soft-sediment deformation in the form of load casts and small flame structures. These sandstones

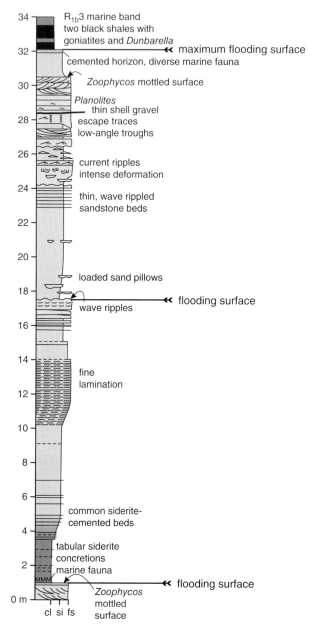

Fig. 9.3.15. Log of the transgressive sediments that lie above the Tullig Sandstone at Stop 4, Trusklieve. The top of the Tullig Sandstone is reworked by extensive *Zoophycos* and the transgressive package consists of a broad coarsening-upwards sequence that shows increasing evidence of oscillatory flows towards its top. The uppermost siltstone is capped by a cemented horizon with marine fauna and then overlain by black shales with goniatites and the $R_{1b}3$ marine band.

Fig. 9.3.16. Photograph of bedforms exposed on the reworked top of the Tullig Sandstone at Stop 4, Trusklieve.

appear to represent the presence of combined flows or reversing flows with a long period (hours?), with the predominant currents being introduced from the SW-WSW. Also present are chevron-style escape traces and thin lenses of fragmented brachiopod shells that indicate marine conditions.

This entire succession represents an upward-shallowing marine sandbody with the sediment being increasingly reworked, first by waves and then by unidirectional/combined flows.

A sharp change in lithology from sandstone to shale is found when climbing down from this sandstone ledge and into the gully on the north side of the outcrop (Figs 9.3.14 and 9.3.15), where two marine bands can be traced along the entire gully and represent the maximum flooding surface at the top of the Tullig Cyclothem. The Kilkee Cyclothem is taken to begin above these marine bands. The first of these marine bands is picked out by a surface with abundant slickensides approximately 0.05 m above the marine band (Fig. 9.3.17). The shales around these marine bands contain scattered elongate siderite nodules. The marine bands consist of grey and black shales that contain orthocone nautiloids, *Dunbarella*, *Caneyella*, crinoid ossicles and the $R_{1b}3$ goniatite *Reticuloceras* aff. *stubblefieldi*.

Transfer to Excursion 9.4

The coast road south of Kilkee offers a spectacular series of cliff-top vistas that are best viewed whilst driving from the south-west to the north-east. It is, of course, possible to drive in the opposite direction but this will

(A)

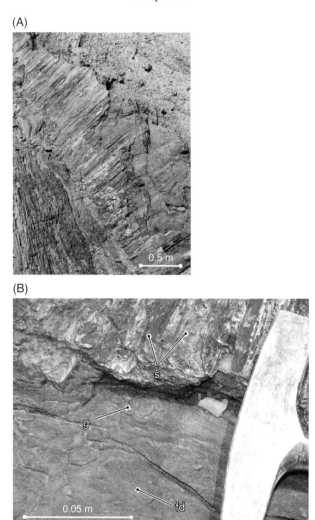

(B)

Fig. 9.3.17. The top of the Tullig Cyclothem exposed at Stop 4, Trusklieve. A) Slickensides developed just above the marine band, which is shown in B). In B), 'g' highlights a large goniatite, 'fd' is fossil debris (including crinoid fragments) and 's' points to the slickensides.

require a lot of looking back over one's shoulder. From the Trusklieve outcrop parking spot, head back to Trusklieve and turn left at the junction with the tight bend, heading up hill towards the hamlet of Oldtown. Continue on this road and then drive back downhill where you will come to a T-junction. Turn left at the junction and follow the road, where after a few tight turns the road emerges to run parallel to the cliff edge and heads towards Kilkee. Although this chapter focuses on the Tullig and Kilkee cyclothems, this coast road affords superb overviews through parts of the Doonlicky Cyclothem and some of the most spectacular coastal scenery in County Clare. Fully viewing these sections will take approximately 2 hours.

9.4 The Coast Road South of Kilkee

Part of the Doonlicky Cyclothem is seen in the cliffs of the small island of Illaunonearaun near to the start of this coastal drive (UTM 450665 m E, 5833458 m N; Map 63, 083014 m E, 156654 m N; Fig. 9.4.1). The lower part of the cyclothem provides a broad array of often large-scale deformation features, including slumps and growth faults. A prominent sandstone bed, probably the sandstone at the top of a major parasequence, can be clearly seen to have been displaced and foundered into the underlying siltstones (white arrow on Fig. 9.4.1). The siltstones overlying this prominent sandstone also possess sandstone blocks and deformed siltstones (yellow arrows on Fig. 9.4.1) and again bear witness to the substantial soft-sediment deformation in this upper cyclothem. This part of the Doonlicky Cyclothem records prograding slope sediment that was frequently prone to failure and extensive soft-sediment deformation. Like the older cyclothems, the upper part of the Doonlicky Cyclothem also possesses a major, sharp-based, fluvio-deltaic sandstone – the Doonlicky Sandstone. This sandstone is exposed facing the Atlantic Ocean on the west side of Illaunonearaun and thus cannot be seen from this cliff-top viewpoint. However, the thick Doonlicky Sandstone can be seen from a viewpoint at Goleen Bay, 0.8 km to the south (UTM 450120 m E, 5833130 m N; Map 63, 082464 m E, 156333 m N; Fig. 9.4.2). Here, the Doonlicky Sandstone forms approximately the upper one third of the cliffs and possesses a sharp contact with the underlying siltstones (arrowed on Fig. 9.4.2). The lower part of the cyclothem is siltstone and again dominated by an array of large-scale deformation features, including slumps and growth faults.

Continue driving north along this road, until the road takes a right bend and you can see a stretch of cliffs at Foohagh Point in front of you (viewpoint at UTM 453083 m E, 5835424 m N; Map 63, 085461 m E, 158586 m N; Foohagh Point at UTM 452950 m E, 5835860 m N; Map 63, 085334 m E, 159025 m N; Fig. 9.4.3; see online *GigaPan* image). These cliffs reveal another spectacular series of syn-sedimentary faults within the Doonlicky Cyclothem, including an orthogonal section through a pair of back-to-back listric faults at the left (north-western) edge of the cliffs (labelled f1

Fig. 9.4.1. The island of Illaunonearaun as viewed from the coastal road. The cliffs here are *c.* 48m high. A prominent sandstone bed is displaced and has foundered into the underlying siltstones (white arrow), whilst the overlying siltstones possess sandstone blocks and deformed siltstones (yellow arrows). See online *GigaPan* image.

Fig. 9.4.2. The island of Illaunonearaun as viewed from Goleen Bay. The Doonlicky Sandstone outcrops on the Atlantic side of the island, forming the upper one third of the cliffs and possesses a sharp contact (arrowed) with the underlying siltstones. The cliffs are *c.* 45 m high.

and f2, Fig. 9.4.3), where the right-hand fault shows clear displacement of the interbedded sandstones and siltstones by several metres, but with little deformation within these beds. On the left side, the fault scarp is infilled by a series of siltstones and thinly bedded sandstones, with no trace of the interbedded sandstones and siltstones. A thin sandstone is seen to overly these two fault zones a little further up the cliff (labelled s1, Fig. 9.4.3), demonstrating the syn-sedimentary nature of the faulting. Another fault is present just to the left side of the sea arch (labelled f3, Fig 9.4.3) but the sense of movement here is reversed as the beds to the left are uptilted to the left. The same sandstone bed noted to the left (labelled s1, Fig. 9.4.3) also drapes over this deformed region, although this sand thins to the right (south-east) side of the cliffs. Finally, two more normal faults are present to the right side of the cliffs (labelled f4 and f5, Fig. 9.4.3), with f4 showing sandstone beds thinning and terminating against this fault on both the hangingwall (labelled s2 and s3, Fig. 9.4.3) and footwall (labelled s4 and s5, Fig. 9.4.3) sides of the fault. A rafted block of sandstone is also present on the down thrown side of this fault (labelled b1, Fig. 9.4.3); and deformed and tilted strata can be seen between f3 and f4 (labelled d1, Fig. 9.4.3). This superb cliff exposure also provides views of a range of smaller-scale deformation associated with these large features described above, developed during progradation of a highly unstable delta slope.

Fig. 9.4.3. Photomontage of the spectacular cliffs at Foohagh Point, which reveal a series of syn-sedimentary faults within the Doonlicky Cyclothem. The cliffs at the far left of the image are c. 57 m high. Five faults are depicted on the photomontage (labelled f1 to f5), together with several sandstone marker beds (labelled s1 to s5), a block of rafted sandstone (labelled b1) and an area of deformed strata (labelled d1). See text for explanation. See online *GigaPan* image.

Transfer to Excursion 9.5

From the coast road south of Kilkee, continue to drive north into Kilkee and head towards the West End car park on the south-west side of Moore Bay (UTM 455115 m E, 5837090 m N; Map 63, 087517 m E, 160225 m N; Fig. 9.5.1).

9.5 Moore Bay, Pollack Holes, Diamond Rocks and George's Head, Kilkee

Moore Bay, Pollack Holes and Diamond Rocks

The upper parts of the Tullig Cyclothem are exposed to the north and south sides of Moore Bay, Kilkee (Fig. 9.5.1). These outcrops afford an opportunity to view the uppermost transgressive sediments of the Tullig Cyclothem, including some truly superb exposures of large-scale, soft-sediment deformation. The two localities on either side of Moore Bay are easily reached by road and visitors should use the car parks near West End on the south-west side of Moore Bay (UTM 455115 m E, 5837090 m N; Map 63, 087517 m E, 160225 m N) and at East End near the golf course to the north-east side of the bay (UTM 455990 m E, 5837580 m N; Map 63, 088399 m E, 160704 m N). West End is reached by driving along the road to the west side of Kilkee Bay and parking near the café at Diamond Rocks.

The walk along the south side of Moore Bay from West End towards Knockroe Point Beach allows examination of readily accessible rocks that begin in the upper part of the Tullig Cyclothem and finish in the overlying Kilkee Cyclothem. These exposures to the south side of Moore Bay will take half a day to complete.

Stop 1 (UTM 455035 m E, 5837118 m N; Map 63, 087437 m E, 160255 m N)

The lowest strata consist of a sandstone body, termed the Moore Bay Sandstone (Wignall & Best, 2000) that is exposed close to high-tide level over an extensive stretch of coastline from the Pollack Holes (Gill, 1979) to Duggerna Rocks and to Knockroe Point. Duggerna Rocks are at the mouth to Moore Bay and witnessed the wrecking of the *Edmond*, a ship taking emigrants to the USA during the Great Irish Famine, on November 19th 1850, with the loss of 98 lives. Access to this section and onto the rocks near the Pollack Holes (a series of natural holes in the rocks with steps into them, where bathing is possible at low tide) is easy from near the café at the car-park, but **care should be taken both to observe tide times and to watch one's footing as the rocks become extremely slippery and hazardous when wet.**

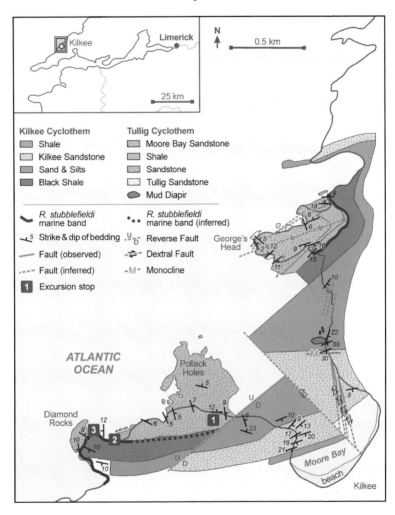

Fig. 9.5.1. Geological map of Moore Bay, Kilkee.

The low cliff sections on the foreshore (UTM 455035 m E, 5837118 m N; Map 63, 087437 m E, 160255 m N) adjacent to the café display a series of mouth-bar deposits (Fig. 9.5.2). The cliff trends north-west to south-east and shows a series of decimetre-thick sandstone beds that are arranged in large accretion surfaces that dip and thin markedly to the SE, with the lower surfaces possessing metre-scale swaley/hummocky cross-stratification and with the upper domal surfaces well preserved (yellow arrows on Fig. 9.5.2), often with superimposed oscillatory flow ripples. The outcrops

Fig. 9.5.2. Photomontage of outcrop by Diamond Rocks café (Stop 1) showing sandstone with large accretion surfaces that dip and thin markedly to the SE. Undular bedding surfaces possessing HCS are indicated by yellow arrows. Rucksack (height 0.45 m) is circled in yellow for scale. See online *GigaPan* image.

Fig. 9.5.3. HCS exposed on bedding planes above, and to the north-west of, those shown in Fig. 9.5.2.

that lie slightly to the NW just above this exposure also show a series of well-bedded sandstones that possess hummocky surfaces (Fig. 9.5.3) and again bear witness to the combined flows in this mouth-bar/shoreface depositional environment. The exposure of the foreshore at Pollack Holes displays a wide range of large-scale bedforms and planform exposure that Gill (1979) documents as showing a series of thin sheet slumps that show movement directions to the SE and ENE (Fig. 9.5.4), comprising three phases of thin sheet slumping and displaying four 'discontinuity' (erosion) surfaces (Fig. 9.5.4). However, the exposure is heavily covered with barnacles and these details are difficult to discern.

Evidence for such slumping is more easily seen by walking carefully along the foreshore (only accessible at low tide) to UTM 454917 m E, 5837150 m N (Map 63, 087319 m E, 160288 m N), to view the outcrop that is a westerly extension of the last exposure. Here, a series of deformational and erosional (?) features are visible in the cliff face, with bedding planes (Fig. 9.5.5) that can be easily accessed. This exposure was described in Gill (1979) and shows a series of small slumps/slides in the upper cliff face with movement directions to the east/south-east, which Gill (1979) speculated could be influenced by the slope generated by nearby mud diapirs (see Stop 3 below). The exposure shows a series of small, 1 to 1.5 m high listric fault surfaces (Figs 9.5.5), above what Gill (1979) termed Discontinuity 2, which dip to the east/south-east and show a series of massive and deformed sandstones between them. Below this, a series of rippled, massive and deformed sandstones infill a depression (labelled 'D', Fig. 9.5.5) that is draped by rippled sandstone beds. Gill (1979) documents this as being part of a slump complex that is linked to deformed horizons to the west (Fig. 9.5.4).

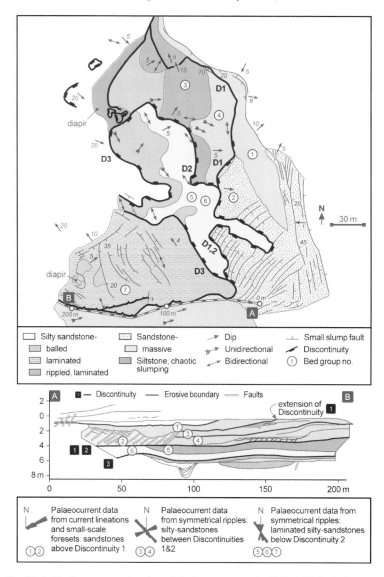

Fig. 9.5.4. Planform map and section of the features exposed at Pollack Holes (redrawn from Gill, 1979; copyright Geological Society of Ireland).

It is also noticeable here that the western and eastern margins of the depression are very sharp, that the overlying beds drape onto this surface and that the planform expression of the depression is rather curved. Although occurring in association with the deformed slump sheets in which the sediments possessed some shear strength, it may be that this

Fig. 9.5.5. Photomontage of the growth fault complex, Stop 1, Kilkee. The yellow arrows point to two, 1 to 1.5 m high listric faults, with the depression discussed in the text being labelled 'D'. See online *GigaPan* image.

depression is partly erosional in nature, defining a shallow channel associated with topography, due to slumping. It is worthy of note here that the unidirectional palaeocurrents at this site are consistently to the north-east (Gill, 1979; Obrock, 2011) with oscillatory flows orientated north-west–south-east and with some south-west–north-east.

Stop 2 (UTM 454460 m E, 5836920 m N; Map 63, 086859 m E, 160064 m N)

From Stop 1, rejoin the cliff-top pathway and walk up the hill to a viewpoint where the cliff to the north-east displays some large-scale soft-sediment deformation features. The Moore Bay Sandstone at Kilkee is best known for its series of mud diapirs and a spectacular view of one example is exposed in the cliffs at Knockroe (Fig. 9.5.6), which can be viewed best from the cliff path and by looking back to the east. This cliff, some 30 m in height, shows a cross-section through a domal structure that does not appear to have penetrated to the palaeo-seafloor, except perhaps at one point. The beds at the top of the sequence can be seen to be deformed but drape the diapir and on the right-hand side a chaotic assemblage of blocks is present that bears witness to the entrainment of surrounding partly-lithified sediments as the diapir was rising. The change in dip of the beds on the right-hand side of the diapir also suggests probable deflation of the diapir, which caused drag on the surrounding beds and tilted them downwards at the diapiric contact. If the tide is low, it is possible to walk onto Duggerna Rocks to examine a series of small-scale, syn-sedimentary graben structures in the upper part of the Moore Bay Sandstone. These can be interpreted as the response of the sea bed to doming by non-penetrative diapirs.

Stop 3 (UTM 454337 m E, 5836953 m N; Map 63, 086737 m E, 160099 m N)

From Stop 2, proceed slightly further up the path and look for the steps that descend to Diamond Rocks. After descending these steps, a sequence can be examined that shows a section through the very uppermost Moore Bay Sandstone and the overlying *Reticuloceras* aff. *stubblefieldi* marine band. Several features can be observed here:

i) Another diapiric intrusion low in the outcrop (Fig. 9.5.7).

ii) A series of broad, swaley, hummocky and large-scale cross-stratified units, which together with parallel-laminated upper-stage plane beds, show sedimentation on a wave-influenced and storm-influenced coast (Fig. 9.5.8). Some of these features have also undergone later soft-sediment deformation, as evidenced by overlying load structures (Fig. 9.5.8A).

iii) The topmost metre of the Moore Bay Sandstone at this locality rests abruptly on a surface covered in small, spherical calcite concretions interpreted to be reworked from a palaeosol (Fig. 9.5.9). Large logs showing

Fig. 9.5.6. Mud/siltstone diapir in the upper Tullig Cyclothem, Stop 2, Kilkee. Note the tilting of beds down towards the diapir on the right hand side of the diapir (blue arrow). The yellow arrow depicts the top of the diapir penetrating into the sandstones above, whilst the white arrow highlights the chaotic assemblage of blocks on the right side of the diapir. See online *GigaPan* image.

Fig. 9.5.7. Mud/siltstone diapir in the upper Tullig Cyclothem at Diamond Rocks, Stop 3, Kilkee, with white arrow pointing to the top of the diapir.

(A)

(B)

Fig. 9.5.8. Hummocky cross-stratified beds exposed at Diamond Rocks, Stop 3, Kilkee. A) Small HCS bedforms with wave-rippled tops, lying below a bed with extensive soft-sediment deformation. Person's feet for scale. B) Undular tops (white arrows) to HCS and low-angle dipping internal accretion surfaces (yellow arrow).

Fig. 9.5.9. Spherical, calcite-cemented concretions (rhizoconcretions?) concentrated on a ravinement surface in the upper part of the Moore Bay Sandstone, Tullig Cyclothem, Diamond Rocks, Stop 3, Kilkee.

Lepidodendron bark are also associated with this surface, which is considered to have its origin in ravinement erosion, marking the passage of a zone of shoreface erosion at the start of transgression.

iv) The topmost Moore Bay Sandstone fines rapidly upward and the succeeding retrogradational package culminates with marine shale. In this transgressive section, wave-rippled sandstones are replaced with intensely bioturbated, muddy sandstone (*Zoophycos* dominated), then calcareous siltstones with a diverse marine fauna (crinoids, brachiopods, bivalves, gastropods) and finally a thin black shale with the goniatite *R.* aff. *stubblefieldi*. This rapid transgression, culminating in deep-water anoxic deposition, marks the boundary between the Tullig and Kilkee Cyclothems.

v) A maximum flooding surface represented by the *R.* aff. *stubblefieldi* marine band that contains abundant goniatites, *Dunbarella*, crinoid debris and rounded phosphate clasts.

vi) The lower shales of the Kilkee Cyclothem show evidence of later tectonic deformation, with thrusting and production of a duplex structure at this locality that is responsible for repeating the lower several metres of the strata.

From this section, a view to the south-west also shows the coarsening-upwards sequence of the Kilkee Cyclothem that terminates in the thick and erosively based Kilkee Sandstone. The lower Kilkee Cyclothem again records delta progradation and this can be seen in the cliffs beneath West End, Kilkee (UTM 455680 m E, 5837005 m N; Map 63, 088081 m E,

Fig. 9.5.10. Large, stacked unidirectional dune cross-sets in the upper part of the Kilkee Sandstone near Kilkee Beach. Rucksack for scale.

160133 m N), which can be reached by retracing your steps from Diamond Rocks, past the café and car park and heading towards Kilkee town. The erosive basal contact of the Kilkee Sandstone can be examined adjacent to Kilkee beach – take the small concrete steps on your left just before a sharp, right-hand bend in the road. These take you onto a rocky platform that varies in size according to the tide, but plenty can be seen even at high tide. The base of the Kilkee Sandstone is marked by an erosive surface draped in large intraclasts, mostly rafts of shale, some approaching a half-metre in size. This contact is interpreted to be a sequence boundary associated with incised valley generation and subsequent infill by a fluvial system (Hampson *et al.*, 1997). The fluvial nature of the Kilkee Sandstone is suggested by the lenticular geometry of the stacked sandbodies and the local presence of trough cross-stratification and unidirectional palaeoflow indicators (Fig. 9.5.10). However, many of the sandstones appear massive.

The Kilkee Sandstone is overlain by two packages of mudstone-dominated strata that have well-developed palaeosols at their tops. These exhibit a mature soil profile, siderite concretions, abundant rootlets and stigmarian root systems. By this stage in basin development, aggradation had clearly reached base-level. Higher levels in the stratigraphy are often hidden beneath the sandy beach, but in some years a thick succession of black shales (basal Doonlicky Cyclothem strata) can be seen in a narrow outcrop at the foot of the seawall.

Transfer to Stop 4 (UTM 455990 m E, 5837580 m N;
Map 63, 088399 m E, 160704 m N)

From West End, head back into Kilkee and take the road around the bay,
turning left onto George's Head Road, where you will pass the Kilkee
Aquatic Centre on your right-hand side. Proceed along this road for
c. 0.6 km and then park at East End car park, near the golf course on the
north side of Moore Bay. These exposures at George's Head to the north
side of Moore Bay will take *c.* 2 hours to complete.

Stop 4: George's Head (UTM 455445 m E, 5838238 m N;
Map 63, 087863 m E, 161369 m N)

From East End car park, walk along the coastal path to George's Head,
where another diapir at the top of the Tullig Cyclothem is beautifully
exposed in both section and planform. At this locality, a mud diapir is seen
to intrude through the overlying siltstones and sandstones and breaks
through these beds (Fig. 9.5.11). The internal structure of the siltstone is
very blocky and structureless, demonstrating the substantial dewatering
that has occurred. Various blocks of the surrounding lithologies can be
found rafted within this diapiric mud/silt. Some of the sands overlying the
diapiric siltstone have also undergone soft-sediment deformation (labelled
'd', Fig. 9.5.11). If you continue to walk around George's Head, the sediments
in the upper part of the Tullig Cyclothem can be examined, consisting of

Fig. 9.5.11. Mud diapir in the upper Tullig Cyclothem, George's Head, north side of
Moore Bay, Kilkee. Soft-sediment deformation in the sandstones overlying the diapiric
siltstone is labelled 'd'.

a series of sideritic siltstones and deformed sandstones (UTM 455810 m E, 5838480 m N; Map 63, 088231 m E, 161606 m N), with the *R.* aff. *stubblefieldi* marine band being located approximately 500 m to the east of the mud diapir in a small bay at UTM 455903 m E, 5838583 m N (Map 63, 088326 m E, 161708 m N).

Transfer to Excursion 9.6

From Kilkee, take the N67 north and after *c.* 10 km you will reach the village of Doonbeg. As you enter Doonbeg, turn left (north) just before a large Catholic church and proceed along this road, around the south-western edge of Doonbeg Bay, then follow the road up the hill for *c.* 2 km towards Killard. At the top of the hill, turn right at the T-junction and follow the road downhill towards the beach at White Strand. Park at the car park at White Strand (UTM 462780 m E, 5844385 m N; Map 63, 095287 m E, 167417 m N), where there is a monument dedicated to Pat McDonald, who was born in the nearby village of Doonbeg and won an Olympic Gold Medal in 1912 representing the USA in the men's shot put.

9.6 The Tullig and Kilkee Cyclothems at Killard and Carrowmore Point

The Tullig Cyclothem is exposed in extensive sections of low outcrop on either side of Doonbeg Bay, around Killard to the south and at Carrowmore Point to the north. These outcrops afford an opportunity to examine sections showing both the Tullig Cyclothem and the transition into the overlying Kilkee Cyclothem.

Killard (UTM 462780 m E, 5844385 m N; Map 63, 095287 m E, 167417 m N)

This excursion will begin at the White Strand car park, with a walk to the north-east and north, to examine the wave-cut platforms along this stretch of coast. Some of the exposure, especially that displaying the lower sediments of the Tullig Cyclothem, becomes covered at high tide, and thus it is best to time field study for low tides in this part of the outcrop. The sediments are exposed in a broad syncline in this excursion area and so the guide first describes walking up through the succession and into the core of the syncline, and then walking back down through the repeated stratigraphy on the other limb of the syncline (Figs 9.6.1 and 9.6.2).

The whole section affords a superb exposure through a mouth-bar to fluvial channel sequence, as well as some of the sediments from the

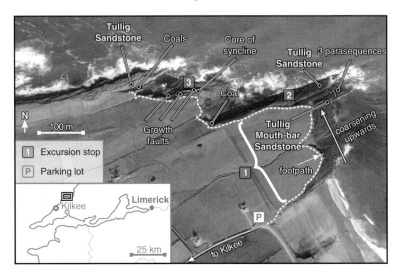

Fig. 9.6.1. Image of the Killard area with the major features detailed in the field excursion marked. Image courtesy GoogleEarth; © DigitalGlobe.

end-Tullig transgression at the centre of the syncline. Three broad exposures will be described: 1) the first coarsening-upwards sequence and three overlying parasequences; 2) the Tullig Sandstone; and 3) the overlying sediments that represent the transgression at the top of the Tullig Cyclothem. These three stops at Killard can take a day to view fully.

Stop 1: The first coarsening-upwards sequence and three overlying parasequences (UTM 455445 m E, 5838238 m N; Map 63, 087863 m E, 161369 m N)

From the car park, walk along the footpath that runs along the low cliff top on the south-west side of Doonbeg Bay. The sediments exposed in the adjacent wave-cut platforms (Figs 9.6.1 and 9.6.2) largely consist of a series of massive and bedded siltstones that have occasional thin sandstones within them. As you proceed up-section to the north-east and north, the succession generally coarsens with an increasing abundance of fine sandstones, which eventually display oscillatory flow ripples, marking shallowing of the sequence from sub-wave-base to between fair-weather and storm wave-base (Pulham, 1989). This coarsening-upwards sequence ends in a laterally extensive thick sandstone, herein termed the Killard Mouth-bar Sandstone (UTM 463000 m E, 5844748 m N; Map 63, 095512 m E, 167777 m N; Stirling, 2003; Figs 9.6.3 and 9.6.4). This sandstone shows extensive soft-sediment deformation at its base and possesses a wide range

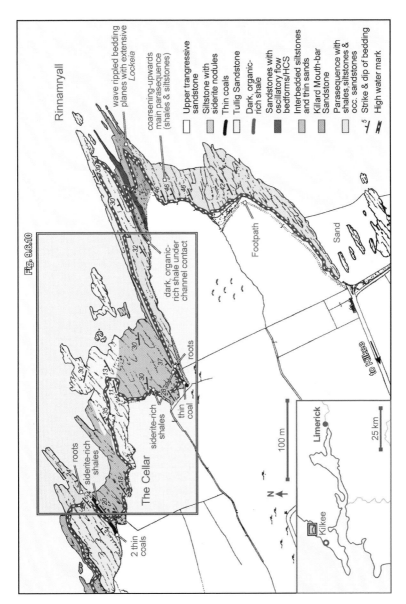

Fig. 9.6.2. Simplified geological map of the Killard area. Inset shows location of Fig. 9.6.10.

Fig. 9.6.3. The Killard Mouth-bar Sandstone exposed on the south side of the Killard syncline. The figures refer to the heights in the sedimentary log shown in Fig. 9.6.4.

of sedimentary structures and trace fossils, including *Lockeia* resting traces, on wave-rippled surfaces.

Above the Killard Mouth-bar Sandstone, a small inlet (Figs 9.6.1 and 9.6.2) exposes three smaller parasequences (Fig. 9.6.5; Pulham, 1989) that have been interpreted as recording mouth-bar progradation and relative sea-level fall, but it is unclear if each successive parasequence records a deepening- or shallowing- upward trend. These parasequences contain a wide range of wave and combined-flow bedforms (including HCS) and a variety of shallow marine/shelfal ichnofacies dominated by *Asterophycus* and *Planolites*, but also with rare evidence for limulid resting traces. These parasequences represent both mouth-bar progradation within a largely wave-influenced environment and perhaps interdistributary bay-fill sequences in a marine-influenced setting. It is worthy of note that the upper parasequence ends in a dark, organic-rich, shale that contains wood fragments, which suggests this may thus have formed more proximal to the Tullig river channels, perhaps within an interdistributary bay or mud-filled abandoned channel. If this is the case, then the upper parasequence demonstrates a continued shallowing-up in the Tullig Cyclothem.

Stop 2: The Tullig Sandstone (UTM 462955 E, 5844780 N; Map 63, 095467 m E, 167851 m N)

The third parasequence is sharply overlain by the Tullig Sandstone that forms the distinct promontory at Killard (UTM 462955 E, 5844780 N; Map 63, 095467 m E, 167851 m N; Figs 9.6.1 and 9.6.2). Stirling (2003),

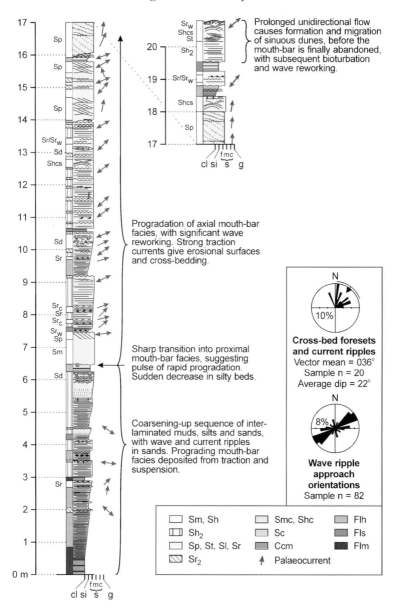

Fig. 9.6.4. Sedimentary log of the Killard Mouth-bar Sandstone exposed on the south side of the Killard syncline. Arrows indicate palaeocurrent directions. See Table 9.3.1 for facies codes and descriptions.

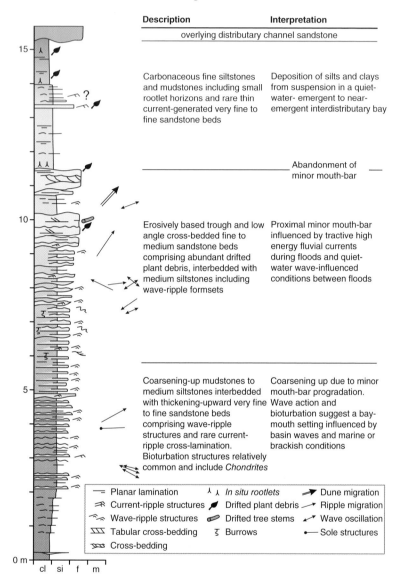

Fig. 9.6.5. Sedimentary log of wave-influenced bay-mouth sequence, Tullig Cyclothem, Killard (redrawn from Pulham, 1989, with permission from the Geological Society of London).

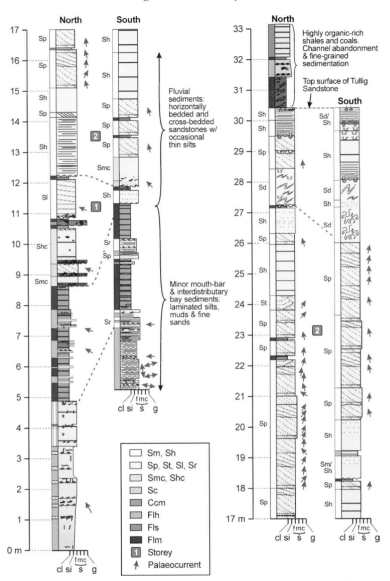

Fig. 9.6.6. Sedimentary log of the Tullig Sandstone exposed on the south and north side of the Killard syncline, detailing sediments in the Killard Mouth-bar Sandstone and Tullig Sandstone. Arrows indicate palaeocurrent directions. See Table 9.3.1 for facies codes and descriptions.

from detailed study of both sides of the syncline, recognised two storeys (Fig. 9.6.6) within the Tullig Sandstone at Killard:

1 Storey 1 shows well-developed horizontal and low-angle cross-bedding, with the palaeocurrents being consistently towards the NNW. Lateral variation within this storey shows the well-bedded facies to be occasionally interrupted by internal erosion surfaces and conglomeratic lags, which are rich in plant material and deformed mudstone rip-up clasts. Stirling (2003) interpreted these scours and coarser facies to represent flood-related erosion and deposition and perhaps individual channel bases within the storey. Storey 1 possesses less than 0.25 m of erosion at its basal surface and reaches a maximum thickness of 5.1 m, before being abandoned and draped by a laterally persistent, laminated silty mudstone that averages 0.45 m thick (Fig. 9.6.7).

2 Storey 2 erodes into the first storey with a relief of up to 4 m and reaches a maximum thickness of 18.2 m on the north limb of the syncline (see below). Internally, Storey 2 is similar to Storey 1, with vertical accretion of facies Sp, Sh and St (Table 9.3.1). The main differences between the two storeys are the lack of conglomeratic facies and reduced abundance of fine-grained facies in Storey 2, compared with Storey 1. Flow depths estimated from the cross-set sizes are approximately 5 to 6 m in Storey 2, with unidirectional palaeocurrents in this storey predominantly to the north-north-east (Fig. 9.6.7). Towards the top of Storey 2 in the Tullig Sandstone at this outcrop, an extensive large-scale deformation horizon is present, with 0.5 to 0.8 m high flame and load structures being present solely within the sandstone. This horizon probably represents seismically-induced liquefaction. The top surface of Storey 2 (Figs 9.6.1 and 9.6.2) represents an abandonment surface, with extensive large-scale stigmarian rootlets, with the fragile tap roots often still visible *in situ* (Fig. 9.6.8) and shaley coal above (UTM 462644 m E, 5844678 m N; Map 63, 095155 m E, 167712 m N). A thin, bioturbated sandstone and sideritic mudstones lie above this low-quality coal and mark a flooding surface (see Stop 3 below). Fragments of small, pyritized brachiopods are found in the base of these mudstones.

Stop 3: The transgressive Tullig at Killard (UTM 462630 m E, 5844730 m N; Map 63, 095141 m E, 167765 m N)

Above the thin coal and shale that overly the Tullig Sandstone lies a shale (Figs 9.6.1 and 9.6.2) that in its lower several metres contains abundant sideritic nodules, which are often of sufficient number and density to form thin, persistent sideritic bands. Above these beds, the shale loses this siderite and contains occasional thin siltstones and sandstones, which in places possess rare *Psammichnites plumeri* traces (resembling *Scolicia*). As you progress up-section to the north, these siltstones are overlain by a sandstone

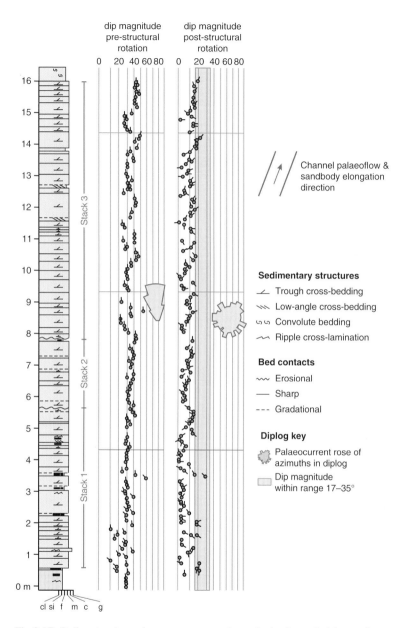

Fig. 9.6.7. Sedimentary log, palaeocurrent roses and magnitude of cross-bed foreset dips from the stacked fluvial channel deposits of the Tullig Sandstone at Killard. Redrawn from Williams & Soek (1993), who demonstrated that only dips in the range of 17 to 35° should be used to reconstruct palaeoflow, as these were in the axial zone of the bedform leeside and in the direction of the palaeocurrent and sandbody elongation. Dips of lower magnitude were associated with the arms of more lunate, three-dimensional dunes and were thus at an oblique angle to the principal flow direction.

Fig. 9.6.8. *In situ Stigmaria* roots preserved at the top of the Tullig Sandstone, Killard.

bed that has an extensively loaded base and that passes upwards into a section with abundant soft-sediment deformation (Fig. 9.6.9). This zone also corresponds to the hinge of a Variscan syncline (UTM 462595 m E, 5844790 m N; Map 63, 095107 m E, 167825 m N), with the centre of this NE-plunging syncline being occupied by a complex series of sandstones and siltstones that show a series of growth faults in both planform and section. A map and log of this succession are given in Figs 9.6.10 and 9.6.11, respectively. The throw on these growth faults is between approximately 2 and 10 m. We will examine three principal localities in this area:

1 A view from the top of the outcrop looks down onto a large growth fault (UTM 462603 m E, 5844800 m N; Map 63, 095115 m E, 167835 m N; Fault 1, Figs 9.6.10 and 9.6.12) that down-throws to the N/NE and in which the side-view rotational contact and arcuate planform shape can be easily

Fig. 9.6.9. Intense soft-sediment deformation at the base of a sandstone in the transgressive Tullig sediments, Killard, with large-scale load structures indicated by yellow arrows.

viewed. The beds in the hangingwall of the fault can be seen to be back-tilted towards the footwall. A section through the side of this fault at this locality (Fig. 9.6.12) shows that the sediment was soft when deformation occurred, with sediment being squeezed along the fault planes; and that the surface depression created by the growth fault was infilled by a massive deformed sandstone, which is extensively dewatered with abundant fluid injection features.

2 A section through Fault 2 (UTM 462558 m E, 5844786 m N; Map 63, 095070 m E, 167822 m N, Fig. 9.6.13; see Fig. 9.6.10 for location) shows that the beds thicken towards the fault, typical of syn-depositional fault growth.

3 Two viewing points at the north-west side of this deformed region (UTM 462540 m E, 5844780 m N; Map 63, 095052 m E, 167816 m N) show a section through the growth fault complex (Fig. 9.6.14). These again illustrate the complex interplay between extension at the back (south) of the fault region (that also develops east-west trending conjugate shear fractures in the siltstones to the south of the southerly growth fault) and a progressive increase in soft-sediment deformation towards the north, where the sequence eventually develops into a series of growth faults. It is also evident that later quartz-veining, associated with the Variscan compression, has exploited these earlier soft-sediment growth faults, with thick quartz veins now lining the original growth fault surfaces.

This complex zone of soft-sediment deformation has been caught up in the centre of the later Variscan syncline and these fault blocks appear to

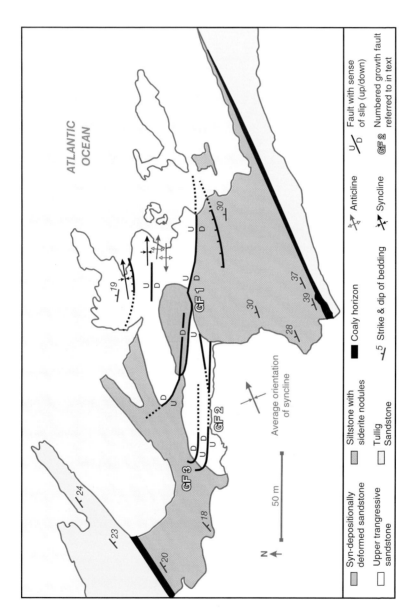

Fig. 9.6.10. Map of major lithological units and structural features in the centre of the syncline, Tullig Cyclothem, Killard. U and D denote upthrown and downthrown sides to growth faults respectively.

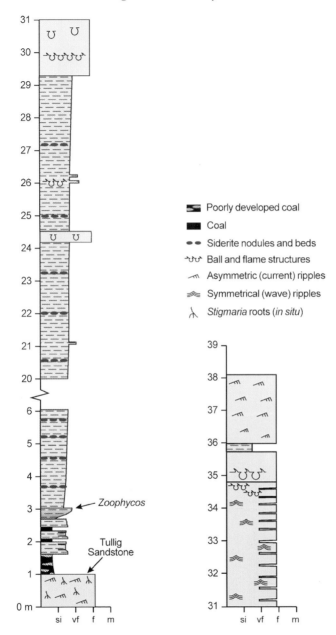

Poorly developed coal
Coal
Siderite nodules and beds
Ball and flame structures
Asymmetric (current) ripples
Symmetrical (wave) ripples
Stigmaria roots (*in situ*)

Fig. 9.6.11. Sedimentary log through sediments above the Tullig Sandstone and in the centre of the syncline, Tullig Cyclothem, Killard.

deformed sandstone

soft sediment intruded along fault plane

side view of growth fault with downthrow to left

arcuate planform of growth fault

~2 m

Fig. 9.6.12. Growth fault in transgressive Tullig Cyclothem at Killard. See online *GigaPan* image.

Fig. 9.6.13. Section through Growth Fault 2 (see Fig. 9.6.10 for location) in transgressive Tullig sediments, Killard. Note thickening of beds towards the fault (yellow arrow), with downthrow to the left (white arrow).

Fig. 9.6.14. Section through Growth Fault 3 (see Fig. 9.6.10 for location) in transgressive Tullig sediments, Killard. Downthrow to the left (white arrows). Person for scale.

have reoriented slightly in this later deformation, such that some of the beds show some divergence from the major plunging synclinal axis. A conceptual model for the structural features at Killard (Figs 9.6.15 and 9.6.16) suggests they can be grouped into two broad categories: 1) syn-depositional growth faulting with associated structures; and 2) post-lithification Variscan deformation. Syn-depositional structural features include large

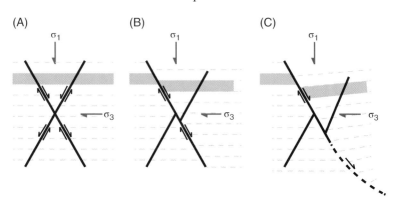

Fig. 9.6.15. Schematic cross-section of conjugate shear fractures, showing the inception and development of growth faults. A) Conjugate shear fractures are initiated, with the principal stresses σ_1 (maximum) and σ_3 (minimum) oriented vertically and horizontally respectively. Any offset of beds is minimal. B) Offset occurs predominantly along the basinward dipping fracture (fault) and marks the initiation of slumping. C) Progressive deformation leads to fault plane rotation, which defines the listric normal faults found at Killard.

listric growth faults, fluid-escape structures and conjugate shear fractures in semi-consolidated mouth-bar sands. Post-lithification structural features include pressure-solution cleavage, pencil cleavage, shear bands with *en echelon* (sigmoidal) veins, broad asymmetric folding with small-scale roof-thrusts, high-angle reverse faults and brecciated fault zones with prevalent slickensides.

Sedimentation associated with mouth-bars of the Tullig delta resulted in local, gravitational extension that was accommodated by the development of conjugate shear fractures and listric normal faults (Figs 9.6.15A to C and 9.6.16A). Steeply dipping fracture sets (42 to 60°) are most prominent in the upper interbedded silts and sandstones, displaying undular surfaces due to stress refraction between the two lithologies. Fluid-escape conduits follow these fractures, with offset along the shear fractures increasing to the north, ultimately giving way to larger-scale slope failures and gravitational slides (Fig. 9.6.16B). This illustrates the genetic relationship between syndepositional slope failures and the conjugate fractures present at Killard.

During later Variscan times, the principal stress orientation shifted from vertical to in-plane (planform) as compression began from the south. Structural evidence for this compression at Killard is seen in the weak pressure-solution cleavage and broad folds that characterise the area. At the centre of the syncline, former growth faults appear to have served as planes for later mineral-rich fluid movement during Variscan compression, and potentially became re-activated in the form of high-angle reverse faults

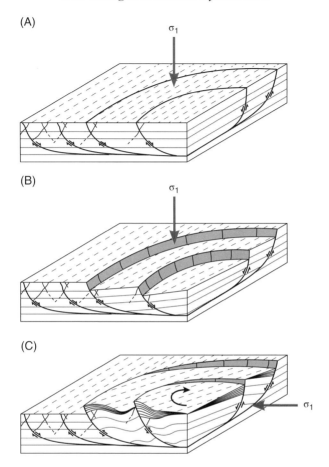

Fig. 9.6.16. Schematic cartoon of the structural evolution of the Killard syncline. A) Development of conjugate shear fractures and initiation of listric faulting during the Namurian. Shear sets and growth faults formed under extension, probably from over-steepening of the distributary mouth-bars. Fluid-escape features along the fracture planes indicate these features are syn-depositional with only minimal compaction. Dashed lines in planform represent strike lines at the time of deposition. B) Further development of faulting (slumping) during the Namurian. C) Wrench faulting and reactivation of listric growth faults during Variscan compression. The black arrow shows the ~25° hangingwall rotation due to differential throw along the major reactivated growth faults. Growth strata are evident in the hangingwalls of downthrown (upthrown) fault blocks.

(see Fig. 9.6.16C). Evidence for post-lithification fluid flow is shown by secondary mineralization along upthrown sections of the hangingwall blocks.

At the north-west side of the outcrop, you walk out through the other limb of the syncline and thus back down section (Figs 9.6.1 and 9.6.2).

As we progress back down section, we pass firstly through shales that become increasingly siderite-rich down-section and then into two coals that are separated by a sandstone — this contrasts with the single coal horizon found on the south-east side of the syncline. Below these coals is the thick Tullig Sandstone (UTM 462400 m E, 5844805 m N; Map 63, 094912 m E, 167843 m N) that again displays an array of large, dune-scale cross-sets with a palaeoflow to the north/north-east. Large *in situ Stigmaria* roots are abundant on the upper surface of the Tullig Sandstone, where they gently curve across the bedding surface for many metres. The top sandstone surface was thus extensively vegetated. The zone of abundant large-scale soft-sediment deformation present at the top of the Tullig Sandstone on the south-east side of the syncline is again present here. The very north-west side of the outcrop (UTM 462300 m E, 5844815 m N; Map 63, 094812 m E, 167854 m N) affords another excellent view of the erosive contact between the Tullig Sandstone and the underlying parase-quences of mouth-bar/interdistributary bay facies and the exposure allows close examination of these features (Fig. 9.6.17).

Transfer to Carrowmore Point (UTM 466415 m E, 5846512 m N; Map 63, 098953 m E, 169495 m N)

This locality lies across Doughmore Bay from Killard and allows inspection of another section through the upper Tullig Cyclothem, the Tullig Sandstone and into the Kilkee Cyclothem. From the Killard parking area, proceed back to the N67 at Doonbeg and turn north. Follow the N67 for *c.* 6 km to where the road takes a tight right-hand turn to the east and a small road forms a junction. Take this small road at this junction (the smaller road heads off straight from the main road at the right-hand turn) and follow the smaller road for *c.* 1 km and then turn left along another small road towards Doughmore and Carrowmore North. After *c.* 1 km at Doughmore, the road takes a tight turn to the right, but here you should drive straight on towards the coast, where access to the beach and outcrops can be gained across the fields. Parking in the farmyard at the end of this road may be possible after seeking permission from the farmer, from whom permission for access across the fields should also be gained. The locality is best visited at low tide, although the upper parts of the succession here are accessible at all tides. The two stops at Carrowmore Pont will take *c.* 3 hours to visit.

Stop 1: Carrowmore Point

These Tullig Cyclothem sediments outcrop on the limbs of an ENE-WSW-trending anticline (UTM 466415 m E, 5846512 m N; Map 63, 098953 m E, 169495 m N), the core of which is exposed at Carrowmore Point (Fig. 9.6.18A). Much of the progradational sequence of the Tullig mouth-

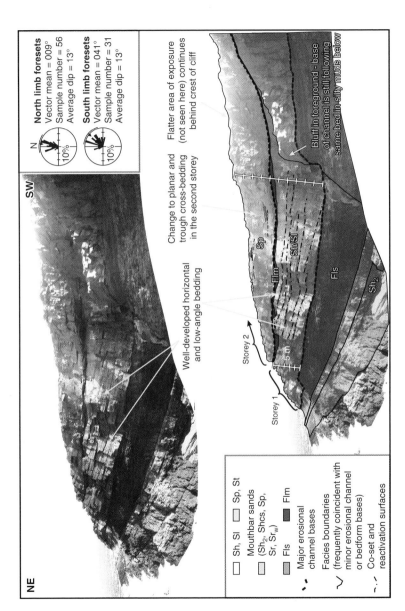

Fig. 9.6.17. Photomontage of the Tullig Sandstone and underlying mouth-bar sediments on the northern limb of the syncline, Killard. Lower diagram shows facies interpretation. Note the erosive base to the Tullig Sandstone in foreground of image. See Table 9.3.1 for facies codes and descriptions.

(A)

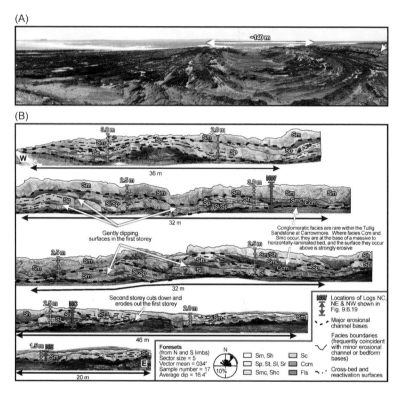

(B)

Fig. 9.6.18. A) Anticlinal exposure of the Tullig Cyclothem at Carrowmore Point. The person at the right of the photomontage (arrowed) is standing at the base of the Tullig Sandstone that is detailed in Fig. 9.6.18B. B) Photomontage of the Tullig Sandstone at Carrowmore Point, showing facies interpretations. Panels run from west to east, top to bottom in the figure. Note erosional base of Storey 2 as it cuts into facies Sp and Sh of Storey 1. See Table 9.3.1 for facies codes and descriptions.

bar is covered by beach sands, but the erosional contact with the Tullig Sandstone is well-exposed at this location (UTM 466318 m E, 5846519 m N; Map 63, 098856 m E, 169503 m N). The maximum total thickness of the Tullig sandbody here is *c.* 22 m and its erosive base cuts down into underlying deltaic facies. A photomontage of this locality on the northern limb of the anticline (Fig. 9.6.18B) shows an outcrop *c.* 150 m in length. Like the Killard section, Stirling (2003) recognised two storeys in the Tullig Sandstone at Carrowmore Point that are separated by a strongly erosive surface (see logs in Fig. 9.6.19).

Storey 1 contains very well-developed low-angle and planar cross-bedding with north-easterly dipping internal erosion surfaces and co-set surfaces, denoting flow towards the north-east. Rare conglomeratic lags

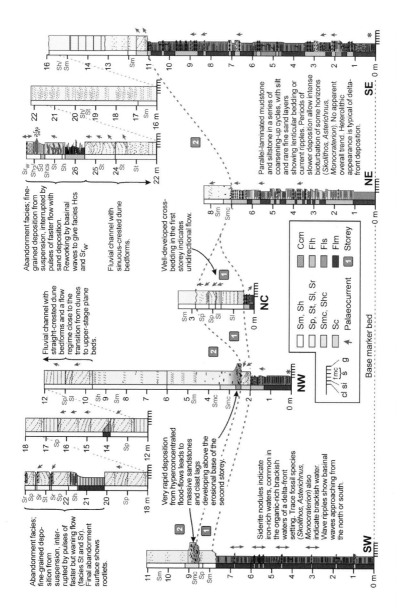

Fig. 9.6.19. Correlation panel of sedimentary logs at Carrowmore Point, showing logs from northern (NE, NC, NW) and southern (SE, SW) limbs of the fold. See Fig. 9.6.18B for log locations and Table 9.3.1 for facies codes and descriptions. Only logs marked with asterisks reach the top of the sandbody. Arrows indicate palaeocurrent directions.

line the base of this storey, which has a basal erosional relief of *c.* 1.5 m. Correlation of this basal surface across the two limbs of the anticline suggests it has a relief up to 4 m. This storey reaches a maximum thickness of 2.5 m before being eroded by Storey 2. Flow depths estimated from cross-set thickness here are a maximum of *c.* 5 m.

Storey 2 is far thicker than Storey 1 and reaches a maximum thickness of 21.7 m, with a basal erosional relief of up to 4 m. It completely erodes out Storey 1 at the east of the outcrop. Storey 2 possesses a range of facies, including massive sandstone with intraformational conglomerates, well-bedded sandstones with horizontal bedding surfaces and finer-grained facies that are interbedded with trough and planar cross-stratified sandstones, climbing unidirectional ripples and some wave ripples. The unidirectional flow indicators show flow to the north-east, as seen in most Tullig Sandstone outcrops (Fig. 9.6.20). Channel depth, as calculated from cross-set thicknesses, is *c.* 7 m. Within the top 4 m of the sandbody on the

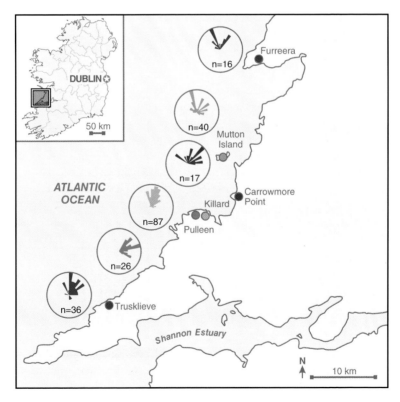

Fig. 9.6.20. Summary of palaeocurrent data for the Tullig Sandstone from various localities in County Clare.

northern limb of the anticline there is a return to fine-grained facies, with the occurrence of a silty mud bed ~1.5 m thick, which is laterally continuous across the outcrop (~80 m). This bed coarsens up from silty mud to fine sandstone and is succeeded by fine-to-medium planar cross-bedded sandstone, which has a thickness between 1.3 and 2.6 m. This final sandy interval is interrupted by thin beds of silty sand (<20 mm thick), which are characterised by thin layers of current ripples within the sandy intervals, showing current directions to the north and north-east. The top surface of Storey 2 again possesses *in situ* rootlets and large *Stigmaria*.

The Tullig Sandstone at Carrowmore Point has some similarities to the outcrops examined at Trusklieve and Killard, but also significant differences. Storey 1, with its downflow dipping accretion surfaces and well-developed cross-bedding, is similar to Storey 2 at Trusklieve. Stirling (2003) interprets this storey to represent the downstream migration of a bar form within a fluvial channel. The second storey at Carrowmore Point is initially very similar to Storey 4 at Trusklieve, with a strongly erosional base and an upward transition from massive to planar and trough cross-bedded sandstone. The principal differences between the Tullig Sandstone at Carrowmore Point when compared to Trusklieve and Killard are at the base and top of the sandbody. Stirling (2003) reports that the basal contact is locally more erosive at Carrowmore Point (4 m of scour relief) than at Killard (0.25 m) and Trusklieve (2.5 m). Additionally, the top of the Tullig Sandstone at Carrowmore Point shows a more gradual transition to abandonment than at Killard, with interbedded muds, silts and fine sands (facies Flh) interrupting sand deposition. These finer-grained facies are evidence of prolonged periods of abandonment during the later stages of channel evolution, prior to its permanent abandonment. The wave reworking seen within the sandy and silty facies near the top of the sandbody at Carrowmore Point is further evidence of abandonment and could indicate the influence of basinal waves (Stirling, 2003; Obrock, 2011). A 0.5 m thick, low-grade, shaley, pyritic coal is present above this sandstone, which marks the initial flooding surface at the top of the Tullig Sandstone; this thus offers a different character to the commencement of the Tullig transgression to that seen at Trusklieve, but it is similar to that at Killard.

Stop 2

Above the Tullig Sandstone, the sequence shows three distinct parasequences that record an overall transgression and the transition to the Kilkee Cyclothem (Fig. 9.6.21; Obrock, 2011). The first parasequence is 9.5 m thick (Fig. 9.6.21) and commences with a siderite-rich laminated siltstone capped by a thoroughly bioturbated sandstone. The second parasequence is *c.* 28 m thick (Fig. 9.6.21) and again consists firstly of laminated

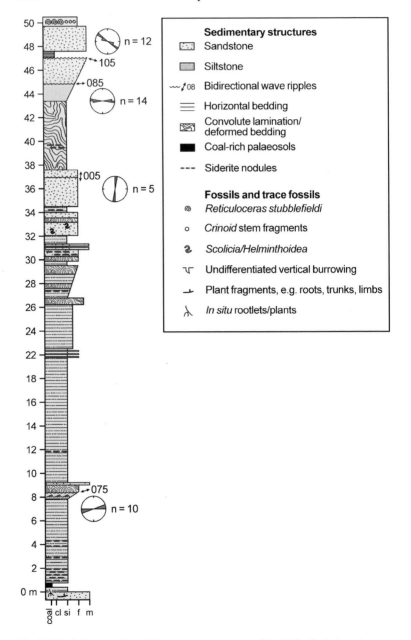

Fig. 9.6.21. Sedimentary log of the upper parasequences of the Tullig Cyclothem at Carrowmore Point.

siltstones, which possess several distinct small slumps. A series of thicker, well-bedded sands lie above these siltstones and show either a massive character, probably due to bioturbation, or well-developed lamination. Parasequence 3, which is *c.* 11 m thick, begins with 6 m of deformed silt-stones (Fig. 9.6.21) that coarsen upward into sandstones that possess wave ripples and *Helminthoidea* burrows, as well as a range of undifferentiated vertical and horizontal burrows. A dark shale lies directly on top of these sands, with a fossil assemblage including *Reticuloceras* aff. *stubblefieldi*, crinoid ossicles and rare orthocone nautiloids. This marks the maximum flooding surface and top of the Tullig Cyclothem.

Transfer to Excursion 9.7

From Carrowmore Point head back to the N67 and turn north, where after *c.* 4 km you will meet the junction with the R483. Turn left (north) at this junction and continue for *c.* 3.5 km until you arrive at the town of Quilty. Seafield Harbour can be reached by turning left at the bend in the road as you approach the seafront at Quilty. Permission should be sought to go onto Mutton Island since it is privately owned (details should be sought in Quilty). The island can be reached in calm seas from boats that can be rented at Seafield Harbour. Information about local fishermen who may be willing to undertake this journey can be obtained in the Post Office or the Quilty Tavern. The journey takes approximately 20 to 30 minutes and it is possible to land on the shingle beaches. As there are now no piers or landing ramps on the island, you may need to resort to wading in shallow water to get onshore. We have successfully and easily landed on the south-ern shore of the island (at high tide it is easier) near the abandoned houses at UTM 465510 m E, 5851626 m N (Map 57, 098119 m E, 174624 m N; Fig. 9.7.1). Despite these difficulties of access, if the weather is good, you will be rewarded with spectacular exposures in a part of the succession that has received virtually no study, except for the work of the Irish Geological Survey and the map displayed in Gill (1979). **This excursion takes a full day and due to boat access should only be attempted in good weather and avoiding windy weather.**

9.7 Mutton Island

Mutton Island (Fig. 9.7.1) is located 1.5 km offshore from the town of Quilty (Figs 1.2.1 and 9.7.1) and was inhabited until the mid-1940s, latterly being used to harvest seaweed that was processed at the nearby Seafield Harbour at Quilty. The remains of several houses are still present near the south-east coast of the island and there is a Spanish graveyard in the

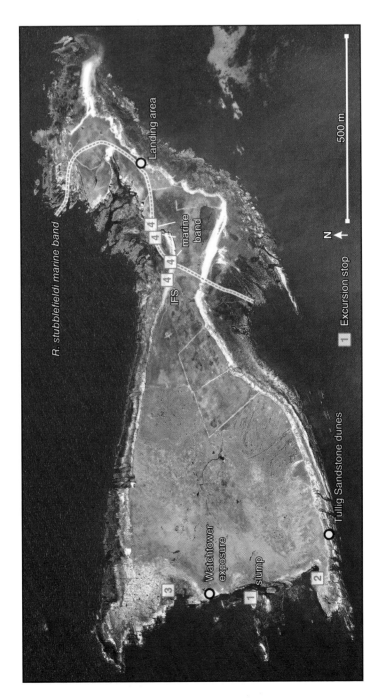

Fig. 9.7.1. Map of Mutton Island showing the stops detailed in the field guide. Image courtesy GoogleEarth; © DigitalGlobe

middle of the island. Here, sailors from the ship *Sao Marcos* are buried, a ship that was part of the Spanish Armada and ran aground off Mutton Island in 1588 (Ancient Origins, 2015). The island allegedly derived its name from sheep carried aboard this ship and brought ashore, although it is uncanny how the planform shape of the island also resembles a leg of mutton!

Mutton Island comprises an easterly plunging anticline, with its axis running *c.* W-E along the centre of the island. The beds on the south of the island dip towards the south, whilst those on the northern Atlantic shores dip gently to the north (Fig. 9.7.1; Gill, 1979). The plunge results in the youngest sediments (of the Kilkee Cyclothem) being present at the east end of the island. The outcrops on Mutton Island display the upper part of the Tullig Cyclothem, with extensive exposures of the Tullig Sandstone, the marine band at the top of the Tullig Cyclothem and the deformed sediments of the lower Kilkee Cyclothem. Four locations are detailed below but the perimeter of the island has excellent exposure and displays many sections, principally through Tullig mouth-bar sequences and Tullig Sandstone on the south, west and north-west shores and especially around the higher cliffs on the west and north-west sides.

Stop 1 (*UTM 464345 m E, 5850995 m N; Map 57, 096944 m E, 174009 m N*)

After landing on the foreshore, proceed towards the 19th Century watch-tower at the western end of the island. The southern edge of the island provides an excellent walk along a series of sandstones that lie within the Tullig Cyclothem, with exposed sandstones around UTM 464345 m E, 5850995 m N (Map 57, 096944 m E, 174009 m N; Fig. 9.7.1) displaying large-scale trough cross-stratification within the Tullig Sandstone and beautiful planform exposures of dunes with steep planar foresets (Fig. 9.7.2A) and sometimes foresets that have leeside ripple fans (Fig. 9.7.2B).

Head towards the watchtower at the NW of the island until you reach a semi-circular-shaped outcrop, approximately 100 m long, located at UTM 464167 m E, 5851280 m N (Map 57, 096770 m E, 174296 m N; Fig. 9.7.1). This exposure shows a 15 m thick sequence (Fig. 9.7.3) characteristic of an uppermost Tullig mouth-bar and the basal surface of the Tullig Sandstone eroding down into these sediments. The mouth-bar sediments exposed here show a series of large bedding planes that possess a range of sedimentary structures, including: i) small- and large- scale oscillatory flow ripples (Fig. 9.7.4A,B; with flow directions being W-E), with occasional beds showing reversals in flow direction as revealed by opposed foreset directions (towards 90° and 325° (Fig. 9.7.3 at 9.55 m on the log) illustrating longer period reversals of flow); ii) beds that have been extensively deformed by liquefaction (Fig. 9.7.4C); iii) surfaces that possess *Planolites*,

(A)

(B)

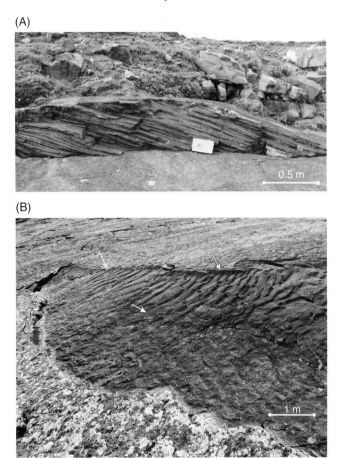

Fig. 9.7.2. Photographs of unidirectional flow dunes, Mutton Island, showing: A) planar dune foresets, flow to the right; and B) planform exposure of dune (crest arrowed in yellow) with ripple fans in its leeside. Flow towards observer with ripple migration direction arrowed in white.

Lockeia and *Asterophycus*; and iv) on some beds accumulations of large, spherical calcite concretions (Fig. 9.7.4D) that are interpreted to be reworked from palaeosols. The Tullig Sandstone cuts erosively down into these sediments at the northern end of this exposure (Fig. 9.7.5), with shale clasts sometimes occurring on the foresets of cross-stratification in the Tullig Sandstone. The Tullig Sandstone here comprises a series of large cross-sets (Fig. 9.7.6) that show a flow direction towards the north-east.

   The sediments at this location thus reveal a section through a strongly wave-influenced mouth-bar, which bears many similarities to that at

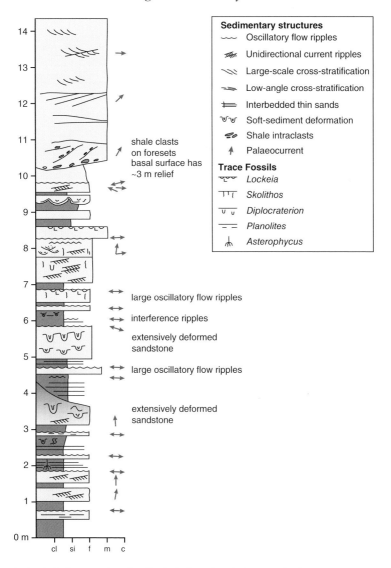

Fig. 9.7.3. Sedimentary log of Tullig Cyclothem, Stop 1, Mutton Island.

Killard (Chapter 9.6). From this locality, the small island that can be seen to the west of the outcrop has a large upper bedding plane that possesses abundant large-scale rafts of sandstone, sandstone balls and distorted strata set in a mudstone matrix (Fig. 9.7.7). The size of this exposure (*c.* 80 × 40 m) suggests this deformed horizon probably represents part of a

(A)

(B)

(C)

(D)

Fig. 9.7.4. Photographs of outcrop at Stop 1, Mutton Island. A) large symmetrical ripples; B) small symmetrical ripples with Lockeia traces (some examples indicated by white arrows); C) soft-sediment deformed horizon (arrowed) and D) spherical calcite concretions.

growth fault and associated slump sheet that has been emplaced in this part of the mouth-bar sequence, possibly towards the top of an underlying parasequence, as the sediments above this horizon in the cliff face are noticeably finer grained and silty. The abundance of soft-sediment deformation in these interbedded sands and silts is shown by both large-scale slumps such as this and smaller-scale bed liquefaction and sandstone balls in the outcrop (Fig. 9.7.4C).

Fig. 9.7.5. Photograph of the erosive contact of the Tullig Sandstone, Stop 1, Mutton Island. The view is towards the NNW, with the white arrows indicating the erosive base of the Tullig Sandstone and the yellow arrow pointing to the deformed bed shown in Fig. 9.7.4C.

Fig. 9.7.6. Large-scale cross-stratification in the Tullig Sandstone, Stop 1, Mutton Island.

Stop 2 (UTM 464215 m E, 5851032 m N; Map 57, 096815 m E, 174047 m N)

From Stop 1, walk approximately 250 m to the south, around an inlet and onto the promontory of rock at the SW corner of the island (Fig. 9.7.1), where one can gain excellent views over several distinctive features in this

Fig. 9.7.7. Slump sheet exposed on the small island near Stop 1, Mutton Island (see location on Fig. 9.7.1). The yellow arrows point to blocks of rafted sandstone within the slump. The width of the field of view is approximately 40 m.

part of the section. If one looks to the north towards the watchtower, two features are of note (Fig. 9.7.8). Firstly, the erosive base of the Tullig Sandstone can be seen in the exposure that lies towards the watchtower (white arrows on Fig. 9.7.8), with the depth of erosion here being *c.* 2 to 3 m. The lateral equivalent of this erosion surface was viewed in the outcrop at Stop 1 (Fig. 9.7.5). Second, the island directly to the north and in front of you contains a clear erosion surface towards its top, with more planar laminated sands cutting down into underlying siltstones (Fig. 9.7.8). These sands are within the parasequences underlying the Tullig Sandstone and bear witness to the erosive surfaces that can occur in the emplacement of mouth-bar sands and may be associated with growth faults and slumps. To further view this erosion, if the viewer turns to look at the main cliff face to the north-east of the viewpoint, these mouth-bar sands can again be seen in splendid cross-section.

Lastly, the Tullig Sandstone is exposed in the upper cliff face in the corner of the inlet at this location (UTM 464250 m E, 5851050 m N; Map 57, 096850 m E, 174065 m N). Here, the erosive base of the Tullig Sandstone cuts down into a series of interbedded siltstones of the underlying parasequence. Below these siltstones, a thick mouth-bar sand is present and this can be traced laterally along the outcrop (Fig. 9.7.9), with distinct internal erosion surfaces and beds that thicken and thin appreciably across the outcrop. The siltstones underlying this mouth-bar sand are laterally equivalent to

Fig. 9.7.8. Photograph taken from Stop 2, Mutton Island, towards the north and the watchtower. The white arrows point to the erosive base of the Tullig Sandstone, whilst the yellow arrow highlights an erosive surface in the underlying mouth-bar sands.

Fig. 9.7.9. Erosive surface (indicated by white arrow) in mouth-bar sands viewed from Stop 2, Mutton Island. The cliffs are approximately 25 m high.

~5 m

Fig. 9.7.10. Mudclast-filled small channels in the Tullig Sandstone, Stop 2, Mutton Island.

those exposed further north and which form the island that possesses the large-scale slump (see Stop 2). In the Tullig Sandstone at this locality in the corner of the inlet, a series of small channels can be viewed in cross-section that are infilled by intraformational mudclast conglomerates (Fig. 9.7.10) and then overlain by sands.

Stop 3 (UTM 464105 m E, 5851395 m N; Map 57, 096710 m E, 1744412 m N)

From Stop 2, walk back north around the exposure of Stop 1 and towards the watchtower. The outcrop immediately to the west of the watchtower (Fig. 9.7.1) displays some large, three-dimensional trough cross-stratification produced by large sinuous-crested dunes and rare casts of large tree trunks. The entire outcrop to the N and NNW of the watchtower is within the Tullig Sandstone and allows access to three-dimensional exposure within these fluvial sands.

Walk north from the watchtower for approximately 80 m and descend the outcrop to a section of *c.* 13 m in height, which runs north-north-west to south-south-east (UTM 464090 m E, 5851540 m N; Map 57, 096697 m E, 174557 m N). This section exhibits three stacked fluvial channels with large-scale, low-angle, accretion surfaces in the lower and upper parts of the outcrop that dip to the west/north-west (Fig. 9.7.11).

Fig. 9.7.11. Photomontage of channels and accretion surfaces in the Tullig Sandstone, Stop 3, Mutton Island. Labels a to f are discussed in the text, with a sedimentary log given in Fig. 9.7.12. See online *GigaPan* image.

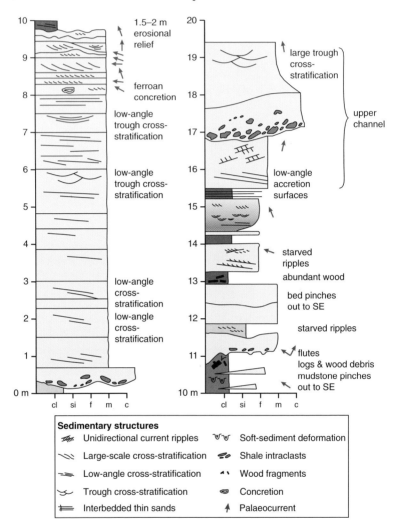

Fig. 9.7.12. Sedimentary log of Tullig Sandstone, Stop 3, Mutton Island.

Walk onto the bedding plane, taking especial care if it is wet, and then down another *c.* 6 m drop onto a bedding plane that can be accessed at low tide (Fig. 9.7.11). At the base of this section, a marked erosion surface within the Tullig Sandstone can be found, with a mud clast conglomerate lining a basal scour (Figs 9.7.12 and 9.7.13; labelled 'a' on Fig. 9.7.11). Above this basal scour, a series of sands possess very low-angle laminae and trough cross-stratification (labelled 'b' on Figs 9.7.11 and 9.7.13), probably produced

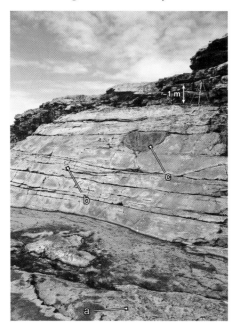

Fig. 9.7.13. Photograph of lower section, Stop 3, Mutton Island (see sedimentary log in Fig. 9.7.12). Labels a to c are discussed in the text, with 'c' highlighting a large ferroan concretion.

by large, low-angle dunes that eventually (in beds above a marked, large ferroan carbonate concretion; labelled 'c' on Figs 9.7.11 and 9.7.13) pass into large dune-scale cross-sets, with flow towards the north-west, north and north-east. The top of these sandstones (at 7 m in the log in Fig. 9.7.12) is the same surface as the notable bedding plane in the outcrop at the base of the main exposure (labelled 'd' on Fig. 9.7.11) and shows an erosive relief across the outcrop of several metres, with erosion becoming greater to the NW with incision of *c.* 3 to 4 m. Low-angle accretion surfaces in the sands above this erosive surface dip to the W/WNW with flow also towards the NW. These sandstones are overlain by a series of siltstones with occasional thin rippled sands (Fig. 9.7.12), with the topmost mudstones and siltstones containing abundant wood fragments, occasional ferroan carbonate concretions and pinching out laterally to the south-east. Above these finer beds, a second series of channel sands erode down into these underlying siltstones and sandstones (*c.* 11 m on log Fig. 9.7.12), with the erosion surface being very evident in the outcrop (Figs 9.7.11 (labelled 'e') and 9.7.12). The base of this sandstone shows flutes and large grooves, with flow direction towards 320°, and the first bed above this surface contains mudstone intraclasts. The beds above show a series of sandstones, which are often discontinuous across the outcrop, and siltstones/mudstones with starved ripples and abundant wood fragments. These beds again gently dip to the NW, with flow direction towards the NE-NW. At the right (southern) end of the outcrop, a third channel is present (Fig. 9.7.11, labelled 'f') eroding into these interbedded sandstones and siltstones, with the depth of erosion to the left (north) being *c.* 2 m. The bed above the erosion surface comprises a basal channel lag of intraformational conglomerate to the south of the outcrop, with large-scale trough cross-stratification in the sandstones above this showing flow towards 350°. Planform exposure of these sandstones displays large-scale trough cross-stratification clearly. The basal conglomerate is eroded into by the overlying sandstones and, to the left (NW), is eventually replaced by these sands.

This exposure thus reveals a series of three stacked channels within the Tullig Sandstone, with the thickness of the lower two storeys being *c.* 8.5 m and 6 m. The channels display large, low-angle accretion surfaces that dip broadly to the north-west and with palaeoflow directions from ripples and dunes giving a mean of 325°. Erosion on the basal contact of these channels at this outcrop is *c.* 3 to 4 m and the channel-fills generally fine upwards, with abundant wood fragments and occasional thin siderite concretions. These characteristics match well those of the Tullig Sandstone at both Trusklieve and Killard and are interpreted to indicate fluvial channels of relatively low sinuosity, with larger bedforms (dunes) migrating either in the channel thalweg or along the low-angle surfaces of larger bar macroforms. The channels and associated finer-grained sediments do not show any evidence for subaerial exposure or *in situ* vegetation here, but the abundant wood fragments in these outcrops attest to the nearby presence of large trees. Importantly, these fluvial channels bear witness to the presence of major feeder channels transporting sediment broadly from south to north at this locality, which is the last exposure of the Tullig Sandstone before the outcrops at Liscannor and Furreera Bay to the north. It appears that the Tullig delta at this point still comprised major trunk channels that were feeding a delta fringe to the north.

Stop 4

From Stop 3, walk east along the northern shore of Mutton Island (Fig. 9.7.1), which displays further exposure within the Tullig Sandstone. As you walk east, you are walking up-section, although the outcrop becomes progressively more poorly exposed as the strata dip gently to the east. The boundary between the Tullig Sandstone and the initial flooding surface of the upper Tullig Cyclothem, as also seen at Trusklieve (Chapter 9.3, Stop 4), is difficult to locate on the outcrop due to cover by

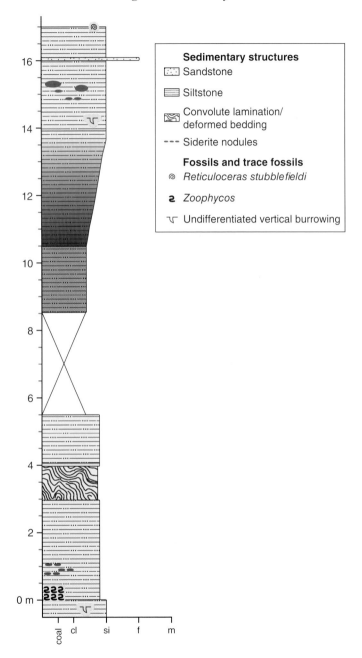

Fig. 9.7.14. Sedimentary log of succession within the upper Tullig Cyclothem, Stop 4, Mutton Island.

boulders on the foreshore. However, as you walk east around UTM 465020 m E, 5851585 m N (Map 57, 097628 m E, 174590 m N), the upper 4 to 5 m of the Tullig Sandstone comprises fine sandstones that are interbedded with mudstones and siltstones. The very top portion of the Tullig Sandstone is intensely folded in some areas due to soft-sediment deformation (similar to that at Killard and Diamond Rocks), is bioturbated and contains thick, metre-scale, sandstone beds that contain occasional symmetrical wave-rippled surfaces indicating 060° north-west to south-east flow oscillations, wood fragments, and 0.05 m thick organic-rich beds that contain compressed logs.

The initial flooding surface that marks the beginning of the end-Tullig transgression (UTM 465125 m E, 5851545 m N: Map 57, 097732 m E, 174548 m N) is obscure but is marked by two consecutive, 0.1 m thick siltstone beds that possess *Zoophycos*. Above this surface, a log of the upper Tullig Cyclothem (Fig. 9.7.14) can be pieced together using partially exposed segments between the covering boulders. The silty mudstones above the flooding surface are dark grey and contain sparse siderite nodules that form along poorly defined bedding planes. The silty mudstones above these nodules are mostly barren, with a 1 m thick deformed siltstone bed being present (Fig. 9.7.14), which is capped by a slickenside surface at 4 m. The silty mudstone coarsens upward to a siltstone, which contains some large siderite nodules and that is overlain by a thin, structureless, fine sandstone. The sediment fines upward to a siltstone, and the sequence is capped by a shale bed containing small *R.* aff. *stubblefieldi* goniatites that can be located in the beach (Fig. 9.7.14; UTM 465180 m E, 5851535 m N; Map 57, 097787 m E, 174537 m N) and nearby low cliffs (UTM 465279 m E, 5851592 m N [Map 57, 097887 m E, 174593 m N]; UTM 465301 m E, 5851606 m N; [Map 57, 097909 m E, 174607 m N]). This small succession marks the top of the Tullig Cyclothem, and the sediments to the east of this point lie within the Kilkee Cyclothem.

Chapter 10
The Tullig and Kilkee Cyclothems
of Northern County Clare

PAUL B. WIGNALL, JIM BEST, JEFF PEAKALL
& JESSICA ROSS

10.1 Introduction

The extensive coastal exposures along the south-facing shore of Liscannor Bay and the spectacular, giant cliffs that stretch from Hag's Head to the Cliffs of Moher provide a series of superb exposures of the northern-most outcrops of the Tullig and Kilkee cyclothems in County Clare. The Cliffs of Moher are also one of the main tourist sites in Ireland but, away from the intense concentration of visitors at this tourist mecca, the majority of locations described here have the same remote, quiet ambience of most County Clare coastal sites. Despite the world-famous nature of the Cliffs of Moher, this chapter provides the first detailed description of the geology of this location and the surrounding area.

The northern sections in the Tullig Cyclothem are somewhat thinner than those seen in the south, although the thickness differential is much less than that seen in the lower parts of the Namurian basin fill (Fig. 1.7.3). The key difference is seen at the level of the Tullig Sandstone. In southern County Clare this is a major sandbody but in the north only isolated channel sandbodies are encountered, which are embedded in finer, delta-top sediments. In contrast, the Kilkee Cyclothem shows no south-to-north thickness variations and comprises a near uniform-thickness blanket of sediment across the basin (Wignall & Best, 2000). Similar facies to those seen in the south are also developed, with a deep-water delta front succession dominating the central part of the Kilkee Cyclothem. This includes one of the largest and best-exposed examples of a growth-fault set seen in any cliff section in the world (Wignall & Best, 2004). These strata can be examined in closer detail at Hag's Head, immediately to the south of the

*A Field Guide to the Carboniferous Sediments of the Shannon Basin,
Western Ireland*, First Edition. Edited by James L. Best and Paul B. Wignall.
© 2016 International Association of Sedimentologists.
Published 2016 by John Wiley & Sons, Ltd.
Companion website: www.wiley.com/go/best/shannonbasin

Cliffs of Moher and on the south side of Liscannor Bay at Freagh Point. Spectacular growth faults and an impressive array of fluidization and liquefaction structures greet the visitor to these locations.

10.2 Tullig and Kilkee Cyclothems between Liscannor and the Cliffs of Moher

Directions to Stop 1

The first stop can be reached by taking the main road (R478) westwards from Liscannor towards the Cliffs of Moher. After 4 km, the road makes a sharp bend to the right and a minor road branches off to the left at the corner. Turn on to this minor road and drive straight ahead. After 150 m, turn left onto a narrow road that takes you down to Furreera Bay and park alongside the seawall on the right. Hop over this wall, cross over a small stream that makes its way through the beach cobbles to the sea and head towards the low sea cliffs to the west. Note that access to this outcrop becomes **difficult at high tide** because the low platform in front of the cliffs is covered by the tide (Fig. 10.2.1). Also, like many rocky surfaces on the County Clare coast, the platform is very slippery when wet.

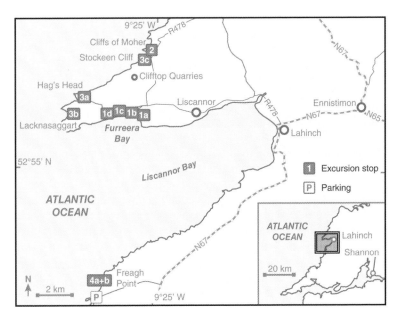

Fig. 10.2.1. Map of northern County Clare showing location of stops described in this chapter.

Stop 1: Furreera Bay (UTM 471153 m E 5865382 m N; Map 57, 103962 m E, 188757 m N)

This bay provides a kilometre-long exposure of continuous low cliff and foreshore in the uppermost part of the Tullig Cyclothem. The rapid lateral facies changes seen at this level provide a fascinating, although somewhat controversial, record of changing conditions in the later stages of this fluvial and wave-influenced delta-top succession.

At Stop 1a (UTM 471153 m E 5865382 m N; Map 57, 103962 m E, 188757 m N), the low cliffs can be clearly divided into two distinct facies - a lower heterolithic package consisting of alternating sandstone and siltstone beds and an upper package of trough cross-bedded sandstones (Fig. 10.2.2). The main sandstone bed of the lower package (Fig. 10.2.3) beautifully displays steeply climbing ripples recording a NNE flow direction. The topmost few centimetres of this bed have also been reworked by oscillatory flows, of variable orientation, and further, thin wave-rippled sandstone interbeds occur in the overlying metre of sandstone. Around 1.5 m above the climbing ripple bed, *in-situ* stigmarian roots, complete with their rootlet fan, are developed in a siltstone layer that also contains abundant siderite concretions (Fig. 10.2.4). This is an incipient palaeosol and it is erosively overlain by a thin sandstone of variable thickness that contains a lag of reworked

Fig. 10.2.2. Sharp-based Tullig channel sandstone resting on heterolithic package of sideritic mudstones and rippled sandstone beds at the eastern end of Stop 1a, Furreera Bay.

Fig. 10.2.3. Climbing ripple bed in the Tullig Cyclothem at Stop 1a, Furreera Bay.

Fig. 10.2.4. Large *Stigmaria* root with rootlets (white arrow), developed in a bed with common siderite nodules, Tullig Cyclothem, Stop 1a, Furreera Bay. Note the brecciated shale of the glacio-tectonic comminution till (see also Figure 10.2.5) at the top of the image (yellow arrow).

Fig. 10.2.5. Brecciated zone in shale bed beneath the Tullig Sandstone, Stop 1a, Furreera Bay. The shale is broken up into either a breccia (white arrows) or is deformed (blue arrow), with some areas of shale remaining undeformed (yellow arrows). This brecciation is interpreted as a glacitectonic comminution till. Photograph courtesy of Drew Phillips.

concretions and woody debris. This, in turn, is overlain by a dark shale that shows thin zones of brecciation, usually at the base and top of the bed, although brecciation occasionally occurs at its centre (Fig. 10.2.5). The breccia layers vary in thickness and often grade laterally into undeformed sediment. This breccia is probably a glacio-tectonic feature referred to as a comminution till, which is generated *in-situ* by fragmentation and rafting of the bedrock by overlying ice, as has also been found elsewhere in Ireland (Hiemstra *et al.*, 2007), or a feature formed by repeated freeze-thaw action of groundwater along bedding surfaces and within joints. The brecciated siltstone bed is sharply overlain by cross-bedded sandstones that form the highest part of the cliff and are the northernmost expression of the Tullig Sandstone. The sharp contact and the occasional presence of siderite concretions in the base of sandstone indicates erosion of the underlying strata.

Pulham (1989, p. 197) has interpreted the upper Tullig Cyclothem outcrops at Liscannor Bay to be a record of "a delta-plain succession dominated by relatively thin single-storey channel sandstones,..... thin fluvial-dominated bay fills [and] emergent, abandonment facies". This scenario is also appropriate for the nearby Furreera Bay section examined here, although with the proviso that there is also significant evidence for wave processes operating at this level (Wignall & Best, 2000). More specifically, a sheltered, inter-distributary bay setting best explains the combination of quiet-water siltstone deposition, wave-reworked sediments and a rooted horizon. The climbing ripple bed could record an episode of crevasse splay from an adjacent distributary, or deposition in a distributary mouth setting. The origin of the overlying sandstone, with its well-developed trough cross-beds, provides a point of debate. A small, distributary channel origin was favoured by Wignall & Best (2000) and by Rider (1969) (whose thesis figures are reproduced in Gill (1979, his fig. 9)). In contrast, Pulham (1989) suggests a crevasse channel developed at 90° to a major, south-easterly flowing, distributary. Although Pulham's idea helps to explain the identical palaeo-currents in the underlying climbing ripple bed because all the sandstone beds in this section become part of a crevasse splay, his notion is unsup-ported by any evidence for a major SE-directed distributary sandbody in the region; in fact, all palaeocurrents show a predominant N/NE palaeo-flow in the Tullig Sandstone here (see Fig. 9.4.20). The Tullig Sandstone is thus interpreted as the smaller, more northerly, distributary channels of the Tullig delta, that have incised down into their underlying distributary mouth-bar and adjacent inter-distributary bay sequences.

   The channel sandbody can be traced by walking westwards for nearly 300 m along the boulder-strewn foreshore to Stop 1b (Fig. 10.2.1; UTM 470895 m E 5865300 m N; Map 57, 103696 m E, 188299 m N). The gentle regional dip brings the top surface of the sandstone down to beach level, where it is overlain by a grey mudstone, with a well-developed pedogenic fabric and a shaly coal. In the parlance of coal geology, these are a fireclay and a cannel coal respectively. The coal provides a valuable marker horizon between Stops 1b and 1c because the foreshore exposure is covered in boulders at this point and the coal horizon, seen in the low cliff at the back of the beach, is the only level that can be traced continuously between these locations. It is fortunate that this is the case because the next section, at Stop 1c (UTM 470590 m E 5865370 m N; Map 57, 103392 m E, 188303 m N), is otherwise difficult to correlate due to rapid lateral change. A sandbody is developed in the low foreshore at the same stratigraphic level as the channel sandbody seen to the east but it is of distinctly different facies. Pulham (1989, his fig. 15) depicts the channel sandbody cutting up between these two locations, which seems probable, although the key evi-dence is hidden beneath beach boulders.

   The first outcrop encountered of the westerly sandbody consists of sandstone beds folded into a broad, elongate, gently sinuous, synclinal fea-ture trending N-S. This is erosive-based with the beds infilling and par-tially draping the erosive surface. Similar, stacked, erosive-based, broad sandstone lenses compose much of the outcrop, as best seen in the upper foreshore sections below the low cliff line. This combination of erosion

Fig. 10.2.6. Outcrop at Stop 1c showing dipping beds of sandstone forming the western limb of a synclinal structure. These beds are truncated by the overlying sandstones that pass laterally (to the right) into dipping foresets that infill the axis of the syncline.

and deposition is also seen in the centre of the foreshore platform. A series of trough-crossed bedded sandstones have been partially eroded, with the resultant surface displaying metre-scale hollows and wave ripples. The infill of these hollows, by lateral accretion, includes south-westerly and southerly prograding foresets, unusual directions in the overall context of this location (Fig. 10.2.6). However, the dominant palaeoflow, recorded by both dunes and linguoid ripples, is to the north and in the same direction as seen in the easterly channel sandbody at Furreera Bay. An undulatory erosion surface truncates the mouth-bar sandstone, with the result that the overlying fireclay varies in thickness as it infills the hollows. This surface can be seen again in the Cliffs of Moher, where it locally cuts out a significant amount of the underlying strata.

The central and western part of these mouth-bar sandstone beds have been folded into a gentle, asymmetric anticline with its steepest limb on the south. The cause of this folding is enigmatic. The fold is perhaps related to underlying deformation, such as rollover in the hangingwall of a growth fault, but conclusive evidence is lacking. More probably, it could represent contemporary syn-sedimentary deformation caused by compression during foreland basin development, similar to that recorded by Tanner *et al.* (2011) in southern County Clare.

The topmost 12 m of the Tullig Cyclothem forms the western end of the Furreera section and shows a range of facies exhibiting considerable marine influence. The fireclay noted above is overlain by the thin coal (also seen previously) and the topmost surface displays some remarkable, *in-situ*, hollow, tree stumps up to 0.5 m tall (Fig. 10.2.7). These are composed of thin, anthracitic walls and possess a 'dished' mudstone infill that records greater compaction than the surrounding mudstones. A thin layer of brachiopod shell hash is developed immediately above the coal, the local development of the Killard Marine Band, and indicates that a gentle marine transgression drowned the Tullig forest to the point where tree stumps were exposed on a marine sea floor. The inner part of the tree subsequently decayed leaving a tube of resistant bark: a remarkable occurrence.

The overlying rocks are beautifully displayed in a series of natural arches and cliffs and show numerous, thin beds projecting from the mudstone faces. In their lower part, these beds are persistent layers of siderite, but at

Fig. 10.2.7. Stop 1c at Furreera Bay showing an *in-situ* tree stump (a hollow cylinder of anthracitic coal; edges arrowed) developed on top of a thin coal, which in turn rests on a palaeosol full of rootlets. The tree stump is infilled with shale that is distinctly compacted only in its upper part. Geological hammer for scale.

higher levels an increasing number of wave-rippled sandstones form the thin ribs. This coarsening-up trend persists for 4 m until the spacing of the sandstone ribs begins to increase. The fining-up trend is then abruptly truncated, after 1 m, by the appearance of a sharp-based, 2 m thick post of sandstone. This is a thin development of the Moore Bay Sandstone, which in its upper half has been intensely bioturbated by a diversity of trace fossils, including *Diplocraterion* and *Teichichnus*. At low tide it is possible to clamber onto the top surface of the sandstone and examine a *Zoophycos*-dominated ichnofabric. The upward change in trace fossil community within this sandstone suggests deepening, a trend that continues higher in the section.

Looking west, whilst standing on the *Zoophycos* surface (Stop 1D, UTM 470150 m E 5865258 m N; Map 57, 102950 m E, 188197 m N), one can see a kilometre-long stretch of strike section that beautifully displays the basal few metres of the Kilkee Cyclothem. Examining these strata presents no problems in dry conditions but the surface is extremely slippery when wet, even by the standards of County Clare sections. The *R.* aff. *stubblefieldi* Marine Band is at the base of the dark shale cliff and contains abundant ferroan carbonate concretions, of various odd shapes, that contain small, uncompacted goniatites.

Transfer to Stop 2

From Furreera Bay, retrace the route back to the main road (R478) and turn left (northwards). Just after this junction on your right you will find yourself driving past 'The Rock Shop', which sells a fascinating array of rocks, fossils and local books. It also has the bonus of a café – a great retreat on a wet day – and an interesting display on the local flagstone industry (*The Story of Liscannor Stone*), which exploits the thin-bedded mouth-bar facies of the Kilkee Cyclothem.

Follow the road northwards and after five minutes you will reach the Cliffs of Moher, with parking for cars on the right hand side of the road. This popular tourist site has ample parking (at a price), a visitors' centre and a well-defined access to O'Brien's Tower, where a spectacular view of the Cliffs of Moher can be gained.

Stop 2. Cliffs of Moher (UTM 471220 m E 5869140 m N; Map 51, 104075 m E, 192066 m N)

Recent changes to the viewing area at the Cliffs of Moher mean that it is no longer possible to visit cliff-top ledge sections and visitors can only view the cliffs from the safety of a stepped viewing area. However, the advantage of these highly restrictive, tourist-control measures is that local stone has been used to pave the paths and line the walls and many truly spectacular examples of the Liscannor Flags (of the Kilkee Cyclothem) can now be examined. These include surfaces intensely burrowed by the tightly sinuous, pascichnial trace *Psammichnites plumeri* (often called *Olivellites* or *Scolicia* in the literature) and others showing beautiful linguoid-rippled surfaces. These beautiful slabs from the Liscannor Flags come from the series of clifftop quarries (UTM 470530 m E, 5867620 m N; Map 51, 103364 m E, 190555 m N) near Stop 3, although permission in advance from the quarry owners is needed to access these sites. Some steps are also made from dark limestones from the Burren area that often display lovely colonies of the rugose coral *Siphonodendron*.

The sandstone ledge in front of the viewing steps displays the same *Zoophycos*-mottled surface at the top of the Moore Bay Sandstone as seen at the west end of the Furreera section of Stop 1. The overlying shale cliff contains the *R.* aff. *stubblefieldi* Marine Band at its base. Recently-constructed walls, together with patrolling security guards, now deny access to these strata. Fortunately, it is still possible to view the upper part of one of the highest sea cliffs in Europe, which shows the upper part of the Tullig Cyclothem and the base of the Kilkee Cyclothem (Fig. 10.2.8). Siltstones

Fig. 10.2.8. Upper part of the Cliffs of Moher. The Moore Bay Sandstone (shown by white arrows) is the main marker bed (its top surface forms the ledge in the foreground) and can be traced in the cliffs to the headland, where it is ~70 m below the tiny figures (black arrow) seen at the cliff top and to the south (white arrows) towards Hag's Head.

and shales belonging to the basal part of the Kilkee Cyclothem occur at the top of the cliff and rest on the sharply-defined top surface of the Moore Bay Sandstone. This 2.5 m thick bed, in turn, sits on a 7 m thick package of predominantly siltstone. Several metres of strata have been removed towards the base of this siltstone, as can be seen by comparing the basal surface in the left of the field of view, where it is in contact with several metres of sandstone beds, to the same level at the right, directly beneath the highest point of the cliff, where a siltstone-on-siltstone contact is developed. The underlying sandstone is cut out over the intervening distance. This erosion surface was seen before, at Furreera Bay, where it capped the top of the mouth-bar sandstone and was clearly undulatory on a decimetre-scale, but not as demonstrably erosive on the metre-scale seen here at the Cliffs of Moher. A further intriguing erosion surface occurs lower in the Tullig Cyclothem beneath O'Brien's Tower but this can only be seen from the viewpoint at Stockeen Cliff that forms part of Stop 3.

Transfer to Stop 3

The Cliffs of Moher is one of the great scenic sites of Ireland but for geologists the most accessible and interesting sections in the area occur at the south-west end of the cliff line around the promontory of Hag's Head. This area can be visited by first parking your vehicle at the end of the lane (UTM 468972 m E, 5866325 m N; Map 57, 101787 m E, 189281 m N) that leads from Liscannor and walking towards Moher Tower that is perched at the cliff edge above Hag's Head. A narrow path to the south of this tower leads down, through a cliff section, to a series of steep grassy slopes developed above 80 m high cliffs – take care!

Stop 3. Hag's Head to Stockeen Cliff

The cliffs around Hag's Head occur at the south-west end of a spectacular development of growth-fault affected strata from the lower Kilkee Cyclothem. A total of 11 growth faults are developed over a 2 km stretch and they display a sequential development from the oldest in the north-east to the youngest in the south-west around Hag's Head. This is perhaps the best-displayed retrogressive growth-fault system seen anywhere in the world (Wignall & Best, 2004). The faulting history is demonstrated by the hangingwall strata of younger faults draping the rotated strata in the footwall of older faults developed down-dip. These relationships are only clearly visible from the sea and it is possible to catch a boat from Liscannor that provides a highly recommended, breathtaking, cruise along the Cliffs of Moher. Fortunately, the cliffs around Hag's Head also provide access to one of the largest growth faults in the section, the ninth to form

(Fig. 10.2.9). It is possible to examine the top termination of this feature at the south-eastward (landward) side of the promontory (Stop 3A; UTM 468536 m E, 5866550 m N; Map 57, 101354 m E, 189512 m N). Here, the fault itself consists of a sharply-defined zone of sandstone dipping at 50° to the north-east. This is slightly more resistant (better cemented) than the surrounding rocks and so it locally forms a low 'wall'. Traced upwards, the fault rapidly flattens out to a near horizontal slip surface (Fig. 10.2.10). Looking back at the cliff section beneath Moher Tower, another growth fault, the tenth, can be seen cutting through the entire height of the cliff, although the associated bed offsets are only minor (less than a metre) because this fault is near its top termination. Several, less extensive, syn-sedimentary faults can also be seen in the footwall strata of this tenth fault.

The downthrow direction of all the growth faults is to the north-east, parallel with the present day cliff line, and the listric geometry of the growth faults is well displayed. In contrast, the south-western cliff face of the Hag's Head promontory is developed at 90° to the main cliff line and thus provides a cross-section normal to the downthrow direction and through several broad, cuspate (convex-down), erosive surfaces that cut up to both the NW and SE. These erosive surfaces are mantled by mudstone or siltstone and often have a dewatered layer just below the erosion surface. They are interpreted to be the expression of minor growth faults, developed in the footwall in close proximity to the ninth growth fault. The overall geometry of at least some of the growth faults was therefore spoon-shaped (Wignall & Best, 2004).

The Hag's Head sections also provide a good opportunity to contrast sedimentation in hangingwall and footwall settings. In the footwall to the south of the fault, coarse siltstones with planar and ripple cross-lamination dominate. Coarser sediment was focused in the hanging-wall section where trough cross-bedded fine sandstones show flow to the north-east, the same direction as the fault downthrow. These sediments are interpreted to be formed in a mouth-bar environment, a typical setting for both high sedimentation rates and growth faulting (Wignall & Best, 2004), as seen elsewhere in County Clare (e.g. Chapter 9.6, Stop 3, Tullig Cyclothem at Killard). In the highest part of the hangingwall stratigraphy, seen in the cliff top 100 m to the east of Moher Tower, a well-developed 'chute fill' is seen. Such features are commonly developed late in the history of growth-fault movement in the County Clare examples. These chute-fills record erosion of elongate scours (tens of metres across), developed orthogonal to the fault line, in the elevated crest of hangingwall anticlines. They are infilled by the downlapping distal portion of subsequent hangingwall strata. In addition, the thin parallel-bedded sandstones and siltstones found in this area, which occasionally possess wave-rippled tops, have

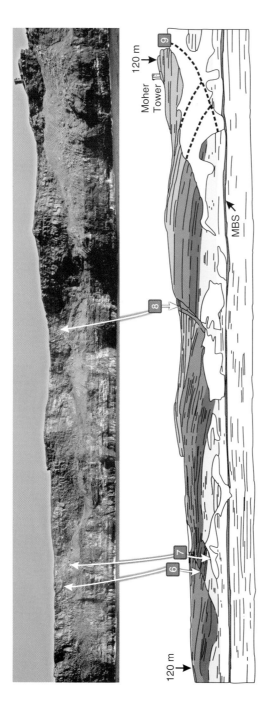

Fig. 10.2.9. View of the Hag's Head cliff section and its associated large growth faults developed above the Moore Bay Sandstone (MBS). The blue numbers refer to the faults (6, 7, 8 and 9) recognised in this sequence of retrogressive faults by Wignall & Best (2004), with the coloured packages depicting strata associated with individual growth faults. Strata not affected by faulting and areas covered by grass and scree slopes have not been coloured. The cliffs are *c.* 120 m high. Modified from Wignall & Best (2004).

Fig. 10.2.10. Geologist standing on the upper part of the growth fault (growth fault 9, Fig 10.2.9) at Hag's Head. As the surface is traced upwards, the dip of the fault rapidly flattens out to near horizontal, as shown by the white arrows.

surfaces that are intensely bioturbated by *Psammichnites plumeri*, as seen in the slabs in the walls at the Cliffs of Moher (Stop 2). These sediments are often beautifully exposed in the clifftop quarries adjacent to the small roads in this area (UTM 470530 m E, 5867620 m N; Map 51, 103364 m E, 190555 m N) and are discussed by Wignall & Best (2004). However, as noted previously, advanced permission from the quarry owners is essential to access these sites.

Having examined the abundance of features to be seen around Hag's Head, the visitor now has a choice of which direction to walk because further intriguing features can be seen both to the north-east and south-west. A short cliff top walk to the south-west will bring you to the south side of Lacknasaggart Bay (Stop 3b; UTM 468330 m E, 5865210 m N; Map 57, 101129 m E, 188175 m N). Looking back to the north, a large cliff section once again displays a considerable portion of the lower Kilkee Cyclothem with abundant soft-sediment deformation. The Moore Bay

Sandstone reaches beach level at this location. The lower part of the cliff shows small-scale, diapiric structures in siltstones with thin sandstones, whereas the upper part of the cliff shows low-angle growth faults defining scoop-shaped slip planes. These are of a smaller scale than the growth faults seen in the Cliffs of Moher and the décollement surfaces are developed at a higher level in the stratigraphy. The profile of the faults is controlled by the orientation of the cliff line and, at this location, it is at a high angle to the NE-directed slip direction. Thus, the section is cut in a near-orthogonal orientation through growth faults seen to have a spoon-shaped profile.

From this location, head back along the cliff line, past Moher Tower and proceed along the spectacular cliff-top path for 2.5 km until you reach the highest point (Stop 3c, UTM 471160 m E, 5868890 m N; Map 51, 104011 m E, 191817 m N). This is at Stockeen Cliff, which approaches 200 m in height, and that provides a spectacular view to the north-east of the stratigraphy seen beneath O'Brien's Tower (Fig. 10.2.11). The lower third of the cliff consists of undisturbed siltstones of the Gull Island Formation

Fig. 10.2.11. The Cliffs of Moher, capped by O'Brien's Tower, as seen from Stockeen Cliff (Stop 3). Thin channel sandbodies of the Tullig Cyclothem dominate the upper half of the cliff and show several internal erosion surfaces (white arrows). The Moore Bay Sandstone is arrowed in yellow and labelled MBS. Goat Island, the elongate cliff line nearest the camera, is capped by a sandstone horizon at the base of the Tullig Cyclothem that is progressively truncated towards the right (blue arrows). The cliffs at O'Brien's Tower are *c*. 177 m high.

(see Chapter 8). The upper 25 m of the cliff shows siltstones and shales from the base of the Kilkee Cyclothem, with the intervening strata belonging to the Tullig Cyclothem and consisting of interbedded sandstones and siltstones. The Moore Bay Sandstone is the most clearly distinguished bed and this tabular sheet can be readily traced along the entire cliff line. A further four, sand-rich, levels occur below this and they display internal erosion surfaces and discontinuous shale layers. It is obviously not possible to make a closer facies inspection of these levels but comparison with strata seen at Furreera Bay (Stop 1), 4 km to the south, suggests a distributary channel or mouth-bar origin. It was noted at the Cliffs of Moher viewpoint (Stop 2) that the topmost sandstone was locally removed by erosion immediately to the south of the O'Brien's Tower cliff. This feature is not visible from Stockeen Cliff but a similar erosive down-cutting and removal of a sandstone bed can be seen to have affected the lowest sand-rich level from this viewpoint. Thus, on Goat Island (a narrow cliff-bound ridge that is not actually an island because it is attached to the main cliff), the fourth sandstone-rich package (the lowest one in the Tullig Cyclothem) is seen to thin from left to right due to erosion of the uppermost sandstone bed. In the cliff section beneath O'Brien's Tower, this same sandstone package has been reduced to a single thin bed. In the cliffs to the north of O'Brien's Tower, the erosive surface cuts up and the fourth sandstone package is again well developed, this time with three beds present. The erosion seen in the Tullig Cyclothem at the western end of the Furreera section was associated with folding and potentially records foreland compression, and a similar origin may be ascribed to the intra-Tullig erosion surfaces at the Cliffs of Moher.

10.3 Kilkee Cyclothem Sediments and their Sand Volcanoes at Freagh Point

Directions to Freagh Point

From the Cliffs of Moher, drive southwards along the R478, which takes you through Liscannor and ultimately reaches a T-junction at the edge of Lahinch. Turn right onto the N67, which passes through the centre of this small town. After nearly 8 km, a small brown signpost indicates a right turn to Freagh and the coast. Take this road and after 1 km you will come to a sharp left hand bend where it is possible to park at the side of the road (UTM 470729 m E, 5858578 m N; Map 57, 103437 m E, 181506 m N). There is a gate into a farmer's field at this bend and adjacent to it a stone wall stile. Cross this stile and head across the field, parallel with the coast, for ~300 m. The coastline now swings to the left as the cliff line becomes higher. Take the small track that leads down to an expanse of wave cut platform (Stop 4A; UTM 470773 m E, 5859045 m N; Map 57, 103487 m E, 181973 m N).

Stop 4. Freagh Point (UTM 470773 m E, 5859045 m N; Map 51, 103487 m E, 181973 m N)

The cliff line at Freagh Point is mostly composed of glacial till that shows a gradual downward transition into undisturbed sediments of the Kilkee Cyclothem. Flaggy beds of sandstone (mouth-bar facies) are the main lithology at this stop but there is also a spectacular range of soft-sedimentation deformation to be seen. On arrival at the main rocky platform, the most obvious features are the abundant small sand volcanoes – 369 in total (Ross *et al.*, 2013; Fig. 10.3.1). The flanks of these volcanoes can be seen to pass into horizontal laminae and the flanks of adjacent volcanoes without break. They therefore all appear to have erupted in one continuous

(A)

(B)

Fig. 10.3.1. A) Photomontage of the extrudite sheet, with B) sand volcanoes marked, at Freagh Point. Syn-sedimentary growth faults are indicated in yellow. Redrawn from Ross *et al.* (2013) with permission from the Geological Society of London.

episode; the entire volcano-studded horizon is an impressive example of an extrudite sheet – a bed sourced by fluidized sediment whose lateral transport is driven by gravity flow from (in this case) multiple vent sites. It seems probable that this level was the product of seismic shaking because there is no sense of the lateral displacement seen in gravity-driven slumps. A thin (2 to 3 cm) rippled bed is partially preserved between the deformed horizon and the extrudite sheet, suggesting a period of normal mouth-bar deposition before the shaking occurred, causing dewatering of the sediment and consequent fluidization of the sand and silt that sourced the extrudite (Ross *et al.*, 2013).

Once the extrudite sheet had formed, before deposition of any overlying sediment, the bed was subject to burrowing by *Psammichnites* trace makers and then minor growth faulting that cut several of the volcanoes (Figs 10.3.2 and 10.3.3). Remarkably, the sand volcanoes do not show any signs of collapse or reworking onto the hangingwall, even though the faulting must have been penecontemporaneous with deposition. The extrudite sediment was thus capable of maintaining slopes of >70° shortly after emplacement, which suggests very rapid sea floor lithification – an observation that probably helps explain the excellent preservation of these features.

Other soft-sediment deformation features seen at this location are mud diapirs, which punctured up through the deformed sediments and are now visible as concentrically-lined pits, where the platform is not covered by the extrudite. Heavily-deformed horizons also overlie the sand volcanoes, which mostly show dewatering features, including superb dish structures and sand pillows (Fig. 10.3.4).

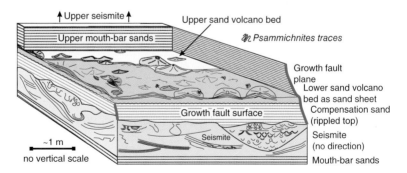

Fig. 10.3.2. Schematic diagram summarising the depositional features and processes recorded at the sand volcano horizon, Kilkee Cyclothem, Freagh Point. Redrawn from Ross *et al.* (2013) with permission from the Geological Society of London.

Fig. 10.3.3. Person standing on the truncated flank of a sand volcano seen on the footwall of a small growth fault at Freagh Point. The hangingwall is in the foreground and has been downthrown by ~0.30 m.

Fig. 10.3.4. Superb dish structures within a large sand pillow boulder, Freagh Point.

Fig. 10.3.5. A ~7 m high cliff section showing the upward transition from dewatered strata, with abundant 'ball and pillow' structures, to undisturbed flaggy sandstones at Freagh Point. At the transition, note the large raft of sandstone to the left that has partially detached from the upper beds.

Fig. 10.3.6. Beautifully exposed growth fault at Freagh Point. The geologists to the left are standing on the edge of the footwall, a pale white boulder sits on hangingwall strata and to the right a minor antithetic fault forms a low wall. The viewpoint is looking to the north-west with Furreera Bay and Hag's Head seen in the far distance.

One hundred metres to the north of this sand volcano surface, a narrow embayment cuts into the rocky platform (Stop 4B; UTM 470773 m E, 5859135 m N; Map 57, 103488 m E, 182062 m N). This exposure provides a superb cross-section through the same level, which shows a beautiful transition from ball-and-pillow style dewatering in the lower beds to undisturbed, flaggy, mouth-bar sandstones at the top (Fig. 10.3.5). The transition also shows large rafts of sandstone that have partially or completely detached from the upper beds and foundered into the dewatered strata. A further 100 m north of this location, the flaggy sandstones are cut by a major growth fault that downthrows to the north-east. Standing at the edge of the footwall, in the same location as the people seen in Fig. 10.3.6, it is possible to look down at the hangingwall sediments and view how a minor antithetic fault has formed a graben. Looking along the length of this fault from the same location, the gentle, arcuate curvature of the fault plane is readily apparent. This growth fault is developed at the same stratigraphic level as the multiple growth faults in the Cliffs of Moher and is of a similar scale. Such delta front collapse features were clearly extremely common in the Kilkee Cyclothem of northern County Clare.

Chapter 11
The Younger Namurian Cyclothems around Spanish Point

PAUL B. WIGNALL & JIM BEST

11.1 Introduction

The Carboniferous rocks of County Clare are arranged into a broad syncline (Figs 2.3.1 and 2.3.4), with the result that the coastal outcrops in the central part of the County expose the youngest strata of the region – three deltaic cyclothems (the Tullig, Kilkee and Doonlicky cyclothems) and the unnamed cyclothems IV and V. The three youngest cyclothems form the focus for this field excursion. Like the older Tullig and Kilkee cyclothems, the younger examples display a spectacular array of sedimentary features seen in a series of low coastal cliffs and extensive, intertidal platforms. For most of the sections described here, access is possible at all but the highest tides. However, it is always wise to keep an eye on the tide, especially at Black Rock. The locations are described along a north to south transect, and at most points it is possible to park cars within a few tens to hundreds of metres of the area to be visited.

11.2 The Area around Spanish Point

Directions to Stop 1

West of Miltown Malbay in central County Clare is a major headland known as Spanish Point (Fig. 11.2.1) that derives its name from the ships that were attempting to return home after the failure of the Spanish Armada expedition in 1588. To its south is a large sandy bay popular with tourists. Curiously, this bay is rarely named on maps and the location is

A Field Guide to the Carboniferous Sediments of the Shannon Basin, Western Ireland, First Edition. Edited by James L. Best and Paul B. Wignall.
© 2016 International Association of Sedimentologists.
Published 2016 by John Wiley & Sons, Ltd.
Companion website: www.wiley.com/go/best/shannonbasin

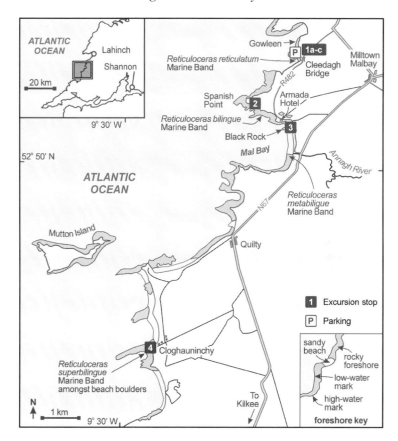

Fig. 11.2.1. Location map of the area around Spanish Point, with the stops visited in this field excursion.

always referred to as Spanish Point, even though this is the headland that defines the north side of the bay. To the north of the point, the coastline is a typically rugged example of the County Clare coast. The first stop occurs at one of the main access points. A small, convenient, car park is found by turning off the R482 immediately north of Cleedagh Bridge (Stop 1a).

Stop 1. Cleedagh Bridge (UTM 471096 m E, 5856825 m N; Map 57, 103779 m E, 179747 m N)

This location provides one of the best and most accessible places to see the topmost beds of the Kilkee Cyclothem and the transition to the overlying Doonlicky Cyclothem. Before examining the uppermost Kilkee sediments

Fig. 11.2.2. Extensive zone of intensely dewatered sandstones and siltstones in the central part of the Kilkee Cyclothem, Stop 1b, Gowleen.

in the small bay in front of you, it is possible to see the lower beds of this cyclothem by taking a short walk to the north-west along the coastal path to the bay at Gowleen (Stop 1b, Fig. 11.2.1: UTM 470980 m E, 5857095 m N; Map 57, 103667 m E, 180019 m N).

At Gowleen, the central part of the Kilkee Cyclothem displays abundant, soft-sediment deformation features that typify this level in all sections north of Kilkee. Here, these consist of a thick zone of dewatered sediment with large, deformed rafts of sandstone (Fig. 11.2.2). The headland that defines the south side of the bay at Gowleen shows the overlying, basal part of the Kilkee Sandstone that consists of a series of sharp-based sandstones with levels of intense bioturbation. The higher parts of the Sandstone are seen by walking part of the way back to Cleedagh Bridge to a small bay to the south of Gowleen (Stop 1c, Fig. 11.2.1; UTM 470885 m E, 5856900 m N; Map 57, 103569 m E, 179825 m N). Here, excellent three-dimensional exposures of dunes record a south-easterly palaeoflow.

Return to the bay at Cleedagh Bridge (Stop 1a) to see the top beds of the Kilkee Cyclothem. These consist of several, metre-thick beds of sandstone with beautifully displayed climbing ripples and bedding surfaces of linguoid ripples that record a palaeoflow to the south-south-west. This direction is roughly orthogonal to the south-easterly flow lower in the section, an observation that may suggest a crevasse splay origin for these rapidly-accumulated beds.

The highest beds of the Kilkee Cyclothem consist of a small-scale shallowing-up unit (parasequence) of probable marine origin. Oscillation ripples and swaley and hummocky cross-stratification dominate the sedimentary structures and suggest a storm-dominated, open-shelf setting. The basal siltstone of this unit is intensely bioturbated by a diverse ichnofauna, including *Zoophycos*, and encloses curious, metre-scale rafts of sandstone that have a circular profile, which have partially sunk into the underlying beds (Fig. 11.2.3). The topmost decimetres of the unit show *Skolithos* burrows and finally a palaeosol (ganister) with stigmarian roots. This marine parasequence can be considered the first of a retrogradational parasequence set. The base of the subsequent parasequences are even more distal in their development because shales and siltstones of the basal Doonlicky Cyclothem dominate the higher strata seen in the cliffs on the south side of the bay. At low tide, the *Reticuloceras reticulatum* Marine Band, a grey shale yielding the zonal goniatite, can be seen at the base of this succession and is overlain by unfossiliferous shales with siderite concretions that are well displayed in the cliffs at the back of the bay.

Fig. 11.2.3. Large raft of sandstone in the upper Kilkee Cyclothem at Stop 1a, Cleedagh Bridge. Geological hammer for scale.

Transfer to Stop 2

Turn right on leaving the car park at Cleedagh Bridge and drive for ~1 km along the R482 and then take a right hand turn down a small road. This brings you to a small, cliff-top car park.

Stop 2. *Spanish Point (UTM 469963 m E, 5855410 m N; Map 57, 102626 m E, 178347 m N)*

From the car park, walk down a short ramp to a broad foreshore outcrop that, even close to high tide, provides excellent exposures of the topmost Doonlicky Cyclothem (Fig. 11.2.1). To the west of this location, at the Spanish Point headland, the cyclothem consists of spectacularly-deformed siltstones and sandstones that are both slumped and faulted. Rider's (1969) unpublished work documents this location in detail with Rider (1978) and Gill (1979, Fig. 11.2.4) presenting the highlights. Syn-sedimentary structures include a broad, half-graben-like, collapse structure suggesting downslope failure to the east, whilst the smaller-scale features include slumps and a series of growth faults in which over-thickened hangingwall sandstones are present.

To the east of this tract of syn-sedimentary deformation, immediately below the car park, the highest beds of the Doonlicky Cyclothem consist of a series of undisturbed, sandstone ribs. These are capped by a distinctive, metre-thick, white sandstone bed that is well cemented in its upper part. This is one of the best examples of an interfluve horizon in the Namurian basin fill and has been studied by Davies & Elliott (1996) and Hampson et al. (1997) who describe and illustrate its pedogenic features (Fig. 11.2.5). Within the sandstone, there is a wispy fabric of rootlets and the top surface is irregular, displaying polygonal patterns typical of well-drained, mature soils. The roots suggest that there was also a water-logged phase in the soil development and it is this complex history that is typical of interfluve surfaces. Such surfaces record an initial phase of emergence during which vegetation can develop, but better drainage ensues because the area is left 'high and dry' as valley cutting occurs during base-level fall.

The succeeding strata record a major fining-up and deepening event, beginning with several beds of well-cemented, fine, dark sandstone. The uppermost bed has been incised by a straight, deep runnel or groove of enigmatic origin (Fig. 11.2.6). Four metres above the palaeosol, a 1.5 m thick black shale, with an abundant fauna of goniatites and the bivalve *Dunbarella*, is well-exposed on the intertidal platform. This is the only exposure of the *Reticuloceras bilingue* Marine Band on the County

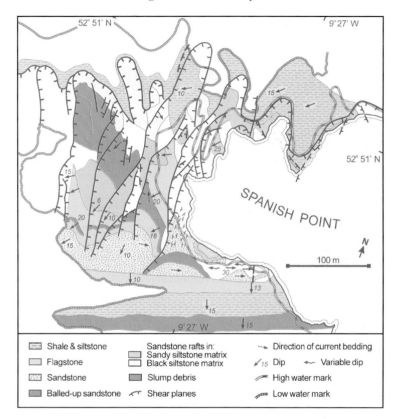

Fig. 11.2.4. Foreshore map of Spanish Point (Stop 2) showing the extensive soft-sediment deformation in the middle of the Doonlicky Cyclothem (redrawn from Gill, 1979; copyright Geological Society of Ireland).

Clare coast (but it is an excellent one) and it marks the base of Cyclothem IV. Much of the fauna has been pyritized and some goniatites are only partially compacted.

Transfer to Stop 3

At low tide it is possible to walk from Stop 2 to the next location at Black Rock along the rocky beach that passes beneath the Armada Hotel. If this foreshore stroll is chosen, then it is possible to examine the intense soft-sediment deformation present in the lower part of Cyclothem IV. The styles of deformation here have been recorded in Rider's (1969) unpublished study and include small graben structures, intense 'balling' of some

Fig. 11.2.5. Uppermost beds of the Doonlicky Cyclothem, Stop 2. The figure in a red jacket is standing on a well-developed, sandy palaeosol (white arrow) that contains abundant rootlets and stigmarian roots.

Fig. 11.2.6. Topmost sandstone bed of the Doonlicky Cyclothem at Stop 2 showing a deeply incisive runnel. In the background upper left, the figure in the red jacket is standing on the *Reticuloceras bilingue* Marine Band (white arrow).

sandstone horizons and numerous examples of small-scale slump structures resting on listric fault surfaces, that Rider (1969) termed 'dish slumps'. This extensive zone of soft-sediment deformation and slope failure would repay further study.

The alternative drive to Stop 3 involves rejoining the main road. Pass the Armada Hotel and turn right at the next junction. After a short distance the main car park at Spanish Point appears on the right.

Stop 3. Black Rock (UTM 470870 m E, 5854895 m N; Map 57, 103526 m E, 177820 m N)

The section at Black Rock begins immediately below the Spanish Point car park and stretches into the northern corner of the bay below the Armada Hotel. Immediately below the hotel, a series of sheet sandstones displays beautiful linguoid ripples on extensive bedding surfaces that record a south-westerly flow (Fig. 11.2.7). Along strike to the west, these beds become progressively deformed and pass into the zone of intense soft-sediment deformation described above. Two, siltstone-dominated, coarsening-up cycles (parasequences) overlie this level and culminate in several beds of wave-rippled, sheet sandstones. These

Fig. 11.2.7. Linguoid rippled surface in upper Cyclothem IV below the Armada Hotel, Stop 3, Spanish Point.

form extensive outcrops, including the two 'finger-like' promontories that are the eponymous 'black rocks' of this location. The higher levels of both parasequences display lovely examples of small-scale load features that Gill (1979, his plates 4 to 6) illustrated as the best example of their kind in County Clare (Fig. 11.2.8). The upper parasequence has dark-grey, pyritic shale at its base, which resembles typical marine band facies, although no marine fossils are present. The topmost sandstone beds of both parasequences are intensely bioturbated by diverse trace fossils including *Zoophycos*, *Monocraterion*, *Palaeophycus* and an ichnotaxon resembling a slender variant of *Thalassinoides* with pellet-filled burrows (Fig. 11.2.9): a rare trace fossil in the County Clare stratigraphy.

The stratigraphy of Spanish Point bay is folded into a syncline with the result that the parasequences seen in the upper part of Cyclothem IV at Black Rock can be seen again in the southern limb in the southern corner of the bay. Reaching there involves a pleasant walk along the sandy beach south of Black Rock, although expect to get your feet wet when crossing the mouth of the Annagh River. At higher tides it becomes very difficult to cross this river and the southern section becomes essentially

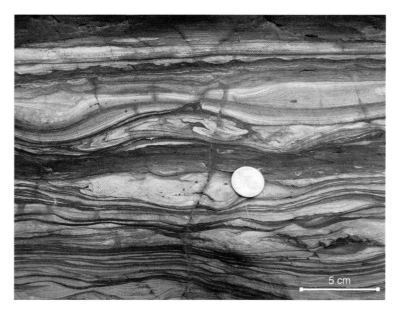

Fig. 11.2.8. Small-scale load structures in wave-rippled, flaser-bedded, strata of upper Cyclothem IV at the Black Rock reef, Stop 3, Spanish Point.

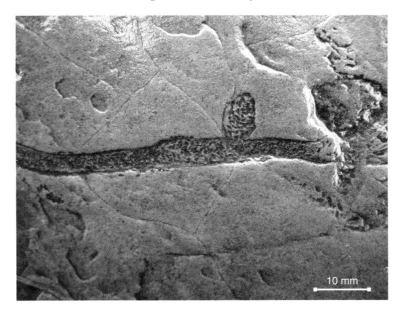

Fig. 11.2.9. Branching burrows (aff. *Thalassinoides*) with pelleted infill and clay lining. This burrow is seen in the highest sandstone of Cyclothem IV at Black Rocks, Stop 3, Spanish Point.

inaccessible. If the section is reached, then the small-scale cycles of Black Rock are seen once again, with the added bonus that an overlying, third parasequence can also be seen. This third parasequence has a well-developed marine band at its base, containing abundant examples of the zone fossil *Reticuloceras wrighti* and, at the top, a sandstone bed with pedogenic features. These are the basal beds of Cyclothem V, and above this basalmost parasequence the sediments pass upwards into a distinctly non-cyclic, thick succession of siltstones that form the core of the syncline seen in the cliffs in the central part of the bay either side of the mouth of the Annagh River. The highest parts of Cyclothem V can be examined at the next location.

Transfer to Stop 4

Turn right from the Spanish Point car park and follow the road round to the T-junction by the Bellbridge House Hotel. Turn right onto the N67 and drive 3 km to the small town of Quilty where you turn right onto a small coast road. After 3 km, park next to the beach at Cloghauninchy.

Stop 4. Cloghauninchy (UTM 467230 m E, 5849075 m N; Map 57, 099804 m E, 172048 m N)

The youngest Namurian rocks of County Clare are seen in the extensive foreshore sections at this location, with plenty to examine even at high tide. A major sandstone body dominates the lower part of the section, which consists of tabular beds that form over a kilometre of strike section at low tide. The upper part of the section forms the southern part of this extensive reef and the headland at Cloghauninchy. The beds here are arranged into small-scale, sandstone-dominated, cycles that record rapidly changing depositional conditions and an overall upward increase in marine influence. Oscillation ripples and linguoid climbing ripples dominate the sedimentary structures and indicate a combination of basinal and fluvial processes, possibly in a distal mouth-bar setting. A single, metre-thick, set of tabular cross-strata is also present. This bed records a south-westerly flow that is broadly in agreement with the other unidirectional ripple flow indicators that are to the west.

The two highest cycles are capped by palaeosols and are overlain by retrogradational strata: dark-grey, bioturbated, fine-grained sandstones with occasional chert nodules. The trace fossil diversity in these beds is considerable and includes *Zoophycos*, *Rhizocorallium* and pyritized *Planolites*. Above the highest parasequence, the outcrop passes southwards into a boulder-covered beach but careful searching amongst the boulders reveals *in-situ* dark shales and, for a height of at least 4 m above the topmost sandstone, these yield abundant specimens of *Reticuloceras superbilingue*. Many of the shale boulders on the beach are derived from this marine band and, more rarely, septarian concretions with beautifully-preserved goniatites are also encountered amongst the beach pebbles.

Appendix: List of *GigaPan* Images

Gigapan images of several outcrops are provided in the companion website to this field guide. A listing of these images, and their corresponding figures in the guide, are provided below.

Figure 8.2.10. The Point of Relief (Chapter 8; Page 213; Stop 1f).

Figure 9.3.2A. Photomontage of the Tullig Cyclothem, Stop 1a at Trusklieve (Chapter 9.3, Page 255).

Figure 9.4.1. The island of Illaunonearaun as viewed from the coastal road (Chapter 9.4, Page 274).

Figure 9.4.3. Photomontage of the spectacular cliffs at Foohagh Point, which reveal a series of syn-sedimentary faults within the Doonlicky Cyclothem (Chapter 9.4, Page 276).

Figure 9.5.2. Photomontage of outcrop by Diamond Rocks café, Kilkee (Chapter 9.5, Page 279, Stop 1).

Figure 9.5.5. Photomontage of the growth fault complex, Kilkee (Chapter 9.5, Page 282, Stop 1).

Figure 9.5.6. Mud/siltstone diapir in the upper Tullig Cyclothem, Kilkee (Chapter 9.5, Page 284, Stop 2).

Figure 9.6.12. Growth fault in transgressive Tullig Cyclothem at Killard (Chapter 9.6, Page 302, Stop 3).

Figure 9.7.11. Photomontage of channels and accretion surfaces in the Tullig Sandstone, Mutton Island (Chapter 9.7, Page 323, Stop 3).

A Field Guide to the Carboniferous Sediments of the Shannon Basin,
Western Ireland, First Edition. Edited by James L. Best and Paul B. Wignall.
© 2016 International Association of Sedimentologists.
Published 2016 by John Wiley & Sons, Ltd.
Companion website: www.wiley.com/go/best/shannonbasin

References

Abreu, V., Sullivan, M., Pirmez, C. and Mohrig, D. (2003) Lateral accretion packages (LAPs): an important reservoir element in deep water sinuous channels. *Mar. Petrol. Geol.*, **20**, 631–648.

Ancient Origins (2015) *In Search of the San Marcos of the Spanish Armada*, http://www.ancient-origins.net/news-history-archaeology/search-san-marcos-spanish-armada-002071, accessed March 30th 2015.

Anderson, T.B. and Oliver, G.J.H. (1996) Xenoliths of Iapetus suture mylonites in County Down lamprophyres, Northern Ireland. *Journal of the Geological Society of London*, **153**, 403–407.

Ashby, D.F. (1939) The geological succession and petrology of the Carboniferous volcanic area of Co. Limerick. *Proc. Geol. Assoc.*, **50**, 324–330.

Austin, R.L. and Husri, S. (1974) Dinantian conodont faunas of County Clare, County Limerick and County Leitrim. In: *International Symposium on Belgian Micropaleontological Limits, from Emsian to Viséan, Namur, 1974*, (Eds. J. Bouckaert and M. Streel), Geol. Surv. Belgium, **3**, 18–69.

Austin, R.L. Conil, R. and Husri, S. (1970) Correlation and age of the Dinantian rocks north and south of the Shannon, Ireland. *Congrès Collection de l'Université de Liège*, **55**, 179–192.

Bakken, B. (1987) Sedimentology and syndepositional deformation of the Ross Slide, western Irish Namurian Basin, Ireland. Unpublished Thesis, University of Bergen, Norway, 182.

Barham, M., Murray, J., Sevastopulo, G.D. and Williams, D.M. (2015) Conodonts of the genus *Lochreia* in Ireland and the recognition of the Viséan-Serpukhovian (Carboniferous) boundary. *Lethaia*, **48**, 151–171, DOI:10.1111/let.12096.

Braithwaite, K. (1993) Stratigraphy of a Mid-Carboniferous section at Inishcorker, Ireland. Annales de la Société géologique de Belgique, **116**, 209–219.

Brennand, R.P. (1965) The Upper Carboniferous (Namurian) stratigraphy north-east of Castleisland, County Kerry, Ireland. *Proc. Roy. Irish Acad.*, **64**, sect. B, 41–63.

Bresser, G. (2000) An integrated structural analysis of the SW Irish Variscides. *Aachener Geowissenschaftliche Beitrage*, Band 35.

Bresser, G. and Walter, R. (1999) A new structural model for the SW Irish Variscides. The Variscan front of the NW European Rhenohercynian. *Tectonophysics*, **309**, 197–209.

Chapin, M.A. (2007) Sheets and incised channels of the Kilcloher cliff section, Ross Formation, Ireland. In: *Atlas of Deep-Water Outcrops*, (Eds T.H. Nelsen, R.D. Shew, G.S. Steffens and J.R.J. Studlick), AAPG Studies in Geology, **56**, 196–200.

Chapin, M.A., Davies, P., Gibson, J. and Pettingill, H.S. (1994) Reservoir architecture of turbidite sheet sandstones in laterally extensive outcrops, Ross Formation, western

Ireland. In: *Deep-Water Reservoirs of the World*, (Eds P. Weimer, R.M. Slatt, J. Coleman, N.C. Rossen, H. Nelson, A.H. Bouma, M.J. Styzen and D.T. Lawrence), SEPM Spec. Publ. Gulf Coast Section, 53–68.

Chew, D.M. (2009) The Grampian Orogeny. In: *The Geology of Ireland*. (Eds C.H. Holland and I.S. Sanders, I.S.), pp. 69–94, Dunedin Academic Press, Edinburgh.

Chew, D.M. and Stillman, C.J. (2009) Acadian Orogeny and late Caledonian magmatism. In: *The Geology of Ireland*. (Eds. C.H.Holland and I.S. Sanders, I.S.), pp. 143–174, Dunedin Academic Press, Edinburgh.

Clayton, G. and Baily, H. (1999) Organic maturation levels of pre-Westphalian Carboniferous rocks in Ireland and in the Irish offshore. *N. Jahrb. Geol. Paläontol.*, **107**, 25–41.

Clayton, G., Johnston, I.S., Sevastopulo, G.D. and Smith, D.G. (1980) Micropalaeontology of a Courceyan (Carboniferous) borehole section from Ballyvergin, County Clare, Ireland. *J. Earth Sci. Royal Dublin Soc.*, **3**, 81–100.

Clayton, G., Haughey, N., Sevastopulo, G.D. and Burnett, R. (1989) *Thermal maturation levels in the Devonian and Carboniferous rocks in Ireland*. Geological Survey of Ireland, 36 pp.

Coller, D.W. (1984) Variscan structures in the Upper Palaeozoic rocks of west-central Ireland. In: *Variscan tectonics of the North Atlantic region*, (Eds D.H.W. Hutton and D.J. Sanderson) Spec. Publ. Geol. Soc., London, **14**, 185–194.

Collinson, J.D., Martinsen, O., Bakken, B. and Kloster, A. (1991) Early fill of the western Irish Namurian basin: a complex relationship between turbidites and deltas. *Basin Res.*, **3**, 223–242.

Cózar, P. and Somerville, I.D. (2005) Stratigraphy of the upper Viséan carbonate platform rocks in the Carlow area, southeast Ireland. *Geological Journal* **40**, 35–64.

Croker, P. F. (1995) The Clare basin: a geological and geophysical outline. In: *The Petroleum Geology of Ireland's Offshore Basins*, (Eds P.F. Croker and P.M. Shannon), Geol. Soc. London Spec. Publ., **93**, 327–339.

D'Arcy, G. and Hayward, J. (1991) *The Natural History of the Burren*, 2nd Ed., Immel Publishing, London, 168 pp.

Davies, S.J. and Elliott, T. (1996) Spectral gamma ray characterization of high resolution sequence stratigraphy: examples from Upper Carboniferous fluvio-deltaic systems, County Clare, Ireland. In: *High Resolution Sequence Stratigraphy: Innovations and Applications*, (Eds J.A. Howell and J.F. Aitken), Spec. Publ. Geol. Soc., London, **104**, 25–35.

Davies, S.G., Guion, P.D. and Gutteridge, P. (2012) Carboniferous sedimentation and volcanism on the Laurussian Margin. In: *Geological History of Britain and Ireland*. Second Edition, (Eds N. Woodcock & R. Strachan), pp. 233–273, Wiley-Blackwell, Oxford.

Debacker, T.N. and De Meester, E. (2009) A regional, S-dipping late Early to Middle Ordovician palaeoslope in the Brabant Masif, as indicated by slump folds (Anglo-Brabant Deformation Belt, Belgium). *Geologica Belgica*, **12**, 145–159.

Deeny, D.E. (1982) Further evidence for Devon-Carboniferous rifting in central Ireland. *Geol. Mijn.*, **61**, 243–252.

Devuyst, F.X. and Lees, A. (2001) The initiation of Waulsortian buildups in Western Ireland. *Sedimentology*, **48**, 1121–1148.

Dewey, J.F. (2005) Orogeny can be very short. *Proc. Nat. Acad. Sci.*, **102**, 15286–15293.

Dewey, J.F. and Strachan, R.A. (2003) Changing Silurian-Devonian relative plate motion in the Caledonides: sinistral transpression to sinistral transtension. *J. Geol. Soc., London*, **160**, 219–229.

Dolan, J.M. (1984) A structural cross-section across through the Carboniferous of north-west Kerry. *Irish J. Earth Sci.*, **6**, 95–108.

Douglas, J.A. (1909) The Carboniferous Limestone of County Clare (Ireland). *Quart. J. Geol. Soc. London*, **65**, 538–586.

Eggenhuisen, J.T., McCaffrey, W.D., Haughton, P.D.W. and Butler, R.W.H. (2011) Shallow erosion beneath turbidity currents and its impact on the architectural development of turbidite sheet systems. *Sedimentology*, **58**, 936–959.

Elliott, T. (2000a) Depositional architecture of a sand-rich, channelized turbidite system: the upper Carboniferous Ross Sandstone Formation, western Ireland. In: *Deep-Water*

Reservoirs of the World, (Eds. P. Weimer, R.M. Slatt, J. Coleman, N.C. Rossen, H. Nelson, A.H. Bouma, M.J. Styzen, M.J. and D.T. Lawrence), SEPM Spec. Publ. Gulf Coast Section, 342–373.

Elliott, T. (2000b) Megaflute erosion surfaces and the initiation of turbidite channels. *Geology*, **28**, 119–122.

Fallon, P. and Murray, J. (2015) Conodont biostratigraphy of the mid-Carboniferous boundary in Western Ireland, *Geological Magazine*, **152**, 1025–1046.

Fisher, R.V. (1983) Flow transformations in sediment gravity flows. *Geology*, **11**, 273–274.

Fitzgerald, E., Feely, M., Johnston, J.D., Clayton, Fitzgerald, L.J. and Sevastopulo, G.D. (1994) The Variscan thermal history of west Clare, Ireland. *Geol. Mag.*, **131**, 545–558.

Ford, M., Brown, C. and Readman, P. (1991) Analysis and tectonic interpretation of gravity data over the Variscides of southwest Ireland. *J. Geol. Soc., London*, **148**, 137–148.

Ford, M., Klemperer, S.L. and Ryan, P.D. (1992) Deep structure of southern Ireland: a new geological synthesis using BIRPS deep reflection profiling. *J. Geol. Soc., London*, **149**, 915–922.

Frey Martinez, J., Cartwright, J. and James, D.M.D. (2006) Frontally confined versus frontally emergent submarine landslides: A 3D seismic characterisation. *Mar. Petrol. Geol.*, **23**, 585–604.

Friend, P.F., Slater, M.J. and Williams, R.C. (1979) Vertical and lateral building of river sandstone bodies, Ebro Basin, Spain. *J. Geol. Soc., London*, **136**, 39–46.

Gallagher, S.J. (1992) Lithostratigraphy, biostratigraphy and palaeoecology of the upper Viséan platform carbonates in parts of southern and western Ireland. Volumes **1 and 2**, 503 pp. *Unpublished Ph. D. thesis*, National University of Ireland.

Gallagher, S.J. (1996) The stratigraphy and cyclicity of the late Dinantian platform carbonates in parts of southern and western Ireland. In: *Recent Advances in Lower Carboniferous Geology*, (Eds. P. Strogen, I.D. Somerville and G. Ll. Jones), Geol. Soc. London Spec. Publ., **107**, 239–251.

Gallagher, S.J. and Somerville I.D. (1997) Late Dinantian (Lower Carboniferous) platform carbonate stratigraphy of the Buttevant area North Co. Cork, Ireland. *Geol. J.*, **32**, 313–335.

Gallagher, S.J. and Somerville, I.D. (2003) Lower Carboniferous (late Viséan) platform development and cyclicity in southern Ireland; foraminiferal biofacies and lithofacies evidence. *Riv. Ital. Paleont. Strat.*, **109**, 159–171.

Gallagher, S.J., MacDermot, C.V., Somerville, I.D., Pracht, M. and Sleeman, A.G. (2006) Biostratigraphy, microfacies and depositional environments of Upper Viséan limestones from the Burren region, Co. Clare, Ireland. *Geol. J.*, **41**, 61–91.

Gill, D.W. (1979) Syndepositional sliding and slumping in the West Clare Namurian Basin, Ireland. *Geol. Surv. Ireland, Spec. Pap.*, **4**, 1–30.

Gill, D.W. and Kuenen, P.H. (1958) Sand volcanoes in slumps in the Carboniferous of County Clare, Eire. *Q. J. Geol. Soc. London*, **113**, 414–460.

Goodhue, R. and Clayton, G. (1999) Organic maturation levels, thermal history and hydrocarbon source rock potential of the Namurian rocks of the Clare Basin, Ireland. *Mar. Petrol. Geol.*, **16**, 667–675.

Graham, J.R. (2009a) Ordovician of the north. In: *The Geology of Ireland*. (Eds. C.H.Holland and I.S. Sanders, I.S.), pp. 43–68, Dunedin Academic Press, Edinburgh.

Graham, J.R. (2009b) Devonian. In: *The Geology of Ireland*. (Eds. C.H.Holland and I.S. Sanders, I.S.), pp. 175–214, Dunedin Academic Press, Edinburgh.

Graham, J.R. (2009c) Variscan deformation and metamorphism. In: *The Geology of Ireland*. (Eds. C.H.Holland and I.S. Sanders, I.S.), pp. 295–310, Dunedin Academic Press, Edinburgh.

Graham, J.R. and Stillman, C.J. (2009) Ordovician south of the Iapetus suture. In: *The Geology of Ireland*. (Eds. C.H.Holland and I.S. Sanders, I.S.), pp. 103–118, Dunedin Academic Press, Edinburgh.

Hampson, G.J., Elliott, T. and Davies, S.J. (1997) The application of sequence stratigraphy to Upper Carboniferous fluvio-deltaic strata of the onshore UK and Ireland: implications for the southern North Sea. *J. Geol. Soc., London*, **154**, 719–733.

Hansen, E. (1971) *Strain Facies*. Springer-Verlag, Berlin, 207 pp.

Haszeldine, R.S. (1984) Carboniferous North Atlantic palaeogeography: stratigraphic evidence for rifting, not megashear or subduction. *Geol. Mag.*, **121**, 443–463.

Haughton, P., Davis, C., McCaffrey, W. and Barker, S. (2009) Hybrid sediment gravity flow deposits: classification, origin and significance. *Mar. Petrol. Geol.*, **26**, 1900–1918.

Heckel, P.H. and Clayton, G. (2006) The Carboniferous system, use of the new official names for the subsystems, series and stages, *Geologica Acta*, **4**, 403–407.

Hiemstra, J. F., Evans, D. J. A. & O´Cofaigh, C. (2007) The role of glacitectonic rafting and comminution in the production of subglacial tills: examples from southwest Ireland and Antarctica. *Boreas*, **36**, 386–399. Oslo.

Higgs, R. (2004) Ross and Bude Formations (Carboniferous, Ireland and England); reinterpreted as lake-shelf deposits. *J. Petrol. Geol.*, **27**, 47–66.

Higgs, R. (2009) Multiscale stratigraphic analysis of a structurally confined submarine fan; Carboniferous Ross Sandstone, Ireland; discussion. *AAPG Bull.*, **93**, 1705–1709.

Higgs, K.T., Clayton, G. and Keegan, J.B. (1988) Stratigraphic palynology of the Lower Tournaisian rocks of Ireland. *Geol. Surv. Ireland, Spec. Pap.*, **7**, 1–93.

Hill, P.J. (1971) Carboniferous conodonts from Southern Ireland. *Geol. Mag.*, **108**, 69–71.

Hodson, F. (1954a) The beds above the Carboniferous limestone in north-west County Clare, Eire. *Q. J. Geol. Soc. London*, **109**, 259–283.

Hodson, F. (1954b) The Carboniferous rocks of Foynes Island, County Limerick. *Geol. Mag.*, **2**, 153–160.

Hodson, F. (1959) The palaeogeography of Namurian times in western Europe. *Bull. Soc. Belge Géol. Paléont. d'Hydrologie*, **68**, 134–150.

Hodson, F. and Lewarne, G.C. (1961) A mid-Carboniferous (Namurian) basin in parts of the counties of Limerick and Clare, Ireland. *Q. J. Geol. Soc. London*, **117**, 307–333.

Holland, C.H. (1981) *A Geology of Ireland*. Scottish Academic Press, Edinburgh, 335 pp.

Holland, C.H. and Sanders, I.S. (Eds.) (2009) *The Geology of Ireland*, 2nd Ed., Dunedin Academic Press, Edinburgh, 568 pp.

Hudson, R.G.S. and Philcox, M.E. (1966) The Lower Carboniferous stratigraphy of the Buttevant Area, Co. Cork. *Proc. Royal Irish Acad. Section B*, **64**, 65–79.

Hudson, R.G.S. and Sevastopulo, G.D. (1966) A borehole section through the Lower Tournaisian and Upper Old Red Sandstone, Ballyvergin, Co. Clare. *Sci. Proc. Royal Dublin Soc.*, **A2**, 287–296.

Hudleston, P.J. (1973) Fold morphology and some geometrical implications of theories of fold development. *Tectonophysics*, **16**, 1–46.

Jones, C. (2004) *The Burren and the Aran Islands, Exploring the Archaeology*, Collins Press, Cork, 268 pp.

Jones, C. (2007) *Temples of Stone: Exploring the Megalithic Tombs of Ireland*, Collins Press, Cork, 334 pp.

Jones, G. Ll. and Somerville, I.D. (1996) Irish Dinantian Biostratigraphy: practical application. In: *Recent Advances in Lower Carboniferous Geology* (Eds Strogen, P., Somerville, I.D. & Jones, G.Ll.), Geol. Soc. London Spec. Publ., **107**, 371–385.

Jones, O.T. (1939) The geology of the Colwyn Bay district: a study of submarine slumping during the Salopian period. *Q. J. Geol. Soc. London*, **380**, 335–382.

Kelk, B. (1960) *Studies in the Carboniferous Stratigraphy of Western Eire*. Ph.D thesis. University of Reading.

Klemperer, S.L., Ryan, P.D. and Snyder, D.B. (1991) A deep reflection transect across the Irish Caledonides. *J. Geol. Soc., London*, **148**, 149–164.

Kneller, B. (1995) Beyond the turbidite paradigm; physical models for deposition of turbidites and their implications for reservoir prediction, In: *Characterization of deep marine clastic systems* (Eds A.J. Hartley and D.J. Prosser), Geol. Soc. London Spec. Publ., **94**, 31–49.

Landes, M., Prodehl, C., Hauser, F., Jacob, A.W.B. and Vermeulen, N.J. (2000) VARNET-96: influence of the Variscan and Caledonian orogenies on crustal structure in SW Ireland. *Geophys. J. Internat.*, **140**, 660–676.

Landes, M., Ritter, J.R.R., Readman, P.W. and O'Reilly, B.M. (2005) A review of the Irish crustal structure and signatures from the Caledonian and Variscan orogenies. *Terra Nova*, **17**, 111–120.

LeClair, S.F. and Bridge, J.S. (2001) Quantitative interpretation of sedimentary structures formed by river dunes, *Journal of Sedimentary Research*, **71**, 713–716.

Le Gall, B. (1991) Crustal evolutionary model for the Variscides of Ireland and Wales from SWAT seismic data. *Journal of the Geological Society of London*, **148**, 759–774.

Lees, A. (1961) The Waulsortian "reefs" of Eire; a carbonate mudbank complex of lower Carboniferous age. *J. Geol.*, **69**, 101–109.

Lees, A. and Miller, J. (1985) Facies variation in Waulsortian buildups: Part 2: Mid Dinantian buildups from England and North America. *Geol. J.*, **20**, 159–180.

Lees, A. and Miller, J. (1995) Waulsortian Banks. In: *Carbonate Mud-mounds, their Origin and Evolution*, (Eds C.V. Monty, D.W.J. Bosence, P.H. Bridges, and B.R. Pratt), Spec. Publ. Int. Assoc. Sedimentologists, **23**, 191–271.

Levine, J.R. and Davis, A. (1984) Optical anisotropy of coals as an indicator of tectonic deformation, Broad Top Coal Field, Pennsylvania. *Bull. Geol. Soc. America*, **95**, 100–108.

Levine, J.R. and Davis, A. (1989) The relationship of coal optic fabrics to Alleghanian tectonic deformation in the central Appalachian fold-and-thrust belt, Pennsylvania. *Bull. Geol. Soc. America*, **101**, 1333–1347.

Lewarne, G.C. (1959) The junction of the Upper and Lower Carboniferous of Co. Clare and Co. Limerick, Eire. Unpublished PhD Thesis, University of Reading, U.K, 155.

Lewis, K.B. (1971) Slumping on a continental slope inclined at 1-4°. *Sedimentology*, **16**, 97–110.

Lien, T., Martinsen, O.J. and Walker, R. (2007) An overview of the Ross Formation, Shannon Basin, western Ireland. In: *Atlas of Deep-Water Outcrops*, (Eds. T.H. Nilsen, R.D. Shew, G.S. Steffens, G.S. and J.R.J. Studlick), AAPG Studies in Geology, **56**, 192–195.

Lien, T., Walker, R.G. and Martinsen, O.J. (2003) Turbidites in the Upper Carboniferous Ross Formation, western Ireland: reconstruction of a channel and spillover system. *Sedimentology*, **50**, 113–148.

Macdonald, H., Peakall, J., Wignall, P.B. and Best, J. (2011) Sedimentation in deep-sea lobe-elements; implications for the origin of thickening-upward sequences. *J. Geol. Soc. London*, **168**, 319–332.

Martinsen, O.J. (1989) Styles of soft-sediment deformation on a Namurian delta slope, Western Irish Namurian Basin, Ireland. In: *Deltas: Sites and Traps for Fossil Fuels*, (Eds. M.K.G. Whateley and K.T. Pickering), Geol. Soc. London Spec. Publ. **41**, pp. 167–177.

Martinsen, O.J. and Bakken, B. (1990) Extensional and compressional zones in slumps and slides in the Namurian of County Clare, Eire. *J.Geol. Soc., London*, **147**, 153–164.

Martinsen, O.J. and Collinson, J.D. (2002) The Western Irish Namurian Basin reassessed - a discussion. *Basin Research*, **14**, 523–531.

Martinsen, O.J., Lien, T. and Walker, R.G. (2000) Upper Carboniferous deep water sediments, Western Ireland: Analogues for passive margin turbidite plays. In: *GCSSEPM Foundation 20th Annual Research Conference Deep-Water Reservoirs of the World*, (Eds. P. Weimer, R.M. Slatt, J. Coleman, N.C. Rosen, H. Nelson, A.H. Bouma, M.J. Styzen and D.T. Lawrence), pp. 533–555. GCSSEPM Publications, Houston.

Martinsen, O.J., Lien, T., Walker, R.G. and Collinson, J.D. (2003) Facies and sequential organization of a mudstone-dominated slope and basin floor succession: the Gull Island Formation, Shannon basin, western Ireland. *Mar. Petrol. Geol.*, **20**, 789–807.

McNamara, M.E. and Hennessy, R.W. (2010) *Stone, Water and Ice: The Geology of the Burren Region, Co. Clare, Ireland*, Burren Connect Project, Ennistymon, 31 pp.

Meere, P., MacCarthy, I., Reavy, J., Allen, A. and Higgs, K. (2013) *Geology of Ireland*, 372 pp., Collins Press, Ireland.

Menning, M., Weyer, D., Drozdzewski, G., Van Amerom, H.W.J. and Wendt, I. (2000) A Carboniferous time scale 2000: Discussion and use of geological parameters as time indicators from central and western Europe. *Geolog. Jahr.*, **A156**, 3–44.

Miall, A.D. (1996) *The Geology of Fluvial Deposits: Sedimentary Facies, Basin Analysis and Petroleum Geology*, Springer-Verlag, Berin Heidelberg, 582 pp.

Middleton, G.V. and Hampton, M.A. (1973) Sediment gravity flows; mechanics of flow and deposition. In: *Turbidites and Deep-Water sedimentation: lecture notes for a Short Course*, (Co-chairmen G.V. Middleton and A.H. Bouma) *SEPM Pac. Sect.*, 1–38.

Morton, W.H. (1965) The Carboniferous stratigraphy of the area north-west of Newmarket, Co. Cork, Ireland. *Sci. Proc. Roy. Dublin Soc.*, **2A**, 47–64.

Nelson, C. (2008) *The Wild Plants of the Burren & the Aran Islands: A Field Guide*, Collins Press, Wilton, Cork, 160 pp.

Nenna, F. and Aydin, A. (2011a) The formation and growth of pressure solution seams in clastic rocks: a field analytical study. *J. Struc. Geol.*, **33**, 633–643.

Nenna, F. and Aydin, A. (2011b) The role of pressure solution seam and joint assemblages in the formation of strike-slip and thrust faults in a compressive tectonic setting; the Variscan of south-western Ireland. *J. Struc. Geol.*, **33**, 1595–1610.

O'Brien, M.V. (1953) Phosphatic horizons in the Upper Carboniferous of Ireland. Proc. 19th Int. Geol. Congress, Algeria, 135–143.

Obrock, E.P. (2011) *Anatomy of a Carboniferous transgression: upper Tullig Cyclothem, County Clare, Ireland,* Unpublished MS thesis, University of Illinois at Urbana-Champaign, USA, 229 pp.

Ordnance Survey of Ireland (2001) *Discovery Series* **63**, County Clare and County Kerry, 2nd ed., 1:50 000.

O'Reilly, B.M., Hauser, F. and Readman, P.W. (2010) The fine-scale structure of upper continental lithosphere from seismic waveform methods: insights into Phanerozoic crustal deformation processes. *Geophys. J. Int.*, **180**, 101–124.

Philcox, M.E. (1961) Namurian shales near Buttevant, north County Cork. *Sci. Proc. Roy. Dublin Soc.*, **A**(1), 205–209.

Philcox, M.E. (1984) Lower Carboniferous lithostratigraphy of the Irish Midlands. Irish Assoc. Econ. Geol. Dublin, 1–89.

Phillips, W.E.A. (2001) Caledonian deformation. In: *The Geology of Ireland.* (Ed. C.H. Holland), pp. 179–199. Dunedin Academic Press, Edinburgh.

Phillips, W.E.A., Stillman, C.J. and Murphy, T. (1976) A Caledonian plate tectonic model. *J. Geol. Soc., London*, **132**, 579–609.

Pointon, M.A., Cliff, R.A. and Chew, E.M., (2012) The provenance of Western Irish Namurian Basin sedimentary strata inferred using detrital zircon U–Pb LA-ICP-MS geochronology, *Geological Journal*, **47**, 77–98.

Pracht, M. (1996) *Geology of Dingle Bay.* Geological Survey of Ireland, Dublin.

Pracht, M., Lees, A., Leake, B., Feely, M., Long, B., Morris, J.H. and McConnell, B. (2004) *Geology of Galway Bay: A Geological Description of the Galway Bay Area*, with accompanying Bedrock Geology 1:100,000 Scale Map, Sheet 14, Geological Survey of Ireland, Dublin.

Prélat, A., Hodgson, D.M. and Flint, S.S. (2009) Evolution, architecture and hierarchy of distributary deep-water deposits: a high-resolution outcrop investigation from the Permian Karoo Basin, South Africa. *Sedimentology*, **56**, 2132–2154.

Price, C.A. and Todd, S.P. (1988) A model for the development of the Irish Variscides, *J. Geol. Soc., London*, **145**, 935–939.

Pringle, J.K., Clark, J.D., Westerman, A.R. and Gardiner, A.R. (2003) The use of GPR to image three-dimensional (3-D) turbidite channel architecture in the Carboniferous Ross Formation, County Clare, western Ireland. In: *Ground Penetrating Radar in Sediments*, (Eds. C.S. Bristow and H.M. Jol), Geol. Soc. London Spec. Publ., **211**, 315–326.

Pulham, A.J. (1989) Controls on internal structure and architecture of sandstone bodies within Upper Carboniferous fluvial-dominated deltas, County Clare, western Ireland. In: *Deltas: Sites and Traps for Fossil Fuels*, (Eds. M.K.G. Whateley and K.T. Pickering), Geol. Soc. London Spec. Publ., **41**, 179–203.

Pyles, D.R. (2004) On the stratigraphic evolution of a structurally confined submarine fan, Carboniferous Ross Sandstone, western Ireland. *Unpublished PhD dissertation*, University of Colorado, USA.

Pyles, D.R. (2007) Architectural elements in a ponded submarine fan, Carboniferous Ross Sandstone, western Ireland. In: *Atlas of Deep-Water Outcrops*, (Eds. T.H. Nelsen, R.D. Shew, G.S. Steffens and J.R.J. Studlick), AAPG Studies in Geology, **56**, CD-ROM, 19.

Pyles, D.R. (2008) Multiscale stratigraphic analysis of a structurally confined submarine fan: Carboniferous Ross Sandstone, Ireland. *AAPG Bull.*, **92**, 557–587.

Pyles, D.R. (2009) Multiscale stratigraphic analysis of a structurally confined submarine fan, Carboniferous Ross Sandstone, Ireland: Reply: *AAPG Bull.*, **93**, 1710–1721.

Pyles, D.R. and Jennette, D. (2009) Geometry and architectural associations of co-genetic debrite-turbidite beds in basin-margin strata, Carboniferous Ross Sandstone (Ireland): applications to reservoirs located on the margins of structurally confined basins. *Mar. Petrol. Geol.*, **26**, 1974–1996.

Pyles, D.R., Strachan, L.J. and Jennette, D.C. (2014) Lateral juxtapositions of channel and lobe elements in distributive submarine fans: three-dimensional outcrop study of the Ross Sandstone and geometric model. *Geosphere*, **10**, 1104–1122.

Pyles, D.R., Syvitski, J.P.M. and Slatt, R.M. (2011) Defining the concept of stratigraphic grade and applying it to stratal (reservoir) architecture and evolution of the slope-to-basin profile: an outcrop perspective. *Mar. Petrol. Geol.*, **28**, 675–697.

Pyles, D.R., Tomasso, M., and Jennette, D. (2012) Flow processes and sedimentation associated with erosion and filling of sinuous submarine channels. *Geology*, **40**, 143–146.

Pyles, D.R., Jennette, D.C., Tomasso, M., Beaubouef, R.T. and Rossen, C. (2010) Concepts learned from a 3D outcrop of a sinuous slope channel complex: Beacon Channel Complex, Brushy Canyon Formation, West Texas, USA. *J. Sed. Res.*, **80**, 67–96.

Ramsbottom, W.H.C., Calver, M.A., Eagar, R.M.C., Hodson, F., Holliday, D.W., Stubblefield, C.J. and Wilson, R.B. (1978) A correlation of Silesian Rocks in the British Isles. *Special Report Geol. Soc. London*, **10**, 81.

Readman, P.W., Hauser, F., O'Reilly, B.M. and Do, V.C. (2009) Crustal anisotropy in southwest Ireland from analysis of controlled source shear-wave data. *Tectonophysics*, **474**, 571–583.

Rider, M.H. (1969) Sedimentological Studies in the West Clare Namurian, Ireland and the Mississippi River Delta. Unpublished PhD, Imperial College, London.

Rider, M.H. (1974) The Namurian of west County Clare. *Proc. Roy. Irish Acad.*, **74B**, 125–142.

Rider, M.H. (1978) Growth faults in the Carboniferous of Western Ireland. *AAPG Bull.*, **62**, 2191–2213.

Riley, N.J., Claoue-Long, J., Higgins, A.C., Owens, B., Spears, A., Taylor, L. and Varker, W.J. (1993) Geochronometry and geochemistry of the European mid-Carboniferous boundary global stratotype proposal, Stonehead Beck, North Yorkshire, UK. *Annales Soc. Geol. Belgique*, **116**, 275–289.

Rodríguez, S. and Somerville, I.D. (2007) Comparisons of rugose corals from the Upper Viséan of SW Spain and Ireland: implications for improved resolutions in late Mississippian coral biostratigraphy. In: Rasser M, Hubmann B. (eds), *9th International Symposium on fossil Cnidaria including Archaeocyatha and Porifera*. Graz, Austria, 3-7th August 2003. Austrian Academy of Sciences. Schriftenreihe der Erdwissenschaftlichen Kommissionen, **17**, 275–305.

Ross, J.A., Peakall, J. and Keevil, G.M. (2013) Sub-aqueous sand extrusion dynamics. *J. Geol. Soc., London*, **170**, 593–602.

Rutter, E.H. (1976) The kinetics of rock deformation by pressure solution. *Phil Trans. Royal Soc. London*, **A283**, 203–219.

Rutter, E.H. (1978) Discussion on pressure solution. *J. Geol. Soc., London*, **135**, 135.

Sevastopulo, G.D. (1981a) Lower Carboniferous, In: *A Geology of Ireland*, (Ed. C.H. Holland), pp. 147–172, Scottish Academic Press, Edinburgh.

Sevastopulo, G.D. (1981b) Upper Carboniferous, In: *A Geology of Ireland*, (Ed. C.H. Holland), pp. 173–199, Scottish Academic Press, Edinburgh.

Sevastopulo, G.D. (1982) The age and depositional setting of Waulsortian Limestones in Ireland. In: *Symposium on the paleoenvironmental setting and distribution of the Waulsortian Facies*. (Ed. Le Mone, D.V.), 65–79. University of Texas at El Paso and El Paso Geological Society.

Sevastopulo, G.D. (2001) Carboniferous (Silesian). In: The Geology of Ireland, (Ed. C.H. Holland), 289–312. Dunedian Academic Press, Edinburgh.

Sevastopulo, G.D. (2009) Carboniferous: Mississippian (Serpukhovian and Pennsylvanian). In: *The Geology of Ireland*. (Eds C.H. Holland and I.S. Sanders, I.S.), 269–294, Dunedin Academic Press, Edinburgh.

Sevastopulo, G.D. and Wyse Jackson, P.N. (2009) Carboniferous: Mississippian (Tournaisian and Visean). In: *The Geology of Ireland*. (Eds. C.H.Holland and I.S. Sanders, I.S.), pp. 231–268, Dunedin Academic Press, Edinburgh.

Shelford, P.H. (1967) The Namurian and Upper Viséan of the Limerick Volcanic Basin, Eire. *Proc. Geol. Assoc.*, **78**, 121–136.

Shephard-Thorn, E. (1963) The Carboniferous Limestone succession in North-West County Limerick, Ireland. *Proc. Royal Irish Acad.*, **62B**, 267–294.

Sibly, T.F. (1908) The faunal succession in the Carboniferous Limestone (Upper Avonian) of the Midland area (north Derbyshire and North Staffordshire). *Quarterly Journal of the Geological Society of London*, **64**, 34–80.

Simms, M. (2006) *Exploring the Limestone Landscapes of the Burren and Gort Lowlands*, The Universities Press, Belfast, 2nd Ed., ISBN 0-9540892-1-9, 64 pp.

Sinclair, H.D. and Cowie, P.A. (2003) Basin-floor topography and the scaling of turbidites. *J. Geol.*, **111**, 277–299.

Sleeman, A.G. and Pracht, M. (1999) Geology of the Shannon Estuary. Geological Survey of Ireland, Dublin.

Somerville, H.E.A. (1999) Conodont biostratigraphy and biofacies of Upper Viséan rocks in parts of Ireland. Ph.D. thesis, University College Dublin (National University of Ireland).

Somerville, I.D. (2003) Review of Irish Lower Carboniferous (Mississippian) mud-mounds: depositional setting, biota, facies and evolution. In: *Permo-Carboniferous Carbonate Platforms and Reefs*. (Eds W. Ahr, A.P. Harris, W.A. Morgan, and I.D. Somerville) SEPM Spec. Publ 78 and AAPG Memoir **83**, 239–252.

Somerville, I.D. and Jones, G.Ll. (1985) The Courceyan stratigraphy of the Pallaskenry Borehole, County Limerick, Ireland. *Geol. J.*, **20**, 377–400.

Somerville, I.D. and Strogen, P. (1992) Ramp sedimentation in the Dinantian limestones of the Shannon Trough, Co. Limerick, Ireland. *Sed. Geol.*, **79**, 59–75.

Somerville, I.D, Strogen, P. and Jones, G.Ll. (1992) Biostratigraphy of Dinantian limestones and associated volcanic rocks of the East Limerick Syncline. *Geol. J.*, **27**, 201–220.

Somerville, I.D, Waters, C.N. and Collinson, J.D. (2011) South Central Ireland. In: *A Revised Correlation of Carboniferous Rocks in the British Isles* (Eds. C.N. Waters, I.D. Somerville, N.S. Jones et al.). Geol. Soc. London, Spec. Rep., **26**, 144–152.

Stirling, E.J. (2003) *Architecture of fluvio-deltaic sandbodies: the Namurian of C. Clare, Ireland, as an analogue for the Plio-Pleistocene of the Nile Delta*, Unpublished PhD thesis, University of Leeds, UK, 357 pp.

Strachan, L.J. (2002a) Geometry to genesis: A comparative field study of slump deposits and their modes of formation. Unpublished PhD thesis, University of Cardiff.

Strachan, L.J. (2002b) Slump-initiated and controlled syndepositional sandstone remobilization: an example from the Namurian of County Clare, Ireland. *Sedimentology*, **49**, 25–41.

Strachan, L.J. (2008) From slump to turbidite: a case study of flow transformation from the Waitemata Basin, New Zealand. *Sedimentology*, **55**, 1311–1332.

Strachan, L.J. and Alsop, G.I. (2006) Slump folds as estimators of palaeoslope: A case study from the Fisherstreet Slump of County Clare, Ireland. *Basin Res*,. **18**, 451–470.

Straub, K. and Pyles, D.R. (2012) Quantifying the hierarchical and spatial organization of compensation in submarine fans using surface statistics. *J. Sed. Res.*, **82**, 889–898.

Stow, D.A.V. (1986) Deep Sea Clastics, In: *Sedimentary Environments and Facies*, (Ed: Reading, H.G.), 2nd Ed., Blackwell Scientific, Oxford, 399–444.

Strogen, P. (1988) The Carboniferous lithostratigraphy of southeast County Limerick, Ireland, and the origin of the Shannon trough. *Geol. J.*, **23**, 121–137.

Strogen, P., Somerville, I.D., Pickard, N.A.H., Jones, G. and Fleming, M. (1996) Controls on ramp, platform and basinal sedimentation in the Dinantian of the Dublin Basin and Shannon Trough, Ireland. In: *Recent Advances in Lower Carboniferous Geology*, (Eds. P. Strogen, I.D. Somerville and G. Jones), Spec. Publ. Geol. Soc. London, **107**, 263–279.

Sullivan, M., Jensen, G., Goulding, F., Jennette, D., Foreman, L. and Stern, D. (2000) Architectural analysis of deep-water outcrops: implications for exploration and development of the Diana sub-basin, western Gulf of Mexico. In: *Deep-Water Reservoirs of the World*, (Eds. P. Weimer, R.M. Slatt, J. Coleman, N.C. Rossen, H. Nelson, A.H. Bouma, M.J. Styzen and D.T. Lawrence), SEPM Spec. Publ. Gulf Coast Section, 1010–1031.

Talling, P.J., Amy, L.A. Wynn, R.B., Peakall, J, and Robinson, M. (2004) Beds comprising debrite sandwiched within co-genetic turbidite; origin and widespread occurrence in distal depositional environments. *Sedimentology*, **51**, 163–194.

Tanner, D.C., Bense F.A. and Ertl, G. (2011) Kinematic retro-modelling of a cross-section through a thrust-and-fold belt: the Western Irish Namrurian Basin. In: *Kinematic Evolution and Structural Styles of Fold-and-Thrust Belts*, (Eds. J. Poblet and R.J. Lisle), Spec. Publ. Geol. Soc. London., **349**, 61–76.

Tate, M.P. and Dobson, M.R. (1989) Pre-Mesozoic geology of the western and north-western Irish continental shelf. *J. Geol. Soc. London* **146**, 229–240.

Tattershall, J.A. (1963) The geological succession of the Carboniferous Limestone of south County Clare Ireland, west of the Fergus River. *Unpublished Ph.D. thesis*, University of Southampton.

Todd, S.P. (2000) Taking the roof off a suture zone: basin setting and provenance of conglomerates in the ORS Dingle Basin of SW Ireland. In: *New perspectives on the Old Red Sandstone*, (Eds P.F. Friend and B.P.J. Williams), Spec. Publ. Geol. Soc. London, **180**, 185–222.

Todd, S.P., Murphy, F.C. and Kennan, P.S. (1991) On the trace of the Iapetus suture in Ireland and Britain. *J. Geol. Soc., London.* **148**, 869–880.

Van den Berg, R. (2005) Granulite-facies lower crustal xenoliths from central Ireland. Unpublished PhD thesis, University College Dublin

Vaughan, A. (1905) The palaeontological succession in the Carboniferous Limestone of the Bristol area. *Q. J. Geol. Soc. London*, **61**, 181–307.

Vaughan, A.P.M. and Johnston, J.D. (1992) Structural constraints on closure geometry across the Iapetus Suture in eastern Ireland. *J. Geol. Soc., London* **149**, 65–74.

Vermeulen, N.J., Shannon, P.M., Landes, M., Masson, F. and the VARNET group. (1999) Seismic evidence for subhorizontal crustal detachments beneath the Irish Variscides. *Irish J. Earth Sci.*, **17**, 1–18.

Walsh, P.T. (1967) Notes on the Namurian Stratigraphy north of Killarney, Co. Kerry. *The Irish Naturalists' Journal*, **15**, 254–258.

Warr, L.N. (2012) The Varsican orogeny: the welding of Pangaea. In: *Geological History of Britain and Ireland*. Second Edition, (Eds N. Woodcock and R. Strachan), pp. 273–298, Wiley-Blackwell, Oxford.

Wignall, P.B. and Best, J.L. (2000) The western Irish Namurian basin reassessed. *Basin Res.*, **12**, 59–78.

Wignall, P.B. and Best, J.L. (2002) The Western Irish Namurian Basin reassessed-reply. *Basin Res.*, **14**, 531–542.

Wignall, P.B. and Best, J.L. (2004) Sedimentology and kinematics of a large, retrogressive growth-fault system in Upper Carboniferous deltaic sediments, western Ireland. *Sedimentology*, **51**, 1343–1358.

Williams, H. and Soek, H.F. (1993) Predicting reservoir sandbody orientation from dipmeter data: the use of sedimentary dip profiles from outcrop studies, In: *The Geological Modelling of Hydrocarbon Reservoirs and Outcrop Analogues*, (Eds S.S. Flint and I.D. Bryant), Special Publication of the International Association of Sedimentologists, **15**, 143–156.

Woodcock, N.H. (1979) The use of slump structures as palaeoslope orientation estimators. *Sedimentology*, **26**, 83–99.

Woodcock, N. and Strachan, R. (2012) *Geological History of Britain and Ireland*, 2nd Ed., 442 pp., Wiley-Blackwell, Chichester, UK.

Index

Note: Page numbers in *Bold italics* refer to Field Guide Stop.

accommodation, 5, 6
accommodation space, sediments, 68, 119, 173, 214, 237
Ailladie, 80, *88*, 89
Aillwee Member, 80–82, 85, 88–92
allocyclic controls, 47
Alportian, 12, 98, 106, 128, 176
anticline, 26, 71, 149, 150, 300, 306, 308, 310, 311, 315, 335, 340
Aran Islands, 2–4, 8, 10, 14, 25, 90, 93
architectural elements, 121, 123, 124, 126–128, 130, 131, 172, 173, 199, 202, 206, 207
Ardaneer, *62*, 64
Arenicolites, 108, 258, 267
Arnsbergian, 12, 98, 102, 104, 108, 176
Asbian, 12, 13, 50, 65, 66, 69, 73, 82, 85, 90–92
Askeaton, 49, 51, 53, 61, 62, 68
Asterophycus, 292, 316, 317
autocyclic controls, 47
autocyclic scour, 256

background, 2–4
Balliny Member, 82
Ballybunnion, 3, 8–10, 14, 175, 181, 184, 192, 193
    Clare Shale, 97, 109, *110*, 111
    Gull Island Formation, 181, 184, 192, 193
    Ross Sandstone Formation, 112, 114, 123, 128, 163, *164*, 164–173
    Viséan sediments, 37, 48–50, 51, 67–70, *71*, 71–78, *75*
Ballybunnion Castle beds, 75
Ballyelly Member, 82

Ballymartin Farm, 55
Ballymartin Formation, 12, 50, 53, 55, 56
Ballymartin Point, 53, 55
Ballynash, *58*, 58–60
Ballynash Farm, 58
Ballynash Member, 50, 58, 60, 61
Ballysteen, 50, 51, 53, 56, 58
Ballysteen Formation, 12, 50, 53, 55–60
Ballysteen Group, 55
Ballysteen limestones, 60
Ballyvergin borehole, 52
Ballyvergin Formation, 50, 56
Ballyvergin Mudstone Formation, 53
basin axis, 118, 196
basin margin, 39, 40, 41, 43, 46, 119, 163, 164, 170–173, 196, 237
basin models, 35–47
Beagh Castle, 51, 53, *56*, 57
Black Head, 80, 83, *85*, 99
Black Head Member, 80, 82, 85, 86
Black Rock, Loop Head Peninsula, 198
Black Rock, Spanish Point, 350, 351, 355, *357*, 357–359
Blake's Bridge, 99, *105*
boudinage, 157, 221
box folds, 23, 26
Bridges of Ross
    Anticline, 26
    Gull Island Formation, 206
    Pressure solution seams, 29
    Ross Sandstone Formation, 114, 115, 123, 128–130, 148, *149*, 150, 153, 154, 163, 170–172
Burren, 2, 9, 14, 70, 79–96
Burren Formation, 12, 50, 80, 82, 85, 91, 92, 94

A Field Guide to the Carboniferous Sediments of the Shannon Basin, Western Ireland, First Edition. Edited by James L. Best and Paul B. Wignall.
© 2016 International Association of Sedimentologists.
Published 2016 by John Wiley & Sons, Ltd.
Companion website: www.wiley.com/go/best/shannonbasin

371

Burren Formation, lower, 65, 69, 82, 87
Burren Formation, upper, 69, 82
Burren, karst, 2
Burren, open fold, 26
Burren Way, 104, 105

calciturbidites, 69, 73, 75
Caneyella, 104, 271
Canon Island, 51, 69
Carrowmore North, 306
Carrowmore Point, 242, 243, 289, *306*, 308, 312, 313
channel hierachy, 127
channels
    Corgrig Lodge Formation, 75, 76
    Gull Island Formation, 39, 188, 199, 202, 205–207, 209, 215, 219, 222, 230, 231, 233, 235–238
    Ross Sandstone Formation, 116, 121, 123–125, 127–131, 135, 136, 138–143, 145–152, 154, 155, 158, 159, 161–163, 166, 170–173
    Tullig Cyclothem, 43, 47, 240, 241, 243–244, 246, 247, 249, 251–252, 255–266, 283, 289, 291, 292, 294, 296, 297, 307–311, 322–326, 329, 331, 333–335, 343, 344, 361
Chokerian, 45
Clare Shale Formation, 10–14, 23, 30, 33, 34, 38, 39, 41, 43, 45–47, 50, 63, 64, 68, 70–72, 78, 80, 81, 94–111, 113, 116, 117, 119, 128, 165, 166, 173, 175–178, 180, 181, 183, 185, 188, 192, 196, 216, 217, 223, 227, 235, 239
cleavage, 23, 26, 27, 31, 59–61, 217, 219, 221
Cleedagh Bridge, *351*, 352–354
Cliffs of Moher, 2, 10, 30, 80, 113, 329, 330, 343
    Gull Island Formation, 174, 175, 197, 216, 217, 233, 235, *236*, 237–238
    Kilkee Cyclothem, *337*, 339, 342, 343, 349
    Tullig Cyclothem, 330, 335, *337*, 339, 343, 344
climate, 4
clockwise minibasin model (Pyles), 41–43, 45, 46
Cloghauninchy, 351, 359, *360*
Coal, 33, 185, 241, 290, 291, 295, 296, 300, 301, 306, 311, 312, 327, 334, 336
Coalification, 33
coast road south of Kilkee, *273–276*
collophane, 100
CoMa (Collinson and Martinsen) model, 39–41, 43, 45–47
comminution till, 332, 333
conchostracan, 264
conjugate shear fractures, 299, 304, 305

coral biostromes, 79, 83
Corgrig, 63, *66*, 67, 73
Corgrig Farm, 64, 66
Corgrig House, 64, 66
Corgrig Lodge, 64, 67, 68
Corgrig Lodge Formation, 12, 37, 50, 63, 66–72, 75–78
Courtbrown Point, 58, 59, *60*

Dangan Gate Member, 85
debris flow, 46, 222, 230
delta progradation, 240, 242, 286
Diamond Rocks, 277–279, *283*, 285–287, 328, 361
Diapir, 241, 278, 280, 281, 283–285, 288, 289, 343, 346, 361
Dibunophyllum, 80
Diplichnites, 268
Doolin, 3, 6, 8, 10, 11, 80, 93, 99, 106, 217
Doolin Quay, 80, *94*, 95, 99, 100
Doonbeg, 10, 289, 306
Doonbeg Bay, 289, 290
Doonbeg Number 1 borehole, 33, 34, 38
Doonlicky Cyclothem, 10, 12, 13, 47, 273, 276, 287, 350, 351, 353–356, 361
Doonlicky Sandstone, 273, 275
Doughmore, 306
Doughmore Bay, 306
Duggerna Rocks, 277, 283
Dunbarella, 103–105, 110, 270, 271, 286, 354
Durnish Formation, 12, 50, 62, 63, 65, 66, 69
Durnish Point, 51, 62, 63
Dykes, sandstone, 126, 150, 156, 157

East End, Kilkee, 277, 288

Fahee North Member, 82
Failure Point, *209*, 210
Fanore, 80, *86*, 99
Fanore Member, 80, 82, 85–87
Finavarra, 83
Finavarra Head, *83*
Finavarra Member, 50, 80, 82–85
Finavarra Point, 80, 99
Fisherstreet Bay, 217, 230, 238
    Clare Shale, 99, *104*
    Gull Island Formation, 174, 175, 177, 179, 182, 188, 192–194, 216, *217*, 219, 220, 223, 230
Fisherstreet slump, 105, 194, 195, 197, 217, 218, 220, 224–227, 229–233, 237
folding, 31, 33, 38, 45, 60, 221, 235, 304, 335, 344
    box, 23, 26
    chaotic, 217
    recumbent, 220, 224
    slump, 220, 221, 227–229
    synsedimentary, 33

Foohagh Point, *273–274*, 361
footwall, 275, 299, 339, 340, 347–349
Foreland basin, 38, 45, 47
Foynes, 3, 8, 10, 14, 25, 49–51, 62, 63, 65, 66, 70
Foynes borehole, 65, 66, 68, 69, 73, 78
Foynes Island, 63, 175, 179, 190
Foynes Shales, 177
Freagh Point, 330, 344, *345*, 346–348
Furreera Bay, 242, 310, 326, 330, *331*, 332, 333, 335–339, 344

general facilities, 5
geological map
  Ballybunnion region, 71
  Burren region, 80
  County Clare, 9
  Fisherstreet, 218
  Killard, 291
  Loop Head Peninsula, 198
  Lower Carboniferous, 49
  Moore Bay, Kilkee, 278
  Ringmoylan-Ballysteen area, 53
  Shannon Estuary region, 10
  Western Ireland, 175
George's Head, 277, 278, *288*
Gigantoproductus, 88, 93
Goat Island, 343, 344
Goleen Bay, 273, 275
Goleen Bridge, 58, 59
Goniatite, 23
  Clare Shales, 97, 98, 102, 105–108
  Gull Island Formation, 177, 178, 180, 182, 184, 185, 204, 216, 229, 235
  Lower Carboniferous, 66, 67, 72, 73
  Ross Sandstone Formation, 112, 115, 117, 120, 128, 153, 166, 168, 170
  Tullig and Kilkee Cyclothems, 247, 270–272, 286, 328, 337
  Younger Namurian cyclothems around Spanish Point, 353–355, 360
Gowlaun River, 94, *95*, 99
Gowleen, 351, *352*
growth fault, 43, 46
  Doonlicky Cyclothem, 273, 276, 354, 361
  Gull Island Formation, 212–214
  Kilkee Cyclothem, 329, 339–343, 345–349
  Tullig Cyclothem, 241, 243, 282, 290, 298–300, 302–305, 318, 320, 335
Gull Island, *197*
Gull Island Formation, 10–14, 23, 30, 39–44, 46–47, 80, 81, 117, 118, 149, 174–239, 343
  biostratigraphy, 180
  contact with Ross Sandstone Formation, 116, 215
  contact with Tullig Cyclothem, 243, 245
  definition, 175, 176

depositional environment, 195–197
lithofacies, 221, 222
lower, 116, 118, 119, 202, 214
middle, 207
sandstone:mudstone ratio, 193
sedimentation rates, 188, 190
sedimentology, 183
slumping, 187, 189
thickness, 186
upper, 212
upward patterns in sedimentology, 215

Hag's Head, 329, 330, 338, *339*, 340–342
hangingwall, 209, 212, 275, 299, 305, 335, 339, 340, 346–349, 354
Helminthoidea, 312, 313
Holkerian, 12, 13, 50, 65, 68, 82, 83
Homoceras, 98, 102, 103

Iapetus Suture, 17–19, 21, 22, 37–39, 41, 45, 62, 113, 118
Illaunlea headland, 58–61
Illaunonearaun Island, 273, *274*, 275, 361
Inishcorker Island, 10, 38, 49, 106, *107*, 108–110, 163, 171, 173, 179
Inishtubbrid beds, 37, 68
Inishtubbrid Island, 49, 51, 68, 69
initial flooding surface, 244, 249, 256, 257, 267, 311, 326, 328
Isohomoceras, 98, 104
Itineraries, suggested, 11

karst, 2, 79, 88
Kilbaha Bay, 10, 28, 29, 114, 115, 121, 122, 128, *138*, 139–142, 153, 170, 172, 198
Kilcloher Cliff, 114, 115, 123, 128, 129, *130*, 132–134, 140, 152, 166, 170, 172, 198
Kilfenora, 6, 9, 96, 97, 99, 106, 179
Kilkee, 3–5, 7, 8, 10, 14, 30, 175, 273, 277, 278, 282–287, 289, 361
  East End, 277, 288
  West End, 277, 286
Kilkee Cyclothem, 10–14, 30, 180, 240, 241, 243, 249, 268–269, 271, 277, 286, 287, 289, 306, 311, 315, 328–330, 337–339, 342, 343, 344–349
Kilkee Sandstone, 241, 278, 286, 287, 352
Killard, 33, 34, 193, 242, 243, *289*, 290–307, 310, 311, 317, 326, 328, 361
Killard marine band, 336
Killard Mouth-bar Sandstone, 290–293, 295
Killard syncline, 292, 293, 295, 305
Killimer Ferry, 3, 8, 14, 70
Kinderscoutian, 12, 98, 104, 105, 128, 176
Knockroe Point, 277

Lackglass, *91*, 92–94
Lacknasaggart Bay, 342
Lahinch, 3, 5, 6, 8, 10, 14, 187, 330
LAPs *see* lateral accretion packages
lateral accretion packages, 125, 135,
    137, 141, 142, 143, 148, 154, 155,
    158–163
Laurentia, 17
Lepidodendron, 264, 286
Limerick Limestone Formation, 59
Limerick Syncline, 68, 69
Liscannor, 326, 329, 330, 333, 339
Liscannor Flags, 338
Liscannor Stone, 337
Lisdoonvarna, 3, 8, 10, 14, 80, 81, 94, 96,
    99, 106, 175, 179, 183, 184, 216
Lismorahaun, 99, *106*
Lissylisheen Member, 82, 95, 100
lobes
    Gull Island Formation, 116, 121,
        123–131, 133, 135, 138, 140, 143, 149,
        151–153, 155, 158, 161, 163, 166,
        170–173
    Ross Sandstone Formation, 188, 189,
        199, 202, 203, 205–208, 210, 212, 214,
        215, 238

Magowna Formation, 95, 100, 101
Magowna Limestone, 100
map coverage, 7, 8
marine bands, 47, 104, 108, 112, 115–117,
    119, 126, 127, 133, 176, 178–185, 187,
    188, 190, 194, 197, 216, 236, 237, 241,
    247, 267–272, 278, 283, 286, 289, 314,
    315, 336–338, 351, 353, 354, 356,
    358–360
Marsdenian, 12, 176
Maumcaha Member, 81, 82, 88–90
Maura's Point, 206, *207*, 208, 209
maximum flooding surface, 241, 247, 270,
    271, 286, 313
megaflute erosional surfaces, 121
megaflutes, 116, 121, 122, 125–127,
    130, 140, 142–145, 147, 154, 155,
    163, 203
Mellon House Formation, 50, 52, 54, 55
Moher Tower, 339–341, 343
Moore Bay, 277, 278, 288
Moore Bay Sandstone, 277, 278, 283, 286,
    337–339, 341, 343, 344
mud/siltstone diapir, 241, 278, 280,
    281, 283–285, 288, 289, 343,
    346, 361
mudstone sheets, 121, 123, 124, 126, 131,
    171–173, 199, 202, 206, 207, 209–212,
    214, 215
Mutton Island, 3, 8, 10, 242, 243, 310, *313*,
    313–326, 361
    watchtower, 314, 315, 320–322

nautiloids, 72, 104, 271, 313

Olivellites, 338 *see also Psammichnites*

Palaeocurrents, 39, 41, 43, 47
    Gull Island Formation, 191, 192, 194,
        199, 201, 202, 219, 224, 229, 230, 235
    Ross Sandstone Formation, 113, 114,
        116, 119, 122, 127, 128, 130, 131, 135,
        138, 139, 145, 146, 150, 154, 155, 159,
        161, 164, 165, 167–169, 171, 172
    Tullig Cyclothem, 241, 246, 262, 266,
        281, 293, 295–297, 309, 310, 317,
        324, 334
palaeokarst, 82, 88–93
palaeosol, 37, 89, 241, 283, 287, 312, 316,
    331, 336, 353, 354, 356, 360
Pallaskenry, 53, 58
Pallaskenry borehole, 52, 55, 56, 58, 61
parasequence, 240, 241, 254–259, 273,
    290–292, 306, 311–313, 318, 320, 353,
    357–360
Parsonage Formation, 37, 66, 67, 69, 73–75
pedogenesis, 90
Pendleian, 12, 98, 102, 108, 176
phosphate, 45, 95, 100–102, 109, 168, 286
Planolites, 270, 292, 315, 317, 360
Point of Relief, 46, 361
    Gull Island Formation, *212*, 213, 361
    Tullig Cyclothem, 244, 245, 361
Pollack Holes, *277*, 278, 280, 281
pressure solution cleavage, 3, 26, 28, 29, 31,
    33, 149, 304
pressure solution seams, 29, 31
progradation, 39, 41, 43, 45–47, 107, 118,
    121, 152, 163, 164, 166, 196, 214,
    237, 240–244, 254, 275, 286,
    292–294, 306
Psammichnites, 296, 338, 342, 346

Quarry Point, *209*, 211, *212*, 245
Quilty, 30, 313, 351, 359

Rathkeale Formation, 62, 65, 68, 69
Rehy Hill, 114, 128, 129, *133*, 136, 137,
    140, 142, 153, 166, 170, 172, 198
Reticuloceras, 98, 105, 176, 181, 247, 271,
    283, 312, 313, 327, 351, 353, 354, 356,
    359, 360
Rinevella Point, 123, 124, 128, 130, *145*,
    146, 147, 171, 172, 198
Ringmoylan Quay, 49, 50, *52*, 53–55
Ringmoylan Shale Formation, 12, 50,
    53, 55
Ross Sandstone Formation, 10, 11, 14,
    26–30, 39–44, 46, 47, 106–110,
    112–175, 181, 194, 196, 238, 244
    lithostratigraphy, 116
    palaeocurrent patterns, 116

Ross slump, 128, 150–154, 156–163, 172, 173

safety, 6
Sand volcano, 150, 153, 156, 160, 218, 221, 225, 226, 344–347, 349
Scolicia, 296, 312, 338 *see also Psammichnites*
Seafield Harbour, 313
sediment provenance, 35, 43, 46, 47, 194
sequence boundary, 249, 256, 257, 287
Shanagolden Formation, 66, 69
Shannon Basin
    Dinantian history, 36
    models of basin infill, 35–47
    Namurian basin history, **38**
    pre-Carboniferous history, 17
    structural setting and evolution, 16–34
    thermal history, **31**
    Variscan deformation and metamorphism, **22**
Shannon Estuary, 2, 3, 7, 9–14, 20, 35–39, 41, 45, 46, 48–53, 56, 62, 97, 98, 106, 109, 113, 118, 127, 164, 196
Shannon Trough, 16, 18, 22, 23, 31, 34, 35, 37, 62, 68, 69, 118
Siderite, 267, 269–271, 287, 291, 296, 300, 301, 306, 309, 311, 312, 326–328, 331–333, 336, 353
Siphonodendron, 65, 83–85, 87, 88, 91, 93, 338
Skolithos, 267, 309, 317, 353
Slieve Elva, 4, 88, 105, 106
Slievenaglasha Formation, 70, 80–82, 91, 92, 94, 95, 100, 102
slumps, 39, 40, 42–44, 46
    Doonlicky Cyclothem, 354, 357
    Gull Island Formation, 174, 178, 181, 183, 187–189, 191–196, 199, 202, 204, 206–215, 223, 233, 238, 239
    Kilkee Cyclothem, 346
    Ross Sandstone Formation, 116, 118, 121, 124, 126, 127, 129–131, 149, 169, 171–173
    Tullig Cyclothem, 241, 273, 280, 313, 318, 320
slump sheet, 194, 226, 281, 318, 320
Southern Uplands-Longford Down Terrane, 17
Spanish Point, 3, 8–11, *350*, 351, **354**, 355, 357–359
St. Brendan's Well, **95**, 97, *100*, 101, 103, 104, 106, 108
St. Brendan's Well phosphate bed, 45, 95, 100–102
Stigmaria, 298, 301, 306, 311, 332
Stigmarian roots, 287, 296, 331, 353, 356
Stockeen Cliff, 330, **339**, 343, 344
Stratigraphy, overview, 9, 12

subsidence, 36–38, 45, 62, 68, 118, 119, 173, 196
syncline, 9, 11, 68, 69, 149, 150, 289, 290, 292, 293, 295, 296, 298–301, 304–307, 335, 350, 358, 359
Syringopora, 56, 83, 85, 87, 93, 94

Tarbert Ferry, 3, 8, 14, 49, 70, 109, 164
Teichichnus, 337
thermal maturation, 32
Tinure Fault, 17
Tournaisian, 12
Transgression, 55, 243, 247, 249, 267–269, 286, 290, 311, 328, 336
travel, xi, 5
Trilobites, 104
Trusklieve, 114, 197, 198, 242, 243, 244, 248, *249*, 250, 310, 311, 326, 361
    amphitheatre, *258*, 258–266
    overview, *254*, 255–257
    top of Tullig Sandstone, *266*, 266–268
    transgressive sediments at top of Tullig Cyclothem, *268*, 268–272
Tubber Formation, 80–86
Tullig Cyclothem, 12, 13, 23, 30, 39, 40, 42–45, 47, 176–178, 180, 181, 183, 188, 190, 212, 214, 235–238, 240–359
    lithofacies, 251–253
    transgression, *268*, 290, 311, 328
Tullig Point, 201, 242, *243*, 243–248
Tullig Sandstone, 11, 178, 201, 241, 242, 244–247, 249, 254–261, 263, 266–271, 278, 290–292, 295–298, 300, 301, 306–308, 310, 311, 314–316, 319–324, 326, 328, 329, 333, 334, 361
Tullig transgression, *268*, 290, 311, 328
turbidites, 1, 17, 37, 39, 41–45, 112–239

Variscan fold-thrust belt, 22
Variscan shortening, 23
VARNET-96 1B seismic traverse, 18, 20
Vigo Cave, 96
Viséan limestones, 11–13, 49, 50, 70, 72, 81, 100, 105
Viséan platform carbonates, 37, 45, 48, 68–70, 72, 79, 85, 90
vitrinite reflectance, 33, 34
Volcanic agglomerate, 65
volcanism, 37, 38, 68

Waulsortian carbonates, 11–13, 36–38, 48–50, 53, 58–63, 68, 70–73
West End, Kilkee, 277, 286
WiBe (Wignall and Best) model, 43–47
WIRE1 and WIRE 1B seismic traverse, 18, 21

Zoophycos, 214, 244, 267–269, 270, 286, 301, 327, 328, 337, 338, 353, 358, 360